Science, Space, Society

Olaf Kühne · Karsten Berr

Science, Space, Society

An Overview of the Social Production of Knowledge

 Springer

Olaf Kühne
Eberhard Karls Universität Tübingen
Tübingen, Germany

Karsten Berr
Eberhard Karls Universität Tübingen
Tübingen, Germany

ISBN 978-3-658-39139-3   ISBN 978-3-658-39140-9   (eBook)
https://doi.org/10.1007/978-3-658-39140-9

Responsible Editor: Cori Antonia Mackrodt
This Springer imprint is published by the registered company Springer Fachmedien Wiesbaden GmbH,
part of Springer Nature.
The registered company address is: Abraham-Lincoln-Str. 46, 65189 Wiesbaden, Germany

Preface

The life of modern humans is characterized in all areas of life by the results of science. In the sciences, conceptual and theoretical abstractions serve to access a phenomenal diversity that is scientifically interpreted, ordered and explained. In geography, for example, these are spatial, landscape or physical phenomena. The study of a space-related science (geography, planning sciences, landscape architecture, etc.) is therefore also associated with the examination of different approaches and methods, sometimes also space theories and science theories. But the theories developed in the sciences are also manifold and in turn require systematic abstractions and ordering proposals as well as theoretical explanations of their genesis and their claim to validity or truth. This is a main task of the philosophy of science. This also includes the reconstruction of the social and historical contexts in which knowledge is produced and validity is claimed. Since scientists cannot have the decision and responsibility for turning to specific theories taken away from them, an informed decision based on basic epistemological and scientific-theoretical knowledge is all the more important. The present book therefore aims to facilitate an introduction to the theoretical examination of knowledge, science and space as well as to scientific-theoretical concepts.

We would like to express our sincere thanks to our colleagues in the "Urban and Regional Development" working group at the University of Tübingen, Corinna Jenal and Timo Sedelmeier, who helped us to make the text more understandable, clearer and more user-friendly with their expert feedback (should any passages be unnecessarily difficult to understand, please address complaints directly to them). In addition, we would like to thank our assistants Christiane Lawrenz and Anna-Kathrin Schneider for their support in formatting the manuscript and the time-consuming citation. We think of Peter Kühn and Steffen Seitz for making themselves available as models for an exhausting photo shoot with Timo Sedelmeier (to whom we are also grateful here). We would also like to thank those who were involved in the re-accreditation of the BSc course "Geography" at the University of Tübingen and gave us the opportunity to finally write this book with the introduction of the module "Philosophy of Science". Finally, we would like to thank Cori Mackrodt and Springer VS for once again establishing a trusting and productive cooperation.

Finally, we would like to thank our wives for accepting and supporting our work on this book, which took place mostly in so-called free time.

Tübingen, Germany   Olaf Kühne
                                                                                    Karsten Berr

Contents

Introduction

Why the whole thing? Why deal with theoretical approaches to science, its social embeddedness and spaces? Let's start with spaces: Whatever we do, we do it in spatial contexts, buildings were erected (why exactly here and exactly like this?), they are used (why differently?), we move on paths, on corridors to places (what drives us?), certain places mean more to us, others less (why does this differ from the attention of others?). In short: We are dealing with material objects, but we are also dealing with socially shared symbolic attributions, but also with quite individual attention, because "geography is everywhere" (Cosgrove, 1989, p. 119). Space sciences (whether geography, urban or spatial sociology, planning sciences, architecture, etc.) are confronted with the most diverse spatial structures and processes that are to be observed and understood through an order: such a way of abstraction is a way to theory formation: "Theory is everywhere, in everything we do. Without theory, life (not just geography) would be chaos" (Cresswell, 2012, p. 5). The theoretical domestication of the 'chaos' produces numerous theories (in this case space). This is also quite confusing and calls for abstraction, that is, a theoretical consideration of the origin and the claim to truth of scientific theories. Here we enter the terrain of the philosophy of science. But theories do not arise from themselves. They are created by people and where people are, there are other people who influence their own ideas and interpretations of What-ever, be it in direct exchange or through their artifacts (not only writings, but buildings, paths, corridors …). This means: The generation of theories and meta-theories (theories of theories) takes place in social contexts, which are on the one hand differentiated, on the other hand in constant change, whereby the question of social conditions, generation of knowledge in general and scientific knowledge becomes relevant. These questions are at the center of the present book.

This book goes beyond other introduced and established textbooks because it integrates science theory, sociology of knowledge and science, as well as space theory, which are otherwise treated separately. This also means that the introductions to the individual topic complexes treated here are relatively brief and offer further introductions for a more detailed study, such as science theory (Bauberger, 2016; Carrier, 2017; Chalmers, 2006 [1996]; Detel, 2018; Hübner, 1978; Janich, 2015; Kornmesser & Büttemeyer, 2020; Pfister, 2016; Poser, 2012; Schülein & Reitze, 2012; Schurz, 2014; Seiffert, 1996; Stegmüller, 1985; Tetens, 2013), sociology of knowledge (Knoblauch, 2006; Maasen, 2015; Schützeichel, 2012; Stehr & Meja, 1980), sociology of science (Kaiser & Maasen, 2010; Maasen et al., 2012; Weingart, 2015) and space theory (Aitken & Valentine, 2015; Egner, 2010; Kirchhoff & Trepl, 2009; Läpple, 1992; Oßenbrügge & Vogelpohl, 2014; Schlottmann & Wintzer, 2019).

1.1   Basic Principles and Structure of the Book

The study of a space-related discipline (geography, planning, landscape architecture, etc.) is associated with the examination of different approaches and methods, sometimes also space theories and science theories. The present book would like to facilitate the entry into the theoretical examination of knowledge, science and space. It focuses on the three meta-sciences of science theory, sociology of science and history of science (Weingart, 2015), which are implemented exemplarily in the case of geography. These meta-sciences are characterized by the fact that they make science itself the subject of their investigation. In addition to these meta-scientific approaches, we also deal with "simple" theory formation, on the one hand with the sociology of knowledge, since it provides a framework not only for the sociology of science, but also for the generation and dissemination of knowledge (and also non-knowledge). In addition, space concepts and -theories are introduced in order to enable students of space sciences a differentiated access to space. This book is therefore not only concerned with the question of how scientific knowledge is generated, how this generation has changed, but also with which theoretical approaches there are to space, and not least with how they can be interpreted in relation to the generation of scientific knowledge. This approach requires interdisciplinary perspectives. Therefore, the present overview of science and space theories as well as sociology of knowledge is mainly written from four perspectives:

1. It deals with philosophical foundations of sciences, here you will find a perspective of philosophy.
2. It also deals with questions of social generation and distribution of knowledge, here you will find a sociological perspective.
3. It refers to concepts of space and space theories, here perspectives of space sciences (especially of sociological space and geography) dominate.
4. It offers a brief insight into the disciplinary history of geography, here the reflective perspective of a science in the confrontation with itself becomes clear.

Science and space theories, especially when their social entanglement is considered, are sometimes in close dependency, sometimes rejection relationships with each other. Thus, theoretical foundations are further developed in different disciplines (for example, the Schützian social phenomenology in the sociology of knowledge in Sect. 5.1, the social constructivist landscape research in Sect. 6.5 and the phenomenological theory of space in Sect. 6.6) and possibly related to each other again (as in the neopragmatic approach to space in Sect. 6.11). This means that certain theoretical approaches can occur in different contexts, which we represent in more detail where we see them most strongly contextually anchored. This also has the consequence that we have built in numerous cross-references into the text in order to facilitate a more in-depth reading of a certain topic, so we have supplemented the linearity of a book text with the possibility of almost parallel reading of different text passages. An alternative to this parallel reading of the book is a repetition of the reading, in order to read the front parts of the book, such as Chaps. 5, 6 and 7 with the knowledge from the back parts of the book, 2, 3 and 4, to be able to confront (for example, scientific theories in relation to the history of geography or actor-network theory with regard to the sociology of science). In the social and cultural sciences, this experience of reception, reading parts of a text in parallel or in loops of repetition, is explained with the so-called "hermeneutic circle" (see Sect. 3.5.2). With a certain, quite vague prior understanding, parts of a text can be read and with the understanding thus acquired other parts can be understood, until gradually a preliminary understanding of the "whole" of the text is formed. And from this understanding thus attained, the "parts" already vaguely understood before can in turn be read anew and often differently, newly grouped, arranged and understood "deeper" or "better".

Since this book is an overview that links together several areas of knowledge and science, many of the explanations are kept so brief that they enable a general understanding, in particular of relationships. This also means that certain theoretical perspectives are only mentioned in passing or individual aspects are highlighted, while others—certainly productive—are not mentioned at all in this book. At the end of the chapters (with the exception of the introduction and conclusion), we have added references to further reading, mostly textbooks on the individual topics, but also references to original literature. The (of course incomplete) references to the original literature include works that, in our opinion, shape the complex of topics treated particularly strongly. This book is aimed primarily—in line with its overview-like and introductory claim—at students of the various space sciences, but also at students of related courses, as well as post-graduate students who would like to update the connections between science and space theories and sociology of knowledge again. Of course, also to all those, inside and outside the (space) sciences, who are simply interested in an abstract perspective on knowledge and science.

In order to understand the rather complex subject matter of knowledge, science and space theories, we make use of text boxes. Three different types of these text boxes, distinguished by different colors, can be fo& in the book:

Interim Conclusion Textboxes
in which an interim conclusion is drawn, are gray. Here the essential statements of a chapter are summarized again.

Continuation Text Boxes
in which further approaches or methods are set out. ◄

Terminology Text Boxes
in which essential basic terminology is explained, are highlighted in blue. They are located at the beginning of each chapter and serve to facilitate understanding of what follows. Since more basic terminology needs to be clarified at the beginning of the book, its scope decreases towards the end of the book.

Visualisations in the form of tables and figures are also intended to contribute to better understanding. The sometimes quite abstract text passages are made more concrete by these visualisations, they likewise support the reproduction of conceptual thinking by pictorially explaining the semantic content of concepts and theories as well as the conceptual and theoretical relationships and contexts. This approach can be supported by a famous formulation of the German philosopher Immanuel Kant, who once said that thoughts without content are empty and intuitions without concepts are blind (Kant, 1959 [1781], B 75). In a narrower sense, however, this also refers to the cooperation of conceptual and visual means for the purpose of explaining the content of concepts and theories at the level of interpretation (semantics). In addition to the use of illustrations and tables, we have translated foreign-language quotations into German and also older German quotations into the new spelling in order to facilitate understanding of the text. Emphasis in quotations is—unless the emphasis was made by the authors and this is accordingly pointed out—from the authors of the quotations.

The present book is a translation of a German-language textbook. This also means that certain German-language traditions (may) have a different meaning and interpretation than in English-language reception (and other traditions tied to languages). However, this also means that texts by non-German-speaking authors are interpreted against the background of German-speaking traditions of thought, and to that extent differ from the reading they have in, say, English- or French-speaking contexts. In this respect, the textbook (implicitly) also provides access to the German-language tradition of dealing with knowledge, science and thinking about science. This also applies to translations of English-language authors into German.

In Chap. 2 an introduction to the logical, linguistic, semantic and pragmatic foundations of the theory of science and the sciences is first carried out. On this basis, in Chap. 3 some philosophical foundations of the theory of science are discussed, then essential stages of the development of the philosophical debate on 'science', 'knowledge' and 'truth' are reconstructed, which extend from the debate between rationalism and empiricism to the debate between logical empiricism and Popper's approach. In Chap. 4 positions are presented which have been developed from the history of science and therefore take into account the factor of 'time'. In Chap. 5 we present positions which have arisen from the tradition of sociology and therefore focus on the factor of 'social embedding'. Chap. 6 presents different understandings of space and corresponding space theories on the basis of Karl Popper's 'Three-Worlds-Theory'. In Chap. 7 the disciplinary history of geography is outlined, in particular the scientifically interesting and relevant development lines and development breaks. In Chap. 8 we draw a conclusion and end this overview with the presentation and discussion of the three currently prevailing basic currents of science. For the understanding of the following, some terms are helpful, which you will find in Textbox 1 shown.

Textbox 1: Essential Terms for Chap. 1

Lebenswelt (in English approximately: lifeworld): As a non-scientific term, it refers to everyday life and everyday life. As a scientific term, it has been used to refer to the total of pre-theoretical practice relationships and experiences that determine the lives of people since Edmund Husserl's work *The Crisis of European Sciences* (Husserl, 1954).

Mythos: (from Greek *myein*, say). The myth is the repetition or retelling of handed-down and believed stories, especially of god stories with Hesiod and Homer. In contrast, the logos (Greek *legein*, put together, interpret) is a reflective reception and analysis and a productive discussion of something handed down or given. The often described development 'from myth to logos' (Nestle, 1940) therefore means a progress from the believed to the reflected and rationally appropriated.

Complexity and Complication: Complexity refers to the totality of possible states in a system, it refers to the degree of networking (a traffic system is complex, for example, when places with different transport carriers are connected). Complication refers to the homogeneity or inhomogeneity of a system (it is uncomplicated when only places of equal centrality need to be connected to each other; Hügin, 1996).

Postulate (from Latin *postulare*, demand): In everyday language, a demand, assumption. In epistemology and philosophy of science, this term is also used as a normative demand—in addition to more specific meanings in logic.

Secularization: Following the seizure of church property by secular rulers, secularization in a more general sense means the process of disempowering the church, the process of emancipation from religious or ecclesiastical tutelage, and the process of secularization of all previously sacral life forms.

Skylla and Charybdis: In Homer's *Odyssey*, Skylla and Charybdis are two sea monsters that wreak havoc on opposite sides of a strait. Ships that want to pass through the strait are either killed by Skylla or by Charybdis. It is a classic dilemma situation in which one must choose between two evils or dangers without coming out unscathed.

System: This is a system of action in which the elements are more closely connected to each other by direct mutual action than to their environment (Sachsse, 1971). There is therefore a complexity gradient between system and environment: In comparison to its environment, a system forms a lower complexity (Luhmann, 1984).

Reality: It is a derivative of the verb 'act' and refers to the Middle High German. 'Work' (work). In philosophy, 'reality' can either be the tangible and perceptible totality of beings or, conversely, the 'essential' or 'ideal' being of things over and against the mere empirically and contingently existing and given. In the natural sciences, the term for what is recognized as objectively given after deduction of subjective conditions of perception and taking into account logically correct inferences.

1.2   The Scientific World View in the Knowledge Society

The life of the modern human being is characterized in all areas of life (cf. Poser, 2009) by the results of science and its particular technical implementation: "We live in the scientific-technical civilization" (Tetens, 2013, p. 9)—and we *believe* in the explanatory power of science and its usefulness for our entire life. It should not only explain phenomena, but also guide our actions in difficult or controversial issues, for example by advising politicians in crises or on socially significant basic decisions. Science can even become the only legitimate form of orientation within the world. Other world views such as myth and religion (cf. Feyerabend, 2010 [1975]; Schülein & Reitze, 2012; Tetens, 2013) or orientations that were previously sought in traditions, customs, conventions and practice-proven experiences, lost their theoretical and practical orientation function successively as a result of the triumphal march of modern science in its alliance with successful technology, progressive secularization and industrialization, overall a 'disenchantment of the world' (Weber, 2011 [1919]). In this way, science has "altered our *world view*" (Poser, 2009, p. 12)—for example, the mythological origin stories of the ancient Greek gods (Hesiod, Homer) and the biblical creation story (Genesis) by the theory of evolution (Darwin). The history of science (e.g. Losee, 1977) investigates the conditions of origin and the development process of the scientific world view, which was first wrested from the 'mythical world view' (Tetens, 2013) by philosophers before Socrates (cf. Diels, 2004–2005), but above all by Aristotle, Socrates and Plato, and occasionally referred to as the development 'from myth to logos' (Nestle, 1940).

During the modern industrialization process, in particular, *technology* was scientized (Schülein & Reitze, 2012). This scientization of technology led to an immense application of scientific knowledge in technical products, goods, and services that greatly improved people's living conditions and contributed to "prosperity assurance" (Carrier, 2017, p. 152) and continue to do so (Schiemann, 2017). The current "scientific-technical civilization" can be characterized by two "postulates": the "Postulate of the exclusive access of the sciences to reality" and the "Postulate of world perfection" (Tetens, 2013, p. 9). The first postulate assumes that it is only the sciences that "let us better understand and understand reality", the second that the world is improved by the technological application of science, "because gradually more and more evils and sufferings disappear from it" (Tetens, 2013, p. 9). These two postulates are already implied by the founding fathers of modern science, the empiricist Francis Bacon and his works 'De Dignitate et augmentis scientiarum' (2006 [1605]) and 'Novum Organon' (1990 [1620]) as well as the rationalist René Descartes and his 'Discours de la méthode' (1990 [1637]). (Natural) sciences should improve the lot of humanity, "progress in science appeared ipso facto as progress in humanity" (Poser, 2009, p. 21) and science operated "from the beginning under the double obligation of knowledge and utility" (Carrier et al., 2007, p. 24).

The flip side of this partly quasi-religious optimism about science (cf. Feyerabend, 2010 [1975] and Sect. 4.5) is a "scientific superstition" (Jaspers, 1975, pp. 183–196; see Tetens, 2013, pp. 103–105), which consists in attributing to science also achievements that it cannot fulfill—for example, the complete explanation and control of all natural and cultural phenomena. This science optimism, which increases to science belief and science superstition, plays, to put it bluntly, "the role of the ruling religion of our time" (von Weizsäcker, 2006, p. 5). For science has "in many cases *taken on the function that religions have had*" (Poser, 2009, p. 11)—not shamans, priests or theologians, but scientifically trained experts are asked for expertise in court, in medicine or in politics. *Disappointed* 'scientific superstition' quickly turns into a 'contempt for science' (Jaspers, 1975) around, possible failures, measured against exaggerated expectations, are contemptuously held up to her as a failure. The Corona crisis that arose in 2020 is a typical example: Scientists are confronted with unrealistic expectations of healing (they are "hyped up" as "experts" in the media and treated like "pop stars" or "prophets"). If these expectations cannot be met in the short or medium term or if scientists announce uncomfortable research results (an example from the Corona crisis: a effective vaccine cannot be developed so quickly or not at all, a lockdown has to be extended or newly introduced from a medical point of view), they are met with disappointment to contempt. While the new (natural) science of the 17th century still "came with promises it could not keep", it was not until the late 19th century that the "promised connection between science and technical development" and the "course of the 20th century that science is again overwhelmed by the external expansion of its explanatory claim and the application contexts. The successes have generated expectations that cannot be met in turn" (Carrier et al., 2007, p. 11). The 'world perfection postulate' therefore probably belongs more to the 'scientific belief' (Tetens, 2013). Such 'limits of science' (Chalmers, 1999) to show and recognize is an important task of every scientific theory.

The successes of the scientized technology in particular since the 19th century thus led to the triumph of the sciences not only in technology, but to their

always growing 'reputation' (Chalmers, 2006 [1996]) and corresponding expectations in other areas of life. This process can be referred to as the "scientization of
everyday life" (Schiemann, 2017, pp. 182–185), for example when the everyday
world-proven handling of street or land maps is replaced by electronic navigation
systems that are based on complex scientific and technical requirements; or when
everyday world experiences are permeated with scientific knowledge or when this
knowledge flows into those experiences, such as scientific climate research results
in everyday knowledge of weather and climate phenomena. In other words, leading scientific concepts, ideas and theories from the world of science have flowed
into the non-scientific 'everyday worlds' of human beings. The German philosopher Edmund Husserl (1859–1938) addressed this phenomenon under the title
'Inflow into Everyday Life' (Husserl, 1954, p. 115 and more often; see Held, 1991,
p. 106). A society that is so shaped by scientific knowledge is referred to as a
'knowledge society' (Carrier et al., 2007; Schülein & Reitze, 2012).

In contrast to the scientification of industrialized technology, one can observe
an 'industrialization of science' (Plessner, 1966; cf. also Wingens, 1998) as
described, among others, by Plessner and Max Weber and Martin Heidegger: Now
'science as a profession' can be understood and practiced by 'experts' (Weber,
2011 [1919]). The sciences develop specific career paths, they are increasingly
being developed into ever more differentiated disciplines, fields and research
areas; this leads to the compulsion of specialization and continuous innovation
(cf. Schülein & Reitze, 2012): "Like weeds and turnips, new scientific disciplines
are constantly springing up" (Tetens, 2013, p. 34), the "complexity and multi-
layeredness of the reality to be researched" corresponds to a "variety of sciences"
that has to "divide the 'reality' into small research-friendly portions" (Tetens,
2013, p. 38). The 'industrialization of science' thus leads to a 'struggle for recognition' (Eisel, 1997; Honneth, 1992), that is, for collegial 'attention', academic
reputation, science policy renown, financial resources and institutional security in
the current 'scientific enterprise' (Franck, 2007). In view of the mentioned 'variety of sciences', the "scientific enterprise" thus becomes a "variety zoo" (Tetens,
2013, p. 77) with rivalries and distribution battles. Martin Heidegger consequently
described the character of modern science as an "operation", spoke of the "operational character of research" (Heidegger, 1963, p. 77) and of the "institutional
character of the sciences" (Heidegger, 1963, p. 78).

The "operational character" of science and research has two side effects. One
makes it clear how much the term "operation" is borrowed from business and
industry: In the "endeavor of the sciences to become a need of society, to *integrate into its earning mechanism*", not only are new "special disciplines (= pro-
fessions, = chances of profit)" (Plessner, 1966, p. 131) found, but it is increasingly
expected that sciences not only work for the economic interests of society but also
organize and evaluate themselves according to economic criteria. This "economization of science" contradicts its self-conception, insofar as "the most important
ingredient of scientific knowledge is lost: the trust in that knowledge which is due
to the exclusive orientation towards truth. It is only available in public institutions" (Weingart, 2008, p. 483; cf. Carrier, 2007). The second side effect concerns

a certain blindness to operation both of the busy scientists and of the societies that profit from the fruits of science and want to continue to do so. The unforeseeable and "most life-threatening consequences of scientific research and scientized technology; examples are superfluous" (Poser, 2009, p. 21) are addressed. In order not to have to choose between the Scylla of a "scientific superstition" and the Charybdis of a "contempt for science", one must again remember the task of philosophy of science, which is to determine the limits of science. Both the operational character and the blindness to operation of science thus refer to "the *legitimation problem of science*" (Poser, 2009, p. 21).

In order to be able to determine the *limits* of science, it is however necessary to determine *science*. But then the question has to be answered, what that actually 'is', 'the science', or which criteria of *scientificity* can be named and justified. We will have to postpone this question for the time being, since we first have to provide the necessary logical and linguistic equipment in the form of a 'Logical Propaedeutic' before we can answer it (see Sect. 3.1.2).

References

Aitken, S. C., & Valentine, G. (Hrsg.). (2015). *Approaches to human geography. Philosophies, theories, people and practices* (2. Aufl.). SAGE.

Bacon, F. (1990 [1620]). *Neues Organon. Herausgegeben und mit einer Einleitung von Woflgang Krohn* (Teilband 1 und Teilband 2). Wissenschaftliche Buchgesellschaft.

Bacon, F. (2006 [1605]). *Über die Würde und die Förderung der Wissenschaften: London 1605/1623. Aus dem Englischen übertragen von Jutta Schlösser. Herausgegeben und mit einem Anhang versehen von Hermann Klenner* (Haufe-Schriftenreihe zur rechtswissenschaftlichen Grundlagenforschung, Bd. 19). Haufe.

Bauberger, S. (2016). *Wissenschaftstheorie. Eine Einführung*. Kohlhammer.

Carrier, M. (2007). Erkenntnisgewinn und Nutzenmehrung. Eine verwickelte Beziehung. In M. Carrier, W. Krohn, & P. Weingart (Hrsg.), *Nachrichten aus der Wissensgesellschaft. Analysen zur Veränderung der Wissenschaft* (S. 93–110). Velbrück-Wissenschaft.

Carrier, M. (2017). *Wissenschaftstheorie zur Einführung* (Zur Einführung, 4., überarb. Aufl., Bd. 353). Junius.

Carrier, M., Krohn, W., & Weingart, P. (Hrsg.). (2007). *Nachrichten aus der Wissensgesellschaft. Analysen zur Veränderung der Wissenschaft*. Velbrück-Wissenschaft.

Chalmers, A. F. (1999). *Grenzen der Wissenschaft*. Springer.

Chalmers, A. F. (2006 [1996]). *Wege der Wissenschaft. Einführung in die Wissenschaftstheorie*. Springer.

Cosgrove, D. (1989). Geography is everywhere: Culture and symbolism in human landscapes. In D. Gregory & R. Walford (Hrsg.), *Horizons in human geography* (S. 118–135). Macmillan Press LTD.

Cresswell, T. (2012). *Geographic thought. A critical introduction*. Wiley-Blackwell.

Descartes, R. (1990 [1637]). *Discours de la méthode. Französisch – Deutsch* (Philosophische Bibliothek, Bd. 261, Unveränderter Nachdruck). Meiner (Von der Methode des richtigen Vernunftgebrauchs und der wissenschaftlichen Forschung. Übersetzt und herausgegeben von Lüder Gäbe).

Detel, W. (2018). *Grundkurs Philosophie* (Erkenntnis- und Wissenschaftstheorie, Bd. 4, 3., vollst. durchgesehene u. erw.). Reclam.

Diels, H. (2004–2005). *Die Fragmente der Vorsokratiker. Griechisch und Deutsch* (Unveränderter Nachdruck der 6. Aufl. 1952). Weidmann (Herausgegeben von Walther Kranz).

Egner, H. (2010). *Theoretische Geographie*. WBG.

Eisel, U. (1997). Unbestimmte Stimmungen und bestimmte Unstimmigkeiten. Über die guten Gründe der deutschen Landschaftsarchitektur für die Abwendung von der Wissenschaft und die schlechten Gründe für ihre intellektuelle Abstinenz – mit Folgerungen für die Ausbildung in diesem Fach. In S. Bernhard & P. Sattler (Hrsg.), *Vor der Tür. Aktuelle Landschaftsarchitekru aus Berlin* (S. 17–33). Callwey. http://www.ueisel.de/fileadmin/dokumente/eisel/ Unbestimmte_Stimmungen/Eisel_Unbestimmte_Stimmungen_fertig.pdf. Zugegriffen: 12. Jan. 2019.

Feyerabend, P. (2010 [1975]). *Against method: Outline of an anarchist theory of knowledge*. Verso.

Franck, G. (2007). *Ökonomie der Aufmerksamkeit. Ein Entwurf*. dtv.

Heidegger, M. (1963). *Holzwege* (4. Aufl.). Klostermann.

Held, K. (1991). Husserls neue Einführung in die Philosophie: der Begriff der Lebenswelt. In C. F. Gethmann (Hrsg.), *Lebenswelt und Wissenschaft. Studien zum Verhältnis von Phänomenologie und Wissenschaftstheorie* (Neuzeit und Gegenwart, Bd. 1, S. 79–113). Bouvier.

Honneth, A. (1992). *Kampf um Anerkennung. Zur moralischen Grammatik sozialer Konflikte*. Suhrkamp.

Hübner, K. (1978). *Kritik der wissenschaftlichen Vernunft*. Alber.

Hügin, U. (1996). *Individuum, Gemeinschaft, Umwelt. Konzeption einer Theorie der Dynamik anthropogener Systeme*. Lang.

Husserl, E. (1954). *Die Krisis der europäischen Wissenschaften und die transzendentale Phänomenologie. Eine Einleitung in die phänomenologische Philosophie* (Husserliana, Bd. 6). Martinus Nijhoff (Herausgegeben von Walter Biemel).

Janich, P. (2015). *Handwerk und Mundwerk. Über das Herstellen von Wissen*. Beck.

Jaspers, K. (1975). *Was ist Philosophie? Ein Lesebuch*. Piper.

Kaiser, M., & Maasen, S. (2010). Wissenschaftssoziologie. In G. Kneer & M. Schroer (Hrsg.), *Handbuch Spezielle Soziologien* (S. 685–705). VS Springer.

Kant, I. (1959 [1781]). *Kritik der reinen Vernunft*. Felix Meiner.

Kirchhoff, T., & Trepl, L. (2009). Landschaft, Wildnis, Ökosystem: zur kulturbedingten Vieldeutigkeit ästhetischer, moralischer und theoretischer Naturauffassungen. Einleitender Überblick. In T. Kirchhoff & L. Trepl (Hrsg.), *Vieldeutige Natur. Landschaft, Wildnis und Ökosystem als kulturgeschichtliche Phänomene* (Sozialtheorie, S. 13–68). transcript.

Knoblauch, H. (2006). *Wissenssoziologie*. UVK-Verlagsgesellschaft/UTB.

Kornmesser, S., & Büttemeyer, W. (2020). *Wissenschaftstheorie. Eine Einführung*. Metzler.

Läpple, D. (1992). Essay über den Raum. Für ein gesellschaftswissenschaftliches Raumkonzept. In H. Häußermann, D. Ipsen, R. Krämer-Badoni, D. Läpple, M. Rodenstein, & W. Siebel (Hrsg.), *Stadt und Raum. Soziologische Analysen* (2. Aufl., S. 157–207). Centaurus.

Losee, J. (1977). *Wissenschaftstheorie. Eine historische Einführung* (Beck'sche Elementarbücher). Beck.

Luhmann, N. (1984). *Soziale Systeme. Grundriß einer allgemeinen Theorie*. Suhrkamp.

Maasen, S. (2015). *Wissenssoziologie* (2., komplett überarb. Aufl.). transcript.

Maasen, S., Kaiser, M., Reinhart, M., & Sutter, B. (Hrsg.). (2012). *Handbuch Wissenschaftssoziologie*. Springer.

Nestle, W. (1940). *Vom Mythos zum Logos. Die Selbstentfaltung des griechischen Denkens von Homer bis auf die Sophistik und Sokrates*. Kröner.

Oßenbrügge, J., & Vogelpohl, A. (Hrsg.). (2014). *Theorien in der Raum- und Stadtforschung. Einführungen*. Westfälisches Dampfboot.

Pfister, J. (Hrsg.). (2016). *Texte zur Wissenschaftstheorie*. Reclam.

Plessner, H. (1966). Zur Soziologie der modernen Forschung und ihrer Organisation in der deutschen Universität. In H. Plessner (Hrsg.), *Diesseits der Utopie. Ausgewählte Beiträge zur Kultursoziologie* (S. 121–142). Suhrkamp.

Poser, H. (2009). *Wissenschaftstheorie. Eine philosophische Einführung*. Reclam.

Poser, H. (2012). *Wissenschaftstheorie. Eine philosophische Einführung* (2., überarb. u. erw. Aufl.). Philipp Reclam jun.

Sachsse, H. (1971). *Einführung in die Kybernetik unter besonderer Berücksichtigung von technischen und biologischen Wirkungsgefügen.* Vieweg + Sohn.

Schiemann, G. (2017). Persistenz der Lebenswelt? Das Verhältnis von Lebenswelt und Wissenschaft in der Moderne. In T. Müller & T. M. Schmidt (Hrsg.), *Abschied von der Lebenswelt? Zur Reichweite naturwissenschaftlicher Erklärungsansätze* (2. Aufl., S. 181–200). Karl Alber.

Schlottmann, A., & Wintzer, J. (2019). *Weltbildwechsel. Ideengeschichten geographischen Denkens und Handelns* (utb Geographie, 1. Aufl.). Haupt.

Schülein, J. A., & Reitze, S. (2012). *Wissenschaftstheorie für Einsteiger* (3., akt. u. erw. Aufl.). facultas wuv.

Schurz, G. (2014). *Einführung in die Wissenschaftstheorie* (4., überarb. Aufl.). WBG.

Schützeichel, R. (2012). Wissenssoziologie. In S. Maasen, M. Kaiser, M. Reinhart, & B. Sutter (Hrsg.), *Handbuch Wissenschaftssoziologie* (S. 17–26). Springer.

Seiffert, H. (1996). *Einführung in die Wissenschaftstheorie 1. Sprachanalyse – Deduktion – Induktion in Natur- und Sozialwissenschaften.* Beck.

Stegmüller, W. (1985). *Probleme und Resultate der Wissenschaftstheorie und analytischen Philosophie* (Theorie und Erfahrung, Bd. 2). Springer (Zweiter Halbband: Theorienstrukturen und Theoriendynamik).

Stehr, N., & Meja, V. (Hrsg.). (1980) Wissenssoziologie [Themenheft]. *Kölner Zeitschrift für Soziologie und Sozialpsychologie* (22). Westdeutscher.

Tetens, H. (2013). *Wissenschaftstheorie. Eine Einführung.* Beck.

Weber, M. (2011 [1919]). *Wissenschaft als Beruf* (11. Aufl.). Duncker & Humblot.

Weingart, P. (2008). Ökonomisierung der Wissenschaft. *N.T.M. Zeitschrift für Geschichte der Wissenschaften, Technik und Medizin, 16*(4), 477–484. https://doi.org/10.1007/s00048-008-0311-4.

Weingart, P. (2015). *Wissenschaftssoziologie.* transcript.

von Weizsäcker, C. F. (2006). *Die Tragweite der Wissenschaft* (7. Aufl.). Hirzel.

Wingens, M. (1998). *Wissensgesellschaft und Industrialisierung der Wissenschaft.* Deutscher Universitätsverlag.

Logical Propaedeutic

2

"Logical propaedeutic" will provide the linguistic means with which we orient ourselves in everyday life and in the sciences by means of words, concepts, sentences, judgments, conclusions and theoretical generalizations, try to recognize and understand something and in everyday life and in the sciences alike build a "world". As a rule, "propaedeutic" (from Greek *pró*, before, and *paideúein*, teach, educate) is understood to mean "the preparatory instruction, the 'preschool', of an art or science" (Gabriel, 2004, p. 361). In philosophy, propaedeutic is often limited to logic (Gabriel, 2004) or understood as "logical propaedeutic" in the sense of an introduction to the logical, linguistic and semantic foundations of philosophy. In the German-speaking world, Wilhelm Kamlah and Paul Lorenzen have designed a "Logical Propaedeutic" as a "preschool of reasonable speech" (Kamlah & Lorenzen, 1967) "for everyone", which "*is not merely a vestibule of formal logic*" (Kamlah & Lorenzen, 1967, p. 13), but the doctrine of "*the building blocks and the rules of any reasonable speech*" (Kamlah & Lorenzen, 1967, p. 13)—both in everyday life and in the sciences. The aim is thus an introduction to "methodical thinking" (cf. Lorenzen, 1968) in the form of a "discipline of reasonable speech" (Kamlah & Lorenzen, 1967, p. 13) and "thinking" (Kamlah & Lorenzen, 1967). This 'disciplining' of thinking and speaking serves the purpose, with regard to sciences, of introducing, in a comprehensible and justified way for all other scientists, the basic concepts of a science, "by means of which individual sciences [...] provide their subject areas for scientific investigation" (Wille, 2011, p. 163). "Logical propaedeutic" thus wants to show how basic concepts can be obtained in everyday life and science and how a "understanding of the world" and the respective field of science can be made possible by the use of language: "An *Subject* in the world *is* something that we designate with a word of our language—that is the principle of Logical Propaedeutic" (Seiffert, 1996, pp. 27–28).

A similar approach is taken by the "Logical-semantic Propaedeutic" (Tugendhat & Wolf, 1986), only that this does not want to give an introduction to "methodical thinking", but an introduction to logic and its connection with semantics (see Sect. 2.1). The "Logical-pragmatic Propaedeutic" by Peter Janich (2001) also pursues a different goal: In contrast to the "Logical" and the "Logical-semantic" Propaedeutic, it does not reconstruct the *linguistic* conditions of an understanding of reality, but reconstructs these conditions taking into account pragmatics (see Sect. 2.1) in methodologically successive steps from everyday human *action* (Hartmann & Janich, 1996; Janich, 2015; see Bauberger, 2016).

In the following, we understand "Logical Propaedeutic" as an introduction to the logical, linguistic, semantic and pragmatic foundations of the theory of science and the sciences. This also takes into account typical logical, semantic or linguistic errors, misunderstandings, fallacies that can occur especially at the beginning of the studies, but even with "established" scientists. For the understanding of the presented in this chapter, some terms are essential, which you will find in Textbox 2.

Textbox 2: Essential Terms for Chap. 2

Analysis, analysans and analysandum: In a concept analysis, the concept to be analyzed 'analysandum' is called, the totality of the analyzing concepts '*analysans*'.

Axiom (from Greek *axióma*, claim, or *axioein*, to consider as valuable, to consider as true): In particular, in mathematics, geometry and logic, a term for the basic, unprovable, but assumed statements of a discipline. They are considered as evident (immediately convincing) and indisputable basis for further, from them following statements. For example, the first axiom according to Euclid reads: "It is required: 1. That one can draw a line from any point to any point" (quoted in: Janich, 2015, p. 26).

Definition, definiens and definiendum: (from Latin *definitio*, delimitation, determination). A scientific definition leads a (new, still unknown) and therefore specially to be defined concept ('definiendum') to other concepts ('definiens'), which are already known.

Epistemology: a term in use in philosophy since the beginning of the 19th century for the science of structures, conditions, origins, types and limits of human knowledge in general.

Explanation, explicat and explicandum: (from Latin *explicare*, to execute, to develop). In a concept explanation, a vague, imprecise concept ('explicandum') is replaced or transformed by a more precise concept ('explicat').

Explicit, implicit: 'Explicit' means 'unfolded, presented in detail' in contrast to 'implicit', which means 'included, included'. It is the difference between an unfolded and an still unfolded meaning.

Fallibilism: (from neulat. *fallibel*, subject to error). Since Popper and his followers the indication of the fundamental and unprovable fallibility of human thinking, recognition and action.

Hybridization: Initially, this was understood to mean a biological cross (Hein, 2006), that is, "the development of new combinations by grafting a plant or fruit onto another" (Nederveen Pieterse, 2005, p. 401). In the last decades of the 20th century, the term 'hybridization' also became popular in the humanities, cultural and social sciences and was extended to a "cultural strategy of mixing and negotiating differences" (Hein, 2006, p. 55).

Intension, Extension; Connotation, Denotation: Terms can be distinguished by intension (lat. *intendere*, to strain, to be aware of) or by their content and extension (lat. *extendere*, to extend) or by their scope. The intension as the content of the concept denotes the meaning of the concept and corresponds to the totality of the characteristics or connotations (lat. *con-*, with-, together, lat. *notatio*, note), which belong to the concept and determine it. The extension as the scope of the concept denotes the totality of all objects or denotates (lat. *denotare*, to designate), to which all characteristics or connotations of the concept apply.

Junctors, logic of Junctors: Statements can be combined by logical junctors (lat. *jungere*, to connect), that is, sentence connections like 'not', 'and', 'or', 'if-then', to complex statements, the truth value of which depends on the truth values of the simple connected statements, the type of their connection and the definition of the junctors. Logic of Junctors is a synonym for the term 'statement logic' and emphasizes the importance of junctors for statement logic.

Logic: (from Greek *logike techne*, art of thinking). 1) In the broadest sense, the doctrine of consistent, logical thinking and reasoning, which is then referred to as 'logical'. 2) For example, in the expression 'logic of capitalism', an actual or supposed system of rules or laws that determine a particular area or context of action. 3) In the narrower sense of the philosophy of science, the formal logic of valid deduction from true premises to true conclusions (cf. Thiel, 2004).

Ontology, ontological: (from Greek *on*, being, entity, and *logos*, doctrine, science) since ancient Greek philosophy, the doctrine of being/entity and the first principles of being.

2.1 Semiotics

So far, with a view to a 'Logical Propaedeutic', rather the *Propaedeutic* of a very general introductory definition of science and scientificity has been provided. The following explanations are devoted to the '*Logical*' in the narrower sense, which in the context of traditional logic is concerned with the doctrine of concept, definition, judgment and conclusion, in modern logic with concept and definition, propositional logic and induction, deduction and conclusion rules, and to which we will turn in more detail accordingly.

With Seiffert (1996) 'Logical Propaedeutic' makes "conscious that we appropriate the world and all its objects through our language." And the "logical building blocks of any language are its concepts" and sentences, which are "composed of concepts" (Schurz, 2014, p. 66). Since science also presents its results linguistically in statements and even scientific (after)thinking and experimentation are also written or conveyed linguistically, the 'language as a starting point' (Poser, 2009) is to be taken below. Since language is also an expression form, "which is bound to sensually perceivable signs" with a particular "communicative function" (Poser, 2009, p. 29), a brief reference to 'semiotics' as "the doctrine of signs" is required—in contrast to natural signs (thunder as a sign of a storm, found pollen as a sign of previous vegetation forms) as artificial signs, "which are made into signs by humans by convention" (Zoglauer, 1998, p. 9), such as traffic signs or the legend on a geographical map.

Semiotics (from Greek *semeion*, sign, signal) is divided into *syntax* (from Greek *syn*, together; *taxis*, order) as grammar or usage theory of the signs in their relationship to each other, *semantics* (from Greek *sema*, sign) as the doctrine of the meaning of the signs, and *pragmatics* as "the doctrine of the use of signs in their context" (Zoglauer, 1998, p. 11), i.e. in their effect on a sign user in a communication or cooperation context. In scientific theory, semantics is mainly relevant (cf. Poser, 2009). Syntax cannot take into account meaning and truth; pragmatics has traditionally been ignored in scientific theory because scientific results are to be obtained and justified context- and person-independently, which led to numerous difficulties and problems and still does. Only the contextualization through pragmatic (see Sect. 3.5.1), historiographic (see Chap. 4) and (social) constructivist approaches (see Sect. 6.5) also take into account pragmatics.

Semantics has to do with the meaning of signs (such as words, concepts and terms). Words, concepts and terms can have different meanings in different linguistic contexts, depending on whether they are used in an object language or a metalanguage. After the 'linguistic turn' as the turning point to language and Language analysis (cf. Sect. 3.1.2) is "always what we can designate with a word of our language" (Seiffert, 1996, p. 30), a 'subject' or, more abstractly, an 'object'. These can be things like trees, stones, chairs, but also 'large stones', 'red' or 'green', 'space', 'time', 'truth', 'town planning', 'geography', 'philosophy of science', 'theory of truth', 'landscape', etc. It also makes no difference whether these words are used as nouns, adjectives, verbs or something else. Questions like whether, for example, 'space' exists 'in itself', i.e. independently of a possible observer, or is only constituted by an observer in the current observation (see Sect. 6.2), "which philosophers have been struggling with for millennia, we simply bypass when we say: An *object* is everything for which we have a *word* in our *language*" (Seiffert, 1996, p. 31). Within this 'linguistic paradigm' (Schnädelbach, 1991; on paradigms see Sect. 4.2) the question is: "What can I understand?" (Schnädelbach, 1991, p. 69) or what is the meaning of a linguistic expression. In this sense, one must proceed from the "non-transcendability of language" (Kamlah & Lorenzen, 1967, p. 24).

The *object* language in this sense deals with words in relation to objects or objects, as in the sentence "Tübingen is a city". *Metalinguistic* one could form the sentence "'Tübingen' has 8 letters" or "The sentence 'Tübingen is a city' is true". Further metalevels could be thought of, such as the metalanguage of truth theories (cf. Sect. 3.1.3) and the meta-meta-language of a philosophical metatheory. If these language levels are not distinguished, logical misunderstandings can occur, for example in the sentence "Tübingen has 8 letters". If 'Tübingen' is not made conspicuous as an object of metalinguistic thematization by quotation marks, the sentence is semantically misleading, if not meaningless. So is the sentence "This landscape is beautiful" semantically misleading because it is assumed that 'beautiful' is an object-level property of the object 'landscape' such as 'spacious', although it is a metalinguistic attribution such as 'truc' or 'just'. The correct sentence is therefore "This 'landscape' is beautiful" (in summary: Fig. 2.1).

2.1.1 Terms

In everyday life and especially in the sciences, we use terms in particular for communication (cf. Schurz, 2014). "Singular terms" refer as "proper names" to an individual object ("Tübingen", "Black Forest") or "characterize" it by means of pointing words ("indicators": "this city here", "this landscape here"). "General terms" include in particular "predicates" (also called "predicators") and "relations". "Predicates" designate characteristics of individuals, either as "property term" (the soil sample is "contaminated") or as "kind term" (this tree is a "beech"). "Relations" designate relationships between individuals (the Black Forest "is larger than" the Eifel). Another distinction is that between descriptive and prescriptive concepts (cf. Schurz, 2014). "Descriptive" terms are descriptive terms, in particular as empirical and theoretical terms. Empirical terms refer as "observation terms" to characteristics (the color of the soil sample is "dark gray", value 4.5 according to the Munsell color chart; Fig. 2.2 or as "disposition terms" to dispositions (the calcium carbonate in the soil sample is "soluble in water"). The meaning of theoretical terms is introduced by scientific theories (e.g. "native normal landscape" by social constructivism; cf. Sect. 6.5). Such scientific terms are also referred to as "terms" ("Termini") (Seiffert, 1996), they are "introduced" explicitly into the scientific language of a theory. *Prescriptive* terms are prescriptive terms, either as 'normative terms' ('is forbidden', 'is required') or as evaluative 'value terms' (a meadow is 'valuable' from an ecological perspective; a rapeseed field is 'beautiful'). Science theory is primarily concerned with descriptive terms that *describe* the 'world' or 'reality' (the meadow in the Bliesgau is a habitat for x 'animal species') but *not evaluate* it (the meadow in the Bliesgau 'is beautiful'; Fig. 2.3).

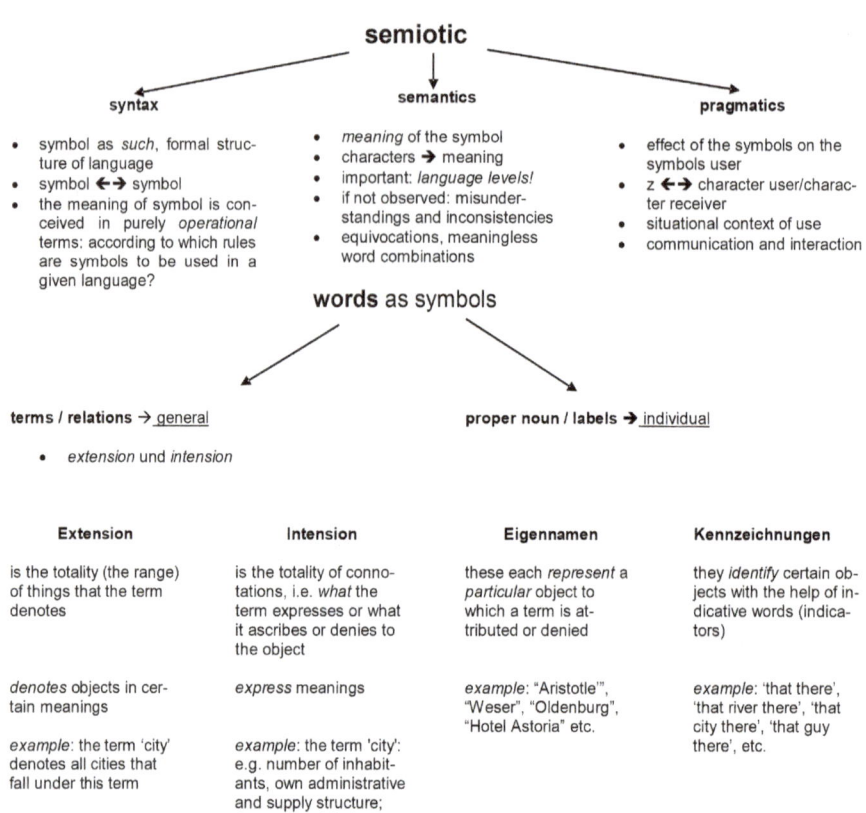

Fig. 2.1 Semiotic, syntax, semantics and pragmatics (own representation)

2.1.2 Definitions

Scientific terms must be defined (lat. *definitio*, delimitation, determination). This means that definitions delimit different meanings of words or concepts from each other. Everyday terms are often introduced and conveyed exemplarily by examples ('the tree in grandma's garden is an apple tree') or by showing ('that is a pear tree'), basically by proven and learned usage. Although the use of everyday terms is also implicitly (and mostly unconsciously) a regulated or standardized use, scientific terms ('termini') must be 'normalized', 'agreed' or 'specified' explicitly for scientific language usage (Seiffert, 1996), so that all scientists know what is meant. Although scientific terms can also be introduced exemplarily at first ("the Harz is a typical example of a 'stereotypical landscape'"), then a scientific definition is required within a theory (here, for example, social constructivism). Although definitions play an important role in the sciences and in the theory of

Fig. 2.2 The determination of soil color with the Munsell color chart can be understood—from a scientific perspective—as a comparison of a specific finding with an empirical term (standardized soil color code). (Photo: Timo Sedelmeier)

science as a meta-level to sciences, "a unified theory that represents all aspects of the concept of definition in a systematic context […] has not yet been developed" (Regenbogen & Meyer, 1998, pp. 136–137). This corresponds to the explanations in handbooks, introductions and other scientific publications on the term 'definition'. Only a few hints are therefore given.

Generally, a scientific definition introduces a (new, still unknown) term to be defined ('Definiendum') in terms of other terms ('Definiens') which are already known (Seiffert, 1996; Zoglauer, 1999). Definiendum and Definiens "can be exchanged in any context without changing the truth value of the sentence. Definiens and Definiendum must therefore be logically equivalent" and 'eliminable' or 'exchangeable' against each other ('substitution rule') (Zoglauer, 1999). Also, the Definiendum may not occur in the Definiens ('requirement of non-circularity') (Schurz, 2014), as in the sentence "Freedom is to be free from constraints"—the term to be defined (freedom) is thus defined in the Definiens (to be free) with the term to be defined. In this sense, the following definition is also circular: "Geography studies geographic spaces" (Duden, 2012, p. 24). The term to be defined (geography) is defined in the Definiens (study of geographic spaces) with the term to be defined.

Fig. 2.3 A view of the Bliesgau, Saarland. (Photo: Olaf Kühne)

In colloquial usage (cf. Zoglauer, 1999), but also in traditional definition theory (Kutschera & Breitkopf, 2007; Pfister, 2019), the definition of a term is given by specifying the next higher general term or the next higher genus (lat. *genus proximum*) and a species-forming difference (lat. *differentia specifica*) which highlights the specific of the term to be defined in contrast to the specifics of other terms (lat. '*Definitio fit per genus proximum et differentiam specificam.*') (cf. Aristotle, 2004 [367-344 BC]; VI, 1) A classic example is Aristotle's definition (Aristotle, 2009, 1253a) of man as a 'rational' (differentia specifica) 'living being' (*genus proximum*). If such a definition is understood in the sense of 'real' or 'essentialist' in the sense that the 'essence' of a thing or an 'object' is thus grasped (see Sect. 3.2), one speaks of 'real definition' (Pfister, 2019; Zoglauer, 1999). A real definition can be understood as an attempt to describe the (often implicitly known) meaning of an already known expression or concept in other words (Seiffert, 1996), which is why they are also called 'lexical definitions' in reference to dictionaries, in which this analysis of an *existing* language usage is carried out. Real definitions are therefore only possible and meaningful in everyday language, while the *scientific* definition explains "basically unknown by already known words" (Seiffert, 1996, p. 66). Such definitions are called 'nominal definitions' or 'stipulative definitions' (lat. *stipulatio*, agreement), they *set* or *legally* establish a *new* language usage 'per definitionem' *conventionally*.

2.1.3 Concept Analysis and Concept Explanation

In a concept *analysis* (cf. Kutschera & Breitkopf, 2007; cf. Pfister, 2019) the concept to be analyzed ('Analysandum') already has a certain meaning which is distinguished from other meanings by itself and in the "ideal case" leads to a "complete definition" (Pfister, 2019, p. 69). The totality of the analyzing concepts is called '*Analysans*'. A concept analysis of 'landscape', which is carried out as a 'linguistic analysis' (Kutschera & Breitkopf, 2007, p. 150), would accordingly have to research the language usage diachronically as well as synchronously, as for example Gerhard Hard (1969, 1970, 1977), Dorothea Hokema (2009, 2013) and Olaf Kühne (Kühne, 2013, 2018; cf. also Berr & Kühne, 2020) have carried out.

In a concept *explication* (lat. *explicare*, carry out, discuss) (cf. Carnap, 1962) a term that is used colloquially 'unclear' (Poser, 2009) or vaguely ('Explicandum') "for its use in a precise theory is determined more precisely in its meaning" (Kutschera & Breitkopf, 2007, p. 150), after previously the "different uses of language" (Poser, 2009, p. 38) were 'collected'. Since in a concept explication an imprecise term is to be replaced or transformed by a more precise term ('Explicat'), one can also "speak of a 'replacing analysis' or 'transforming analysis'" (Pfister, 2019, p. 77). For example, the term 'landscape' can be subjected to a concept analysis in the sense of a concept explication described above in order to supplement a 'fact component' in view of the purposes of theory formation in a scientific theory with a 'determination component' that is distinct from and more precise than this in order to introduce this theory. In this way, imprecise features of a concept can be analytically separated and precisely distinguished and determined for the purposes of the theory. For example, 'landscape' can be broken down into four dimensions, the 'social', the 'individually actualized social', the 'acquired physical' landscape and the 'external space' (Kühne, 2018, pp. 55–69)—namely within the framework of a social constructivist approach with the aim or the "task of providing an overview of essential strands of landscape research" (Kühne, 2018, p. 3).

2.1.4 Intension and Extension

Terms can be distinguished by intension (lat. *intendere*, to strain, to be mindful of) or content and extension (lat. *extendere*, to extend) or scope. Intension as the content of a term denotes the meaning of the term and corresponds to the totality of the characteristics or connotations (lat. *con-*, with-, together, lat. *notatio*, annotation) that apply to the term and determine it. Extension as the scope of a term denotes the totality of all objects or denotates (lat. *denotare*, to designate) to which all characteristics or connotations of the term apply. General terms refer with their extension to a class or set of objects that fall under the term, such as 'geographer' to the geographers Carl Sauer, Winfried Schenk or Gerhard Hard. Proper names refer with their extension to a specific object that is designated by the name, such as 'Carl Sauer' to the geographer Carl Sauer.

The content and extent of a concept are inversely related to each other: The greater or more connotative the content, the smaller the extent, the smaller or less connotative the content, the greater the extent. That is, the more general (extensive) a concept is, the more content-poor or abstract it is. The general concept of 'space' is so abstract that it can refer to very many things, but it is all the more content-poor or meaningless because it can hardly name any concrete features, which is why very different scientific understandings of it have developed, which we will set out in Chap. 6. In philosophy, 'Being' is the most general, but also the most content-poor concept. Everything and nothing is meant by it. The singular concept and proper name 'University of Tübingen', on the other hand, only denotes one object, so it is limited in scope, but rich in content or meaning, because it is associated with very many connotations. Only one object in space is denoted, but at the same time many meanings are connoted. The traditional saying for this wealth of individual 'objects' is '*Individuum est ineffabile*' ('the individual is ineffable') (see Oeing-Hanhoff et al., 2019). However, the abstraction associated with scientific concepts is necessary because concepts stand as general concepts for many 'cases' or 'objects' that fall under these concepts. This is only possible if one 'abstracts' from the innumerable number of possible features, properties or dispositions, if one 'abstracts' a *common* combination of features (Textbox 3).

Textbox 3: Generalization: Generalizing, Typing and Modeling using the Example of Geography

In geography and generally in the sciences, three specific variants of conceptual generalization are common and can be distinguished (cf. for the following and for the examples: Duden, 2012, pp. 27–30): Generalizing, Typing and Modeling. In each case, certain characteristics are abstracted either by omitting or highlighting them.

Generalization (from lat. *generalis*, concerning the genus): Certain characteristics of an individual object are *omitted*, its possible location relationships (in the following example: Germany) remain. As an example, the 'industrial production of Germany' can be mentioned: "The industrial production is based on electronics, mechanical engineering, vehicle construction, the chemical industry as well as the food and beverage industry. These industrial sectors bind together around 60% of industrial jobs and generate the same percentage of industrial sales." It is, for example, omitted to state which electronic products are produced. Too much information would make it difficult to recognize what is common.

Typing (from gr. *typos*, imprint, pattern): Common characteristics of several individual objects are *highlighted* as 'patterns', possible location relationships are omitted. For example, the term 'steep coast' can be defined as a 'type' by disregarding specific location relationships and by specifying common 'typical' characteristics such as "steep, sometimes vertical walls (cliff); relatively solid rock; surf hollow; narrow block beach".

Models (from Latin *modulus*, measure, scale): They represent two- or three-dimensional visualizations of the generalizations mentioned and thus a greatly simplified, but intuitive image of reality. It only agrees with the original in characteristic features. Maps also represent models of the world (Fig. 2.4). ◄

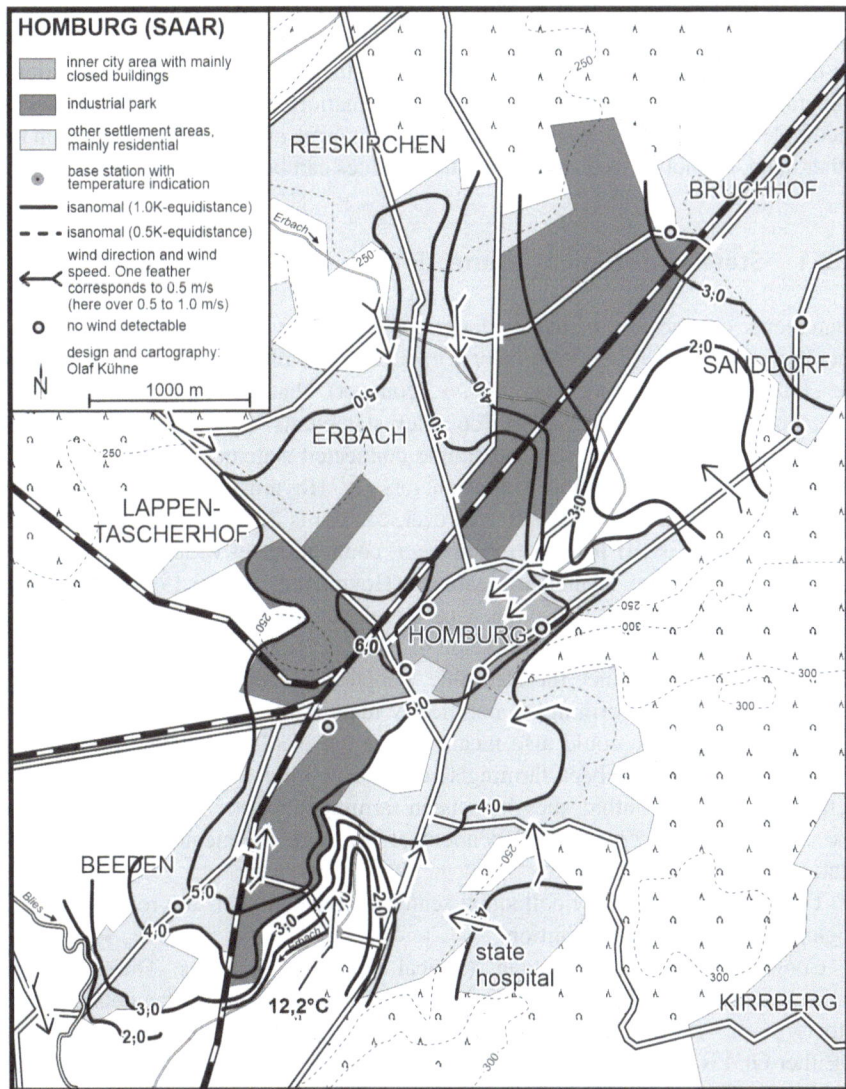

Fig. 2.4 A widely used form of representation in geography is the map, a model of spatial phenomena, in which relevant aspects of the topic are represented in the form of signatures. The example shows the distribution of the equivalent temperature (i.e. the total heat content of the air) in the night of 11 to 12 August 1997 (after: Kühne, 1999)

2.2   Statements

Statements are sentences that refer to facts (Seiffert, 1996, pp. 76–77). In contrast
to descriptive (lat. *describere*, describe) statements, "which state that something
so and so *is*" and describe it accordingly, statements, "which state that something
should be or more generally, so and so is to be evaluated" (Poser, 2009, p. 33),
prescriptive, that is, normative (lat. *norma*, angle measure, guideline, regulation)
or prescriptive (lat. *praescribere*, prescribe). Descriptive statements can therefore
be true or false, truth values can be ascribed to them "true" or "false". Questions,
imperatives, wishes, warnings, requests, exclamations, future-oriented sentences,
incomplete or meaningless sentences as well as norms and value judgments, on the
other hand, cannot be true or false, no truth values can be ascribed to them.

2.2.1   Statement-logical Connections

Statements can be *formalized* by letters as *statement variables* (p, q, etc.) and
thereby *transformed into statement forms*. On the other hand, statements can
be *connected* by junctors (lat. *jungere*, connect), that is, sentence connections
like 'not', 'and', 'or', 'if-then', to complex statements, the truth value of which
depends on the truth values of the simple connected statements, the type of their
connection and the definition of the junctors (cf. Hoyningen-Huene, 1998; Leerhoff et al., 2010; Zoglauer, 1999). However, caution is advised, as this 'truth func-
tionality' only exists if it is an '*extensional*' connection, in which the truth value
is independent of the 'sentence meaning' (Hoyningen-Huene, 1998). For example, if a colloquial 'and' is used as 'and therefore' and thus as 'because', it is an
'*intensional*' connection, the truth value of which depends on the 'sentence meaning'. The statement 'Since the Neolithic, people have settled *and* since the Neolithic they have been farming' is not clearly identifiable as an 'And-Connection',
because this sentence could also mean: 'Since the Neolithic, people have settled
and therefore they have been farming since the Neolithic' or '*Because* people have
settled since the Neolithic, they have been farming since the Neolithic'. But then
the connection is no longer independent of the sentence meaning of the whole
statement.

This ambiguity of many colloquial sentence connections is the reason for their
logical clarification and definition.

Conjunction: The logical 'and' (logical notation: ∧) means: The compound
statement connected by 'and' 'p ∧ q' is then true if both sub-statements are true,
otherwise it is false.*Disjunction*: The colloquial *exclusive* 'Or-Connection'
('Either-Or') is junctionally logical (logical notation: ∨) in the sense of a 'Disjunction' or more precisely an 'Adjunction' (p ∨ q) defined as an *inclusive* 'Or-Connection' so that the compound statement is only false if both sub-statements are false,
otherwise it is true. An *exclusive* 'Or' is called '*Contravalence*' in junctional logic

and, depending on the linguistic context, can also be used in the same way: "The choice of junctional operators used on a formal level is therefore not predetermined by any higher laws, but rather (also) depends on pragmatic considerations" (Leerhoff et al., 2010, p. 19).*Implication*: The colloquial 'if-then' (logical notation: →) in the sense of an implication (p → q) is then false if a false sub-statement is to be concluded from a true sub-statement, otherwise it is true. The implication contradicts the colloquial understanding most clearly, because from a false statement (antecedent) a true statement (consequent) can be concluded and the statement connection is still true. The philosophical tradition of the Middle Ages formulated this fact in the formula '*ex falso sequitur quodlibet*' ('from something false anything can follow'). Conversely, it applies: '*verum sequitur ex quolibet*' ('Truth can be concluded from anything'). In colloquial usage, a causal relationship is implied in the implication, which is not intended in its logical definition (Bucher, 1998). Depending on the linguistic context, therefore, other defined junctional logical conditionals (if-then connections) can be used for the implication. What is more important for understanding the logically defined implication is the following insight: "The deepest reason for the logical thinker's strict adherence to the definition of his implication lies in the usefulness of the logical derivation concept, which is defined so that nothing false can follow from something true" (Bucher, 1998, p. 71).

The *negation* (logical notation: ¬) is the logical specification of the colloquial negation and assigns the opposite truth value to a statement (¬p).

The junctors are usually *defined* by truth tables. In the first two columns, all possible combinations of the two truth values true (w) and false (f) are assigned to the sub-statements p and q. For each possible combination, it is specified (defined) in the other columns which truth value results from the respective combination of the two sub-statements (p and q) and the truth value assignment (w and f) (Tab. 2.1):

Since *negation* only refers to one statement each, the truth table looks different (Tab. 2.2):

The following table shows with *examples* these 4 important statement links (Tab. 2.3):

Since the *negation* only refers to each statement, the example table looks different (Tab. 2.4):

Tab. 2.1 Truth table

p	q	p ∧ q Conjunction	p ∨ q Disjunction	p → q Implication
w	w	w	w	w
w	f	f	w	f
f	w	f	w	w
f	f	f	f	w

Tab. 2.2 Truth table with one statement

p	$\neg$p
w	f
f	w

Tab. 2.3 Conjunction, disjunction and implication

	Conjunction	Disjunction	Implication
First statement	The earth is a planet	The earth is a sun	The earth is a planet
Second statement	The moon is a planet	The moon is a sun	The sun revolves around the earth
Compound statement	Earth and moon are planets	Either the Earth is a sun or the moon is a sun	If the Earth is a planet, then the sun revolves around the Earth
The decisive criterion of truth or falsity of the composite statement	Only if both sub-statements are true, the composite statement is also true	Only if both sub-statements are false, the composite statement is false	Only if a false sub-statement follows from a true sub-statement, the composite statement is false

Tab. 2.4 Negation

	Negation
Statement	The Earth is a planet
Negation of statement	The Earth is *not* a planet

2.2.2 Contradiction, Tautology, Contingency

Some statement forms each have a peculiarity that is to be observed.

A *contradiction* (from Latin *contra*, against, *dicere*, to say) exists when one statement claims a state of affairs and the other denies the same state of affairs—in logical notation: p ∧ ¬p. Aristotle had already formulated the principle of (illegal) contradiction: "for it is not possible that the same thing belong and not belong at the same time to the same thing in the same respect … For it is impossible that someone should take the same thing to be and not to be" (Aristotle, 1991 [348-345 BC], 1005b). A contradiction is therefore always false. However, the falsity only refers to the *relationship* of the two statements, "which says nothing about the factual truth or falsity of one or the other statement. So it is quite possible that two false statements contradict each other" (Hoyningen-Huene, 1998, p. 143)—for example: 'The Black Forest is a mountain range in Africa' and 'The Black Forest is a mountain range in Asia'. Both statements are false and they contradict each other. This is an example of a *contrary* contradiction, in which neither of the two statements can be true at the same time, but both can be false at the same time. The *contradictory* contradiction is "stronger" and corresponds to Aristotle's definition. Here it is *neither* possible that both statements be true *nor* that both be false at the same time: 'The Alps are in Europe' and 'The Alps are not in Europe'. In each case, one statement is the negation of the other (cf. Hoyningen-Huene, 1998).*Tautologies* (gr. *tauto*, the same, *logos*, the word) are statement forms that are always true, "no matter what specific situation may be present. Therefore, the person who utters a tautology does not say something false, but at the same time does not convey any information" (Bucher, 1998, p. 85). For example, the statement conjunction 'Geography is a natural science or it is not' is trivially true (in logical notation: p ∨ ¬p). However, some tautologies may not be immediately recognizable, as in "Hans is sleeping or awake" (Bucher, 1998, p. 86). With very complex statement conjunctions in equally complex arguments, tautologies are often difficult to recognize, so that the information emptiness is overlooked. Only with the help of truth tables, in which the truth values are tabularly assigned to each statement and each conjunction up to the truth value of the overall statement and 'calculated' can be, such tautologies can be traced. This also applies to the contradictions and fallacies already mentioned (Sect. 2.4).A *contingent* (lat. *contingit*, it happens; mittellat. *contingentia*, possibility, chance) statement form can be both true and false, they refer to what is neither necessary nor impossible. Examples are statement forms like 'p'; '¬p'; 'p ∨ q'; 'p ∧ q'; 'p → q'. Contingent statements therefore relate to facts that are *possible* or *accidental* as stated and could therefore be different. The statement 'Mrs. Müller has German citizenship' is contingent, because Mrs. Müller could also have been born in Iraq and have Iraqi citizenship.A truth table may also be used for these special cases of statement forms. Note: Here the junctional logical definitions of disjunction, conjunction and negation apply (Tab. 2.5):

So far, it has been assumed as a matter of course, but not explained in detail, that statements can only be attributed the two truth values 'true' or 'false'. In logic, this attribution of the two truth values is called 'two-valued logic'. In everyday language, this two-valuedness is intuitively clear, in logic it is in need of justification (see Textbox 4).

Tab.2.5 Special cases of statement forms

Statement	Statement	Tautology	Contradiction	Contingency
p	¬p	Example: **p ∨ ¬p**	Example: **p ∧ ¬p**	Example: **¬p**
w	f	w	f	f
f	w	w	f	w
		Always true!	Always false!	Both true and false!

Textbox 4: Two-valued Logic

Aristotle demanded in his "Metaphysics" (1991 [348-345 BC] as "First Philosophy" (gr. *prote philosophia*) the investigation of the axioms (gr. *axiomata*) and principles (gr. *archai*), which are always presupposed by individual sciences with regard to their respective subject area, but which cannot be investigated by these themselves. But since there are axioms and principles that affect all possible subject areas, they would remain unexplored if not the "First Philosophy" took on this task. Aristotle asks whether there are axioms or even a supreme axiom that lies at the bottom of all objects, affects all individual sciences and is the "safest" of all. This "safest" principle or axiom is in the language of contemporary logic the "*theorem of the excluded contradiction*" which is present in an ontological and in a logical formulation at Aristotle:

Ontological: "It is not possible that the same thing should at the same time belong and not belong to the same thing in the same respect" (Aristotle, 1991 [348-345 BC], 1005b).

Logical: "For it is impossible that someone should think that the same thing is and is not" (Aristotle, 1991 [348-345 BC], 1005b).

This means that Aristotle formulates both a logical principle that applies to statements (sentences), and an ontological principle that applies to what is stated (to objects). Aristotle emphasizes that it is a sign of 'lack of education' to demand a direct proof for this axiom, as this would inevitably lead to an infinite regress. However, an *indirect* proof is possible insofar as the rejection of this principle makes any communication impossible: Whoever says something also always says the opposite of what he says. In logical notation of statements, the principle of contradiction therefore reads: ¬ (p ∧ ¬ p) (It is false that something is both and not—or that something is both true and false).

The '*principle of the excluded third*' is inextricably linked to this principle of the excluded contradiction. In ontological as well as logical meaning, it states: There is neither a third truth value nor is there anything third in addition to being and non-being (Aristotle, 1991 [348-345 BC], 1011b-1012a): *Tertium non datur* (a third is not given). In logical notation of statements, this principle therefore reads: p ∨ ¬ p (Either something is or it is not). ◄

2.3   Arguments

Descriptive statements refer to facts. It is natural to assume that such statements are justified by the fact that they are checked against the empirical "reality". The sentence "The Harz is a low mountain range" is then true ("verified": lat. *verifi-*

care, to verify), if the Harz is actually a low mountain range. However, there are also many statements that are justified by other statements—the justifying statement is called "premise" (lat. *praemittere* = to send ahead), the justified statement is called "conclusion" (lat. *concludere* = to conclude, to infer). This type of linking of premise and conclusion is called "argument" (lat. *argumentum* = proof, evidence). By an argument "one wants to convince others or oneself of the truth of a certain statement by *deriving the truth of this statement from other statements* of which one is already convinced" (Tetens, 2014, p. 23). The reason for arguing instead of an empirical proof is that the "conclusion appears to be questionable, but one cannot verify or refute it directly, for example by observations" (Tetens, 2014, p. 24). This function of arguments, to attribute the truth of the conclusion to the truth of the premises, is "the fundamental function of arguments" (Tetens, 2014, p. 36). In addition, there is a second function, namely to "demonstrate that the conclusion follows logically from the premises. A sentence can follow logically from quite different sentences. Often we are surprised which sentences lead to a certain other sentence" (Tetens, 2014, p. 36). For example, the following two sentences, which are largely unproblematic in themselves and independent of their possible embedding in an argument, could lead to a surprising conclusion in the context of an argument: 1) Doing sports is healthy, 2) Boxing is a sport: Conclusion: 3) Boxing is healthy.

2.3.1 Basic Components of an Argument

Since from the premises the conclusion is *drawn*, in logic also of a 'conclusion' is spoken. The premises *support* the conclusion, because the truth of the premises guarantees the truth of the conclusion. Premises and conclusion are *relative* expressions. That means, the same statement can be either a premise or a conclusion depending on the context and argument. The conclusion of an argument can serve as a premise in another argument, a premise in an argument can be a conclusion in another argument. Arguments can, in contrast to premises and conclusions, not be true or false, they can only be valid/invalid or sound/not sound. We will come back to these two distinctions.

The form of an argument can be represented schematically as follows (cf. Zoglauer, 1999, p. 57):

$$
\begin{array}{ll}
\text{Premises:} & p_1 \\
& p_2 \\
& ..... \\
& p_n \\
\hline
\text{Conclusion:} & q
\end{array}
$$

An example (cf. Duden, 2012, pp. 26–27):

Premise 1: The recognition of certain contents of geographic spaces is linked to certain scale ranges.
Premise 2: Geographic spaces have a ranking, a hierarchy.
Premise 3: Taking into account this hierarchy, spatial order and spatial structure are recognized.
Conclusion: Geographic thinking therefore includes the different scale ranges and the hierarchy of geographic phenomena.

Many arguments, especially those dealing with the theory of science, are "descriptive arguments" in which descriptive statements occur that can be true or false. But there are also "normative arguments" in which normative statements (commands, prohibitions) occur that can not be true or false, but can be "right" or "wrong" according to an ethical criterion (Tetens, 2014). With a few exceptions, this overview deals with descriptive arguments.

In arguments and argumentations *theses* play an important or decisive role. Theses are easily confused with hypotheses, they are therefore to be distinguished (Textbox 5).

Textbox 5: Theses and Hypotheses

A thesis is a claim that is to be confirmed or justified by an argumentation in its claim to validity. In the argument structure of premises and a conclusion, the thesis is the conclusion confirmed by the premises in their claim to validity in a logically and factually correct argument. Theses are introduced by typical sentence starters such as:

- 'My thesis is that …'
- 'In my opinion, it is the case that …'
- 'I claim that …'

The use of these typical formulations is not only to be observed in practice, but also recommended. Because a difficulty in argumentation can consist in distinguishing such claims from *judgments* which refer to accepted facts in a descriptive or constative way (from Latin *constare*, to stand, to be) and are not necessarily in need of justification. For example, there is a difference between stating that a soil sample comes from a certain soil horizon, or claiming that it is to be assigned to a specific soil type.

Hypotheses play a decisive role in scientific practice. They are assumptions within and under the assumption of a theory that are formulated tentatively and subject to their correctness. They have to be subjected to a test in experiments or other test arrangements, but also in scientific discourse. In empirical (over) testing procedures, they can be subjected to a corresponding procedure and either (provisionally and subject to) confirmed or refuted. The status of hypotheses, hypothesis formation and hypothesis confirmation or-refutation, which is evaluated differently in tradition and in various scientific theories, is a central topic of the following Chaps. 3 to 5. ◄

Another distinction is that between inductive (from Latin *inducere*, (in) introduce, tempt), deductive (from Latin *deducere*, lead away, derive) and abductive (from Latin abducere, lead away) conclusions (arguments).

2.3.2 Inductive and Deductive Conclusions

'*Inductive* inferences' are "inferences from the particular (individual cases) to the general" and they are "information-expanding" (Zoglauer, 1999, p. 57), because the information content of the conclusion exceeds that of the premises (see also Sect. 3.6). Inductive inferences play an important role in *hypothesis formation*, which occurs in two variants: as a 'forecast' and as a 'generalization' (Zoglauer, 1999). For example, air temperatures have been measured systematically for over two hundred years and it has been found that the average temperature has increased statistically, especially in recent decades. From this, the *forecast* is inductively derived that in the future temperatures will continue to rise. An example of a *generalization* would be the following conclusion: 'The average temperature in Central Europe has risen. The average temperatures in North America, South America and Asia have also risen. Therefore, the average temperature *worldwide* has risen.' In both cases, that of the forecast and that of the generalization, the information content of the conclusion (*forecast*: from n events to a following event; *generalization*: from n events to an underlying rule or law) is more comprehensive than that of the premises (statements about observational data). The knowledge derived from inductive inferences is not logically necessary or compelling, it only has a certain (statistical) *probability*.

'*Deductive* inferences' are "inferences from the general to the particular" and they are "information-preserving" (Zoglauer, 1999, p. 58), because the information content of the premises is transferred to the conclusion and reformulated in another statement (the conclusion). In other words, "the conclusion does not add any new information that would not already be contained in the premises. Deductive inferences are logically compelling" (Zoglauer, 1999, p. 58). Arguments that, as we have explained, pass the truth of the premises on to the conclusion, are deductive inferences in this sense. However, the point of arguing is that this compelling logic of inference does not depend on the *content* (the semantics) of the premises, that is, not on the actual truth of the premises, but only on their logical-syntactic *form*. This is irritating, as the truth of the premises was previously emphasized as guaranteeing the truth of the conclusion. But for the correctness of an argument, it is enough to *initially* see that "the conclusion must be true if the premises are true, without knowing yet whether the conclusion is true" (Tetens,

2014, pp. 24–25). It all depends on the *form* of the argument. Aristotle was the first to discover that arguments have typical forms that always lead to correct conclusions, regardless of the specific content. Such an argument could look like this:

Premise 1: All philosophers are crazy.
Premise 2: Some geography professors are philosophers.
Conclusion: So some geography professors are crazy.

This form of argument—it is a typical conclusion form of Aristotelian syllogism: the 'mode of Darii' (cf. Zoglauer, 1999)—'works' always, for example, if the following content is taken:

Premise 1: All fir forests in Central Europe are endangered by rising temperatures.
Premise 2: The 'Little Citizen Bush' in Oldenburg is mostly a fir forest.
Conclusion: So the 'Little Citizen Bush' is endangered by rising temperatures.

Another well-known and important argument form is the '*modus ponens*' (lat. *modus*, here: conclusion, lat. *ponere*, set: 'concluding set')—in logical notation: $p \rightarrow q$; p; q. We will introduce examples below, first a theoretical derivation and explanation. The first premise is the implication already known to us from its junctional definition. Based on this, the definition of the *modus ponens* can be represented in a truth table as follows (Tab. 2.6):

The *validity* of *modus ponens* as an argument form is shown by the fact that in a distribution in which the two premises are assigned the truth value 'true', the conclusion also has the truth value 'true' *and no* distribution is such that the two premises are assigned the truth value 'true' but the conclusion has the truth value 'false'. The *invalidity* of an argument form can be shown correspondingly by the fact that in a distribution in which the two premises are assigned the truth value 'true', the conclusion has the truth value 'false'—*even if* there is a series with a distribution in which the two premises are assigned the truth value 'true' and the conclusion also has the truth value 'true' (cf. Salmon, 1983).

There is another method of proving the validity (or invalidity) of an argument form: by converting the argument form into a statement form. Here it applies: Every valid argument corresponds to a tautological statement form (which is

Tab. 2.6 Truth table of the modus ponens

Sub-statement	Sub-statement	Premise 1	Premise 2	Conclusion
p	q	$p \rightarrow q$	p	q
w	w	w	w	w
w	f	f	w	f
f	w	w	f	w
f	f	w	f	f

Tab. 2.7 Truth table: The argument form of 'modus ponens' as a tautological statement form.

Conjunct 2 Implication		Conjunct 1 Implication	Antecedent (Conjunction)	Consequence (Conclusion)	**Tautological** Statement Form
p	q	p → q	(p → q) ∧ p	q	[(p → q) ∧ p] → q
w	w	w	w	w	w
w	f	f	f	f	w
f	w	w	f	w	w
f	f	w	f	F	w

always true). The statement form then consists of the conjunction of the premises of the argument form as antecedent (lat. *antecedens*, the preceding, cause) and the conclusion of the argument form as consequence (lat. *consequi*, follow, achieve) of an implication as a statement form: [(p → q) ∧ p] → q. This can also be shown using a truth table (Tab. 2.7):An example of *modus ponens* is the following:

Premise 1: When it rains, the road gets wet.	p → q
Premise 2: It's raining.	p
Conclusion: So the road gets wet.	q

This form 'works' also with any content:

Premise 1: If the global average temperature continues to rise, severe Damage occurs in ecosystems.	p → q
Premise 2: The global average temperature continues to rise.	p
Conclusion: So serious damage will occur in ecosystems.	q

2.3.3 The Hempel-Oppenheim Schema

A prominent application in the practice of science and in scientific theory of the '*modus ponens*' as a deductive conclusion form is the so-called '*Hempel-Oppenheim schema*' (HO-schema), which is also referred to as the 'Deductive-nomological model'. Popper first described the basic structure of this model in 1935 in the 'Logic of Research' (Popper, 2002), without referring to it as such. In 1948, Carl Gustav Hempel and Paul Oppenheim worked out the details of this model, which was later named after them and entered into the history of science (Hempel & Oppenheim, 1948) and further developed by Hempel (1965). It is a model of scientific explanation, namely in response to 'why-questions' seeking causes (Poser, 2009). Hempel and Oppenheim (1948, p. 135) formulate this as follows: "To explain the phenomena in the world of our experience, to answer the question "why?" rather than only the question "what?", is one of the foremost objectives of all rational inquiry". Science is more than just collecting facts, it also has to provide explanations for observed phenomena. However, the concept of explanation itself was in need of explanation or at least explication, so the HO-schema can also be understood as the 'explicate of the concept of explanation' (Poser, 2009). Hempel and Oppenheim now claim that *every* scientific explanation can be traced back to this schema. The schema consists of the *explanans* (lat. *explanare*, to explain) as the explaining, which consists of general hypotheses or laws of nature (premise 1) and of singular sentences as empirical or factual initial conditions (premise 2). From this, the empirical state of affairs to be explained as an 'event statement', that is the *explanandum* as the to be explained, can be deduced purely logically:

$G_1, G_2 \ldots G_n$	Law statements	$p \rightarrow q$
$A_1, A_2 \ldots A_n$	Initial conditions	p
--		
E	Explanandum	q

This model of scientific explanation is fraught with a number of 'adequacy conditions' (Poser, 2009) as well as numerous logical and scientific-theoretical difficulties, which are not to be discussed here in detail (cf. u. v. Bauberger, 2016; Poser, 2009; Schurz, 2014). Two problems should, however, be mentioned briefly.

In the humanities (philosophy, German studies, history, etc.), explanations are given that are based on individual or collective preconceptions and expressed in actions. These explanations cannot be given with the HO schema. Furthermore, *every* (whether natural or humanities) explanation is never finally conclusive, but always based on a previously assumed explanation or interpretation background or on scientific 'preconceptions' (cf. Poser, 2009) as fundamental and commonly shared perspectives, which Ludwik Fleck 'styles of thought' (Fleck, 1980 [1935]) and Thomas S. Kuhn 'paradigms' (Kuhn, 1976) are called (cf. Sects. 5.2 and 4.2). This "connecting horizon of world view" (Poser, 2009, p. 212) is no longer accessible with the methods of the empirical sciences and thus with the HO schema itself, but requires another method, which we will discuss in the section on hermeneutics (Sect. 3.5.2).

The second difficulty is: One of the adequacy conditions requires that the sentences from which the explanans consists must be true (cf. Poser, 2009). Again, it is irritating that, on the one hand, only the logical form of the conclusion is to be responsible for its correctness, independently of its contents, and on the other hand it is now required that the contents of the premises be true. However, the correctness of the argument only requires that, *if* the premises are true, then the conclusion is also true. This specific formal peculiarity of arguments or conclusions is most often referred to as 'validity': "An argument is *valid* exactly when it is *actually* rational to hold the conclusion to be true, if the premises are true" (Beckermann, 2014, p. 21; cf. Pfister, 2019). This means that it is absolutely necessary to "distinguish between the truth of statements in an argument on the one hand and the quality of the transition from premises to conclusion on the other hand [to] take into account" (Pfister, 2019, p. 23).

However, *scientific* conclusions require that the conclusion is *true*. Because even if the conclusion is formally valid, it could transfer the falsity of one of the premises to the conclusion. For example, one could argue:

Geography is a natural science.
All natural sciences use only natural scientific methods.
Therefore, all geographers use only natural scientific methods.

The conclusion is obviously not "in order", because geography also uses social science methods. The reason for the falsity of the conclusion lies in the *first premise*, because geography is a *mixture* of natural and social sciences (cf. Kühne, 2008). This error is transferred to the *formally valid* conclusion about the conclusion. In order for a scientific conclusion to be not only logically formally valid and correct, but also to lead to a true conclusion, it must be ensured in addition to its validity that it is concluded from *true* premises to a true conclusion. Most authors call this additional requirement "coherence" or "soundness": "An argument is *coherent* if and only if it is valid and all of its premises are true" (Beckermann, 2014, p. 22) or: "An argument is sound if it is valid and contains only true premises" (Pfister, 2019, p. 26)—and thus the truth of the premises logically forces the conclusion to the conclusion.

2.3.4 The Abductive Conclusion

In addition to induction and deduction, the American philosopher Charles Sanders Peirce (1839–1914) introduced a new mode of reasoning: *abduction* (from Latin *abducere*, to lead away). Unlike deduction, which infers from a rule and a case to the result (as in the HO schema), and induction, which infers from a case and a result to the rule, abduction infers from a 'surprising fact' (the 'result' in deduction and induction) and a new rule that is assumed to be 'true' and plausible, to the case. Peirce describes this process as follows:

"The surprising fact C is observed; but if A were true, But if A were true, C would be a matter of course, Hence, there is reason to suspect that A is true" (Peirce, 1991, p. 129).

What is decisive is that the assumption ('A') is an 'as-if' assumption that pretends *as if* the assumption is correct and is presented as a *new rule* that did not exist before. Because if it had existed before, fact 'C' would not have been surprising. Abduction is therefore to be characterized by the search for new rules. It tries to suggest something by means of explanatory hypotheses that may be possible. In Peirce's words:

"Abduction is the process of forming an explanatory hypothesis. It is the only logical operation which introduces any new idea; for induction does nothing but determine a value, and deduction merely evolves the necessary consequences of a pure hypothesis."
    "Deduction proves that something *must* be; induction shows that something *actually* is operative; abduction merely suggests that something *may* be."
    [...] "and it needs no reason, since it merely offers suggestions" (Peirce, 1991, p. 115).

Tab. 2.8 Deduction, induction and abduction, a summary

Deduction	Induction	Abduction
(A) Rule: All beans from this bag are white **(B) Case**: These beans are from this bag **(C) Result**: These beans are white	**(B) Case**: These beans are from this bag **(C) Result**: These beans are white **(A) Rule**: All beans from this bag are white	**(C) Result**: These beans are white **(A) Rule**: All beans from this bag are white **(B) Case**: These beans are from this bag
	another example	
Rule: All mice are gray **Case**: 'Mausi' is a mouse **Result**: 'Mausi' is gray	**Case**: 'Mausi' is a mouse **Result**: 'Mausi' is gray **Rule**: All mice are gray	**Result**: 'Mausi' is gray **Rule**: All mice are gray **Case**: 'Mausi' is a mouse
• Conclusion from the general to the particular • Logically compelling conclusion • Information and truth preserving	• Conclusion from the particular (observable regularity) to the general • Probabilistic conclusion • Information and truth expanding	• Conclusion from a singular individual case and a hypothetical rule to an individual case • Possibility conclusion • Information and truth expanding
Logically unproblematic. A mistake could only concern the factual content if, for example, one of the two premises were false	Problem: The conclusion is *logically* invalid, only probabilities can be given (see Sect. 3.3)	Problem: Even grey herons, wolves and raccoons are grey. The conclusion can only be *accidentally* true

Because of this hypotheticity, abduction is also referred to as "inference to the best explanation" (Bartelborth, 1996).

The abduction with the finding of a new rule or hypothesis is the first step in a three-stage 'logic of research'. In a second step, research-leading predictions are deduced from the explanatory hypothesis of the first step. In the third step, it is about the inductive search for observational data that can confirm the hypothesis. This three-stage research logic of abduction, deduction and induction is repeated until hypothesis and observational data match. The abduction is therefore not an algorithm for generating true knowledge, but a heuristic, a search procedure in the sense of an '*ars inveniendi*' (art of finding or inventing). Due to the fact that the 'new' as a 'new rule' is not algorithmically, that is, to (re)find in a rule-recursive way, the abduction also comes close to creativity research (see Sect. 4.5)

Peirce has illustrated the differences between the three syllogistic forms using the famous 'bean example' (Peirce, 1998). The following table illustrates the relationships between deduction, induction and abduction (Tab. 2.8).

The examples of the different types of conclusion presented so far are relatively simple in structure and easy to understand. However, scientific texts are much more complex in their argumentative structure, which makes them more complicated, which is why knowledge of so-called "premise and conclusion indicators" can be helpful (Textbox 6).

> **Textbox 6: Premises and Conclusion Indicators**
>
> It is particularly difficult for beginners in the reading of foreign texts and in the writing of their own scientific texts to identify and decode the corresponding arguments or to formulate them themselves. This difficulty must not be underestimated, as scientific texts are dependent on the logical stringency of the argumentation and its clear, comprehensible formulation and identification. The first hurdle is to identify premises and conclusions in an argumentation. For this purpose, the knowledge of so-called premise and conclusion indicators, that is, words that typically either point to premises or to conclusions, is helpful. Typical *premise* indicators are words or word sequences such as: 'Because', 'Because of', 'Since', 'Follows from', 'How', 'To the extent that', 'As shown by', 'Namely', 'Because', 'The reason is that', 'As indicated by', 'May be inferred from', 'Because of', 'May be traced back to', 'May be derived from', 'To deny this would', 'In view of the fact that' etc. Typical *conclusion* indicators are words or word sequences such as: 'therefore', 'hence', 'thus', 'so that', 'accordingly', 'consequently', 'as a result of', '… proves that', 'So it follows that', 'We may conclude that', 'I infer from this that', 'What shows that', 'What means that', 'What includes that', 'What allows us to conclude that', 'Indicates the conclusion that' etc. The assignment of the indicators to premises and conclusions is always dependent on the respective *context*, so that the knowledge of the indicators does not obviate the sometimes time-consuming and strenuous analysis of the arguments, but only represents a—if also effective—aid. ◄

2.4   Fallacies

Another not to be underestimated problem are fallacies, that is incorrect arguments, which are incorrect in their logical form in deductive arguments and in non-deductive arguments either too weak to support the conclusion or violate the rules of 'reasonable argumentation' (cf. Pfister, 2019; Rosenberg, 2009). Many such fallacies in their function and effect are already known and researched. Part of this tradition of knowledge overlaps with that of rhetoric, which dates back to antiquity and therefore also deals with the strategic instrumentalization of fallacies by means of 'eristic dialectic' (Schopenhauer, 2009), 'manipulation techniques' (Edmüller & Wilhelm, 2009) or 'black rhetoric' (Bredemeier, 2005) in the context of discussions. Arguments play a decisive role not only in texts, but also in discussions. Fallacies are not only committed by beginners, but often enough also by advanced or even 'established' scholars. If they are uncovered, they can cast doubt on an argumentation or discredit it as unsuitable for its purpose. In order to prevent this as far as possible, it is useful to at least know some typical and frequently successful fallacies—not only to avoid them oneself, but also to uncover them in others and not to fall for incorrect arguments.

2.4.1   Formal Fallacies

In the literature, some typical deductive arguments are described which are incorrect due to their logical form. Among these, as a well-known and often committed fallacy, is the *equivocation* (lat. *aequus*, equal and *vocare*, call, sound: 'similarity'). This is a semantic fallacy which is based on the ambiguity of words with the same sound which are used in arguments with different meanings (cf. Pfister, 2019; Rosenberg, 2009). With Rosenberg, the equivocation *also* violates one of the "rules of reasonable thinking": "A word must mean the same thing within an argument, every time it appears. That is the basic rule we are talking about here. To violate it is basically to change the subject in the middle of the argument" (Rosenberg, 2009, p. 89). For example, the term 'space' (see Chap. 6) is often used in arguments with different meanings: for example as a 'container space', as a 'relational space' or as a 'space-time continuum' (for these concepts of space: see Sect. 6.1). If such differences in meaning of the same term *within* an argumentation are not taken into account, an impermissible equivocation results.

In connection with the *mode ponens*, the *'fallacy of the affirmation of consequence'* is also a fallacy that occurs frequently (see Pfister, 2019; Zoglauer, 1999). This is the fallacy from a given effect to a certain cause. For example, it could be claimed:

Premise 1: If there is largely prosperity in a population, then the
Population tolerant of minorities. p (antecedent) $\rightarrow$ q (consequent)
Premise 2: The population is tolerant
towards minorities. q (affirmative consequence)
Conclusion: So there is prosperity. p

The conclusion is false because the tolerance could also have other causes, for example a widespread increase in educational level. Any argument of this form is a fallacy by virtue of its form alone, regardless of its content. This can be easily represented using a truth table (Tab. 2.9):

The *invalidity* can be shown by the fact that in the distribution in the red-marked line, the two premises are assigned the truth value 'true', but the conclusion has the truth value 'false'—namely invalidity, *although* there is also a line in the yellow-marked line with a distribution in which the two premises are assigned the truth value 'true' and the conclusion has the truth value 'true'.

Tab. 2.9 Truth table: education, wealth and tolerance

Partial statement	Partial statement	Premise 1	Premise 2	Conclusion
p	q	$p \rightarrow q$	q	p
w	w	w	w	w
w	f	f	f	w
f	w	w	w	f
f	f	w	f	f

2.4.2 Violation of a Rule of Reasonable Thinking and Arguing

The *petitito principii* (Latin for 'claim of the proof ground') is a logical fallacy that explicitly or implicitly uses the conclusion as a premise (Pfister, 2019; Rosenberg, 2009). This means that what is to be proven is used as proof of what is to be proven. Here, metaphorically bites the cat in its own tail, accordingly this fallacy is also referred to as a 'circular argument'. As an example, a basic pattern of capitalism criticism/socialism criticism (depending on ideological preferences, see Sect. 5.4.1) may serve, which works as follows:

> 'Capitalism/socialism is bad, because the capitalist/socialist economic system is a bad form of economic activity.'

Since capitalism/socialism is a form of economic activity, the conclusion is already more or less obviously contained in the premise. The decisive problem that casts this conclusion into doubt is its argumentative invulnerability. This consists neither in an indisputable formal correctness nor in terms of content with regard to the truth value of the premise (s), but rather in the fact that, if premise and conclusion are identical, at least the key premise is not open to doubt. More precisely: "Since the premise is identical with the conclusion, we have in the questions: 'Is the premise false?' and 'Is the conclusion false?' *the same* question before us" (Rosenberg, 2009, p. 95). This ultimately violates a rule of 'rational thinking' and argumentation, since rational thinking and argumentation require the possibility of doubt and criticism of premises and conclusions.

An *infinite regress* (*regressus ad infinitum*; lat. *regressus*, decline, lat. *infinitus*, infinite) is then given if the premise as justification of the conclusion in turn requires a further justification and this further justification in turn requires a further justification and so on '*ad infinitum*'. For example, the question of a beginning of the world inevitably leads to an infinite questioning—even a '*Big Bang*' *could* be asked about its 'What was before?', if the questioning is not at this point, as the physicists do rightly, *interrupted*. This type of fallacy—if the questioning is *not* interrupted—can also be referred to as a 'fatal infinite regress' (Rosenberg, 2009). The fatal thing is that no real justification can be given, because this justification consists in an infinite series of setbacks in a justification chain that can never give a reliable answer to the justification question. This contradicts a real and serious justification work.

2.4.3 Weak Reasons

There are two typical fallacies known (cf. Pfister, 2019; Salmon, 1983), which either want to base their justificatory force on an appeal to an accepted authority (lat. *argumentum ad verecundiam*, 'proof by reverence'), or, conversely, try to discredit the justificatory force of an argument put forward by questioning the authority of the person arguing (lat. *argumentum ad hominem*, 'argument against the person'). In both forms, the reference is not to factual matters, but to the agreed-upon or not authority of persons. Both fallacies are fundamentally to be objected to that the authority or non-authority (for example, of a professional nature in a particular field of knowledge) *alone* is not sufficient to support or discredit an argument—it is always only a relatively weak form of justification and therefore, if 'good' and thus convincing arguments are to be put forward, to be omitted. An arguer who only has insufficient knowledge in a particular field of knowledge can nevertheless put forward a correct argument just as conversely people who have expert knowledge can put forward false or incorrect arguments depending on the relevant topic. So relying solely on the statements of, for example, Nobel laureates is just as weak and incorrect an argument as the one trying to discredit the statements of non-experts from the outset.

Another typical fallacy in the sense of a weak justification is the so-called 'causal fallacy', which can occur in three variants (Salmon, 1983). The first is traditionally referred to "as *Post hoc ergo propter hoc* (which can be translated as 'after this event', therefore because of this event)" (Salmon, 1983, p. 206). It is also often referred to simply as the '*Post-hoc Fallacy*' (Salmon, 1983). The fallacy lies in "inferring a cause-and-effect relationship from the temporal sequence of events" (Pfister, 2019, p. 109). An example: A student has stomach problems and drinks chamomile tea at the recommendation of a friend. A few days later she feels better, the stomach pain subsides. The chamomile tea is seen as the cause of the improvement. In fact, the cause could be something else, such as a psychological placebo effect, or that she was often outdoors or that the symptoms simply subsided on their own. Sometimes an *occurrence* of events is also used as the cause of a subsequent event. The classic is the example with the storks, on the noticeably increased population of which in a place an equally noticeable increase in the birth rate in this village can be observed. This confirms—so the post-hoc fallacy—that the 'stork' brings the babies. Here, a random correlation (lat. *correlatio*, relationship)—namely, the temporal coincidence of the increased stork population and the increased birth rate—is confused with a causal cause (Fig. 2.5).

The second variant is the "fallacy of confusing cause and effect" (Salmon, 1983, p. 211). This fallacy does indeed require an actually given causal relationship, but then confuses the cause with the effect or vice versa. Salmon (1983) alludes to the following example: In the 19th century, an English reformer

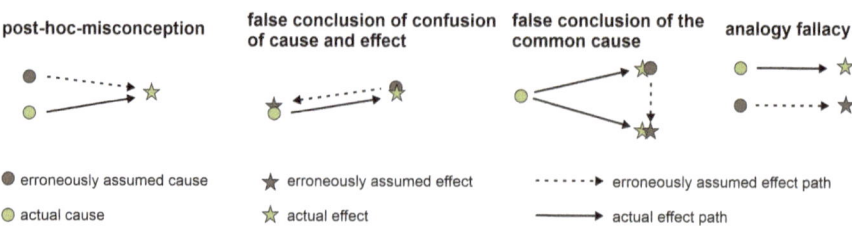

Fig. 2.5 *Graphical illustration of post-hoc fallacy,* fallacy of confusion of cause and effect, fallacy of common cause and analogy (own representation)

observed that hard-working and moderate farmers own at least one or two cows. Lazy or even drunk farmers, on the other hand, have no cow. He therefore concluded that giving a cow to lazy or drunk farmers without cows would make them hard-working and moderate as well. Obviously, here the cause (laziness) and the effect (cowlessness) are confused. With a little imagination, this example can be transferred to many other well-intentioned reform efforts in the present.

The third variant is the "fallacy of the common cause" (Salmon, 1983, p. 212). Here, two events are placed in a causal relationship, with one seeming to be the cause of the other. What is overlooked is that both events can be *effects* of a *common* cause that does not come into view at all. This also has practical consequences, because this fallacy "leads people to confuse symptoms with underlying causes" (Salmon, 1983, p. 213). For example, racism in the USA can be seen as the cause of police brutality against blacks. However, both phenomena may be effects or *symptoms* of a more comprehensive common cause that cannot be seen with the racism diagnosis—such as complicated sociocultural traditions, institutional action logics and sociopolitical customs that lead to *both* racism *and* police violence as *effects*.

An "analogical inference" (from Greek *analogos*, corresponding) is also a weak argument, based on similarities between objects in *one* respect to *another* respect, and is therefore fundamentally dependent on the assessment of "relevant similarities" (Salmon, 1983). Thus, one can distinguish between "a strong and a weak analogy. An analogy is then strong […], if the compared things are equal or similar in relevant respects, and it is then weak, if the compared things are unequal or dissimilar in relevant respects" (Pfister, 2019, p. 100). Famous examples abound in the New Testament with the parables of Jesus, for example the following: "For it is easier for a camel to go through a needle's eye, than for a rich man to enter into the kingdom of God" (Luke 18, 25). Since a camel and a human being do not have many similarities, it is therefore a weak analogy.

2.4.4 Naturalistic and Normativist Fallacy

Geography can be referred to as a 'hybrid' (Kühne, 2008) or 'diffuse' discipline (Hard, 2003; Toulmin, 1978), in which natural, social, humanities and cultural methods and concepts are mixed (Fig. 2.6).

This leads to the fact that, depending on the research question and method, also *normative*, that is moral ('good' vs. 'evil') or evaluative arguments and conclusions come into play (evaluative: with regard to purposes: 'suitable' vs. 'unsuitable'; with regard to values: 'good' in the sense of 'valuable' vs. 'bad'). In connection with the distinction between descriptive and normative statements and thus at the borderline between empiric and the normative, two typical fallacies are to be mentioned which are often committed: on the one hand the 'naturalistic' fallacy, on the other hand the 'normative' fallacy. Both occur in two variants each.

The '*naturalistic fallacy*' is firstly a conclusion from facticity ('Being') to normativity ('Should') and is also referred to as the 'Being-Should Fallacy' (Stuhl-

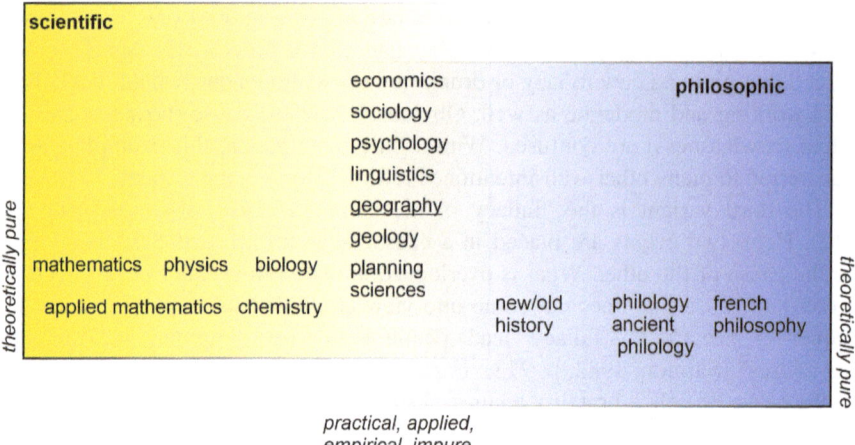

scientific

economics
sociology
psychology
linguistics
geography
geology

philosophic

mathematics   physics   biology   planning
                                        sciences
applied mathematics   chemistry

new/old
history

philology
ancient
philology

french
philosophy

theoretically pure

theoretically pure

practical, applied,
empirical, impure

Fig. 2.6 The position of geography in the canon of sciences (own representation according to: Kühne, 2008; on the basis of: Bourdieu, 1992)

mann-Laeisz, 1983) or 'Hume's Law' (cf. Quante, 2008; Sen, 1966; Hume, 1978) in philosophy. It is a logical-syntactic fallacy in a narrower sense in that a normative conclusion ('Should') is *derived* from a descriptive premise ('Being') purely logically. A complete moral argument (cf. Höffe, 1981; Müller, 2017) requires at least one descriptive and at least one normative premise in order to be able to draw a valid normative conclusion from it. Either a normative premise that is secretly claimed remains hidden or it is strategically ignored. For example, it is often argued that there is the 'beautiful traditional landscape' (descriptive premise), therefore this must be preserved (normative conclusion). The moral argument would be complete if, for example, the (not mentioned) normative conviction were expressed that a 'beautiful landscape' is *valuable* in principle and therefore should be preserved. But then this normative demand would have to be *justified* (cf. Berr & Kühne, 2019).

The naturalistic fallacy occurs secondly in the form of a *semantic* fallacy, whereby the normative or evaluative concept of 'good' shall be defined by empirical, particularly scientific, concepts (Moore, 1996; cf. Schaber, 2011; Quante, 2008). For example, one could argue that a 'historical cultural landscape' is 'good' in the sense of 'valuable', because it is a traditional and empirically observable synthesis of 'land and people', which can be made present to posterity through its protection. But if one asks why the unity of 'land and people' is 'good', a *value criterion* is required, which in turn refers back to the question of what is 'good' (cf. Poser, 2009). "What has been demonstrated here using the example of the concept of 'good', applies to all evaluative predicates, so that one will always have to distinguish between a range of factual statements and a range of normative statements" (Poser, 2009, p. 36).

The "normative fallacy" is, on the one hand, the logical-syntactic fallacy of norms ("should") on facts ("being") and is called the "moralistic fallacy" in this sense ("moralistic fallacy") (Davis, 1978; Moore, 1957). What should be, must also be "of nature" or "natural", can and must not be otherwise. For example, Rousseau concluded from the moral conviction that it is wrong that people are "evil" ("should"), that man is "of nature" ("being") "good" or must be good (Rousseau & Rippel, 1998 [1755]) Christian Morgenstern described the mechanism of this fallacy in his poem "The Impossible Fact" as follows: "Because, he concludes with a razor-sharp/must not be what can not be." (Morgenstern, 1993, p. 105).

The second form of the 'normative fallacy' draws a normative conclusion (should) directly from values or norms ('should') without taking into account the empirical 'reality' ('being') (Höffe, 1981). The incompleteness of the argument lies in the fact that a descriptive or empirical premise is missing or remains hidden or is ignored. "In fact, purely normative considerations only result in general assessment criteria which still have to be mediated by the specific regularities of the respective subject area and by concrete situational factors" (Höffe, 1981, p. 186). For example, from the conviction that the climate *must* be saved from human intervention and therefore harmful substances must be banned or replaced, it is directly concluded that all fossil energy sources *must* be banned or replaced—regardless of the real cultural, technical and social contexts in which this should happen (Berr & Kühne, 2019). The business ethicist Andreas Suchanek speaks of an 'assumption of the ought' (Suchanek, 2004), which sets itself above empirical moral framework conditions, Friedrich von Hayek (1996) of an 'arrogance of knowledge' that claims to be exempt from such conditions.

Textbox 7: Interim Conclusion to: Logical Propaedeutics
In this chapter it was possible to show in general terms how and why people in their everyday lives and in their scientific work use words, concepts, sentences, judgments, conclusions and theoretical generalizations in order to try and understand something and to grasp the "objects" of their everyday or scientific "world". In this way, an introduction was given to the logical, linguistic, semantic and pragmatic foundations of the theory of science and the sciences. In addition to the basics of concept formation, definition theory, statement logic, the correct structure of arguments and the typical conclusion forms in the sciences, typical logical, semantic or linguistic errors, misunderstandings, false conclusions and forms of incorrect argumentation were also considered and presented, which can occur especially at the beginning of their studies, but even among "established" scientists. With this logical-propedeutic toolbox, the decisive stations of philosophical engagement with science, truth, knowledge and understanding can be followed and traced in the following chapter.

Further Reading
Janich (2014). A comprehensive but demanding introduction to the relationships between language and methods and at the same time an excellent introduction to philosophical reflection.

Kruse (2017). A detailed, comprehensive and very illustrative "introduction for students" to the many forms and variants of thinking, in particular critical thinking and argumentation.

Tetens (2014). A very sound, comprehensive and illustrated introduction to the many variants and applications of philosophical argumentation.

Original Literature
Kamlah and Lorenzen (1967). The classic of logical propaedeutics, still worth reading and enlightening today.

Rosenberg (2009). A classic of an introductory and very understandable as entertainingly written book in philosophy—expressly with the subtitle 'A Handbook for Beginners'.

Salmon (1983). A very sound, comprehensive and well-illustrated introduction to logic.

References

Aristoteles. (1991 [348–345 v.u.Z.]). *Metaphysik. Schriften zur ersten Philosophie*. Reclam.
Aristoteles. (2004 [367–344 v.u.Z.]). *Topik*. Reclam.
Aristoteles. (2009). *Politik. Neuausgabe*. Rowohlt (Nach der Übersetzung von Franz Susemihl mit Einleitung, Bibliographie und zusätzlichen Anmerkungen von Wolfgang Kullmann).
Bartelborth, T. (1996). Der Schluß auf die beste Erklärung. In C. Hubig (Hrsg.), *Cognitio humana – Dynamik des Wissens und der Werte* (S. 552–559). Akademie.
Bauberger, S. (2016). *Wissenschaftstheorie. Eine Einführung*. Kohlhammer.
Beckermann, A. (2014). *Einführung in die Logik* (De Gruyter Studium, 4., durchgesehene. Aufl.). de Gruyter.
Berr, K., & Kühne, O. (2019). Moral und Ethik von Landschaft. In O. Kühne, F. Weber, K. Berr & C. Jenal (Hrsg.), *Handbuch Landschaft* (S. 351–365). Springer VS.
Berr, K., & Kühne, O. (2020). *„Und das ungeheure Bild der Landschaft …"*. *The Genesis of Landscape Understanding in the German-speaking Regions*. Springer VS.
Bourdieu, P. (1992). *Homo academicus* (Suhrkamp-Taschenbuch Wissenschaft, Bd. 1002). Suhrkamp (französische Originalausgabe 1984).
Bredemeier, K. (2005). *Schwarze Rhetorik. Macht und Magie der Sprache*. Goldmann.
Bucher, T. G. (1998). *Einführung in die angewandte Logik*. de Gruyter.
Carnap, R. (1962). *Logical foundations of probability*. University of Chicago Press.
Davis, B. D. (1978). The moralistic fallacy. *Nature, 272*, 390. https://doi.org/10.1038/272390a0.
Duden (Hrsg.). (2012). *Basiswissenschule Geografie. 7. Klasse bis Abitur*. Dudenverlag.
Edmüller, A., & Wilhelm, T. (2009). *Manipulationstechniken*. Haufe Lexware.
Fleck, L. (1980 [1935]). *Entstehung und Entwicklung einer wissenschaftlichen Tatsache. Einführung in die Lehre vom Denkstil und Denkkollektiv* (Wissenschaftsforschung). Suhrkamp (Mit einer Einleitung herausgegebn von Lothar Schäfer und Thomas Schnelle).
Gabriel, G. (2004). Propädeutik. In J. Mittelstraß (Hrsg.), *Enzyklopädie Philosophie und Wissenschaftstheorie* (3, P – So, unveränderte Sonderausgabe, S. 361–362). Metzler.
Hard, G. (1969). Das Wort Landschaft und sein semantischer Hof. Zu Methode und Ergebnis eines linguistischen Tests. *Wirkendes Wort, 19*, 3–14.

Hard, G. (1970). *Die „Landschaft" der Sprache und die „Landschaft" der Geographen. Semantische und forschungslogische Studien.* Ferdinand Dümmlers.

Hard, G. (1977). Zu den Landschaftsbegriffen der Geographie. In A. Hartlieb von Wallthor & H. Quirin (Hrsg.), *„Landschaft" als interdisziplinäres Forschungsproblem. Vorträge und Diskussionen des Kolloquiums am 7./8. November 1975 in Münster* (S. 13–24). Aschendorff.

Hard, G. (2003). Studium in einer diffusen Disziplin. In G. Hard (Hrsg.), *Dimensionen geographischen Denkens. Aufsätze zur Theorie der Geographie* (Osnabrücker Studien zur Geographie, Bd. 23, S. 173–230). V & R Unipress.

Hartmann, D., & Janich, P. (Hrsg.). (1996). *Methodischer Kulturalismus. Zwischen Naturalismus und Postmoderne* (Suhrkamp-Taschenbuch Wissenschaft, Bd. 1272). Suhrkamp.

Hayek, F. A. v. (1996). *Die Anmassung von Wissen. Neue Freiburger Studien* (Wirtschaftswissenschaftliche und wirtschaftsrechtliche Untersuchungen, Bd. 32). Tübingen: Mohr.

Hein, K. (2006). *Hybride Identitäten. Bastelbiografien im Spannungsverhältnis zwischen Lateinamerika und Europa.* transcript.

Hempel, C. G. (1965). *Aspects of scientific explanation and other essays in the philosophy of science.* Free Press.

Hempel, C. G., & Oppenheim, P. (1948). Studies in the logic of explanation. *Philosophy of Science, 15*(2), 135–175.

Höffe, O. (1981). *Sittlich-politische Diskurse. Philosophische Grundlagen. Politische Ethik. Biomedizinische Ethik* (Suhrkamp-Taschenbuch Wissenschaft, Bd. 380). Suhrkamp.

Hokema, D. (2009). Die Landschaft der Regionalentwicklung: Wie flexibel ist der Landschaftsbegriff? *Raumforschung und Raumordnung, 67*(3), 239–249.

Hokema, D. (2013). *Landschaft im Wandel? Zeitgenössische Landschaftsbegriffe in Wissenschaft, Planung und Alltag.* Springer VS.

Hoyningen-Huene, P. (1998). *Formale Logik. Eine philosophische Einführung.* Reclam.

Hume, D. (1978). *Ein Traktat über die menschliche Natur. Buch II. Über die Affekte Buch III. Über Moral* (Unveränderter Nachdruck der 1. Aufl. von 1906 (Buch 2 und 3)). Meiner.

Janich, P. (2001). *Logisch-pragmatische Propädeutik. Ein Grundkurs im philosophischen Reflektieren.* Velbrück Wissenschaft.

Janich, P. (2014). *Sprache und Methode. Eine Einführung in philosophische Reflexion.* Francke.

Janich, P. (2015). *Handwerk und Mundwerk. Über das Herstellen von Wissen.* Beck.

Kamlah, W., & Lorenzen, P. (1967). *Logische Propädeutik oder Vorschule des vernünftigen Redens.* Bibliographisches Institut.

Kruse, O. (2017). *Kritisches Denken und Argumentieren. Eine Einführung für Studierende* (UTB, Bd. Nr. 4767). UVK Verlagsgesellschaft mbH; UVK/Lucius.

Kuhn, T. S. (1976). *Die Struktur wissenschaftlicher Revolutionen* (Suhrkamp-Taschenbuch Wissenschaft, Bd. 25, zweite revidierte und um das Postskriptum von 1969 ergänzte Aufl.). Suhrkamp.

Kühne, O. (1999). *Die Wetterlagen-, Tages- und Jahreszeitenabhängigkeit der Verteilung von Lufttemperatur, spezifischer Luftfeuchte, Windfeld, Äquivalenttemperatur und anderer bioklimatisch wirksamer Größen im Lokalklima der Stadt Homburg/Saar.* Dissertation zur Erlangung des akademischen Grades eines Doktors der Philosophie, Universität des Saarlandes, Saarbrücken.

Kühne, O. (2008). *Distinktion – Macht – Landschaft. Zur sozialen Definition von Landschaft.* VS Verlag.

Kühne, O. (2013). *Landschaftstheorie und Landschaftspraxis. Eine Einführung aus sozialkonstruktivistischer Perspektive.* Springer VS.

Kühne, O. (2018). Die Landschaften 1, 2 und 3 und ihr Wandel. Perspektiven für die Landschaftsforschung in der Geographie – 50 Jahre nach Kiel. *Berichte. Geographie und Landeskunde, 92*(3–4), 217–231.

Kutschera, F. v., & Breitkopf, A. (2007). *Einführung in die moderne Logik* (8. Aufl.). Karl Alber.

Leerhoff, H., Rehkämper, K., & Wachtenhofer, T. (2010). *Einführung in die Analytische Philosophie.* Wissenschaftliche Buchgesellschaft.

Lorenzen, P. (1968). *Methodisches Denken* (Theorie, Bd. 2). Suhrkamp.

Moore, E. C. (1957). The moralistic fallacy. *The Journal of Philosophy, 54*(2), 29–42.

Moore, G. E. (Hrsg.). (1996). *Principia Ethica* (erweiterte Ausgabe). Reclam.

Morgenstern, C. (1993). *Gedichte – Verse – Sprüche.* Lechner.

Müller, A. (2017). *Planungsethik. Eine Einführung für Raumplaner, Landschaftsplaner, Stadt-planer und Architekten.* UTB.

Nederveen Pieterse, J. (2005). Hybridität, na und? In L. Allolio-Näcke, B. Kalscheuer, & A. Manzeschke (Hrsg.), *Differenzen anders denken. Bausteine zu einer Kulturtheorie der Trans-differenz* (S. 396–430). Campus.

Oeing-Hanhoff, L., Kobusch, T., & Borsche, T. (2019). Individuum, Individualität. In J. Ritter (Hrsg.), *Historisches Wörterbuch der Philosophie* (Völlig neubearbeitete Ausgabe des „Wör-terbuchs der philosophischen Begriffe" von Rudolf Eisler, Sonderausgabe, I – fvK, Bd. 4, S. 304–310). WBG Academic.

Peirce, C. S. (1991). *Vorlesungen über Pragmatismus. Einleitung, Anmerkung und herausgege-ben von Elisabeth Walther.* Felix Meiner.

Peirce, C. S. (1998). *Elements of logic. Volume 2* (58th Edition, Reprint of the 1931 Edition). Thoemmes.

Pfister, J. (2019). *Werkzeuge des Philosophierens.* Reclam.

Popper, K. R. (2002). *The logic of scientific discovery.* Routledge.

Poser, H. (2009). *Wissenschaftstheorie. Eine philosophische Einführung.* Reclam.

Quante, M. (2008). *Einführung in die allgemeine Ethik* (Einführungen Philosophie, 3. Aufl.). Wissenschaftliche Buchgesellschaft.

Regenbogen, A., & Meyer, U. (Hrsg.). (1998). *Wörterbuch der philosophischen Begriffe.* Meiner.

Rosenberg, J. F. (2009). *Philosophieren. Ein Handbuch für Anfänger.* Vittorio Klostermann.

Rousseau, J.-J., & Rippel, P. (Hrsg.). (1998 [1755]). *Abhandlung über den Ursprung und die Grundlagen der Ungleichheit unter den Menschen.* Stuttgart: Reclam.

Salmon, W. C. (1983). *Logik.* Reclam.

Schaber, P. (2011). Naturalistischer Fehlschluss. In M. Düwell, C. Hübenthal & M. H. Werner (Hrsg.), *Handbuch Ethik* (3., akt. Aufl., S. 454–456). Metzler.

Schnädelbach, H. (1991). Philosophie. In E. Martens & H. Schnädelbach (Hrsg.), *Philosophie. Ein Grundkurs* (S. 37–76). Rowohlt.

Schopenhauer, A. (2009). *Die Kunst, recht zu behalten. In achtunddreißig Kunstgriffen darges-tellt.* Anaconda.

Schurz, G. (2014). *Einführung in die Wissenschaftstheorie* (4., überarb. Aufl.). WBG.

Seiffert, H. (1996). *Einführung in die Wissenschaftstheorie 1. Sprachanalyse – Deduktion – Induktion in Natur- und Sozialwissenschaften.* Beck.

Sen, A. K. (1966). Hume's law and Hare's rule. *Philosophy, 41*(155), 75–79.

Stuhlmann-Laeisz, R. (1983). *Das Sein-Sollen-Problem. Eine modallogische Studie* (Problemata, Bd. 96). Frommann-Holzboog.

Suchanek, A. (2004). Die Rolle empirischer Bedingungen für die Wirtschaftsethik. In P. Ulrich & M. Breuer (Hrsg.), *Wirtschaftsethik im philosophischen Diskurs. Begründung und „Anwend-ung" praktischen Orientierungswissens.* Königshausen und Neumann.

Tetens, H. (2014). *Philosophisches Argumentieren. Eine Einführung.* Beck.

Thiel, C. (2004). Logik. In J. Mittelstraß (Hrsg.), *Enzyklopädie Philosophie und Wissenschaft-stheorie* (2, H – O, unveränderte Sonderausgabe, S. 626–631). Metzler.

Toulmin, S. E. (1978). *Menschliches Erkennen I: Kritik der kollektiven Vernunft.* Suhrkamp.

Tugendhat, E., & Wolf, U. (1986). *Logisch-semantische Propädeutik.* Reclam.

Wille, M. (2011). Die Disziplinierung des Denkens. In B. Pörksen (Hrsg.), *Schlüsselwerke des Konstruktivismus* (S. 160–174). VS Verlag.

Zoglauer, T. (1998). *Geist und Gehirn. Das Leib-Seele-Problem in der aktuellen Diskussion* (UTB für Wissenschaft. Uni-Taschenbücher: Philosophie, Bd. 2066). Vandenhoeck & Rupre-cht.

Zoglauer, T. (1999). *Einführung in die formale Logik für Philosophen.* Vandenhoeck & Ruprecht.

Philosophy of Science— Philosophical Foundations and Positions

3

Theory of science is a philosophical discipline, but is also located in a context of other disciplines dealing with "science" as their object, and has a long tradition of philosophical theories of knowledge. Therefore, we will first discuss some philosophical basics of the theory of science, then reconstruct essential stages of the development of philosophical engagement with "science", "knowledge" and "truth", extending from the debate between rationalism and empiricism to the debate between logical empiricism and Popper's approach. Popper will be considered by us as the high point and culmination of classical theory of science; the science historians (see Chap. 4) and sociologists of science (see Chap. 5) who followed him will be treated in the two following chapters. Terms essential for the following chapter will be explained in Textbox 8.

Textbox 8: Important Terms for Chap. 3

Accidental and essential: (from Latin *accidens*, suddenly occurring, accidental, unimportant; Latin *esse*, to be). In his "Metaphysics" (1991 [348-345 BC]), Aristotle distinguishes between accidental, changing, unimportant properties of an object and its necessary, permanent and essential properties. This distinction determines the entire history of metaphysics; and it continues to determine scientific approaches that can be called "essentialist" (see Sect. 3.2).

Aposteriori and Apriori: (lat. 'from the later one' (*aposteriori*) vs. 'from the former one', from the outset (*apriori*)). Since Kant, this distinction has been used to distinguish experience-based knowledge, which is gained through sensory perception and is therefore empirical and contingent (*aposteriori*), from non-empirical knowledge, which always precedes experience and makes it possible in the first place (*apriori*). The difference can

---

© The Author(s), under exclusive license to Springer Fachmedien Wiesbaden GmbH, part of Springer Nature 2022
O. Kühne and K. Berr, *Science, Space, Society*,
https://doi.org/10.1007/978-3-658-39140-9_3

be understood quite simply as the one between experience-dependent and experience-preceding or experience-independent knowledge.

Conditio humana: (from lat. *conditio*, condition, constitution and *humanus*, human). In general and in philosophical usage, it denotes the basic, distinguishing and determining circumstances, conditions and requirements of human existence (for example, mortality; dependence on food, clothing, housing). This concept is criticized because it is essentialist (see Sect. 3.2), i.e. it postulates something like a fixed essence of man.

Idea and Ideal: (from Greek *idéa*, image, model, archetype—related to Greek *eidos*, form; New Latin *idealis*, exemplary). Plato determined the essence of things *ontologically* as '*Idea*' (Greek *idéa*) and this 'Idea' at the same time as the form or shape (Greek *eidos*) of things. In modern times, starting with Descartes, Locke and Hume (see Sect. 3.3), 'ideas' are interpreted as *content of consciousness*, as 'representations' or 'mental principles'. In everyday language, ideas are something like a 'sudden thought' or an 'invention'. An '*Ideal*', on the other hand, refers to "the exemplary in reality that is never encountered in its pure form, the perfection in the concept in relation to the imperfection in the sensually empirical appearance" and "always aims at a state considered to be of higher standing, thus requiring a knowledge of the possibility of perfection" (Regenbogen & Meyer, 1998, p. 301).

Metaphysics: (from Greek *meta ta physika*, after or next to physics). Regardless of the disputed origin of the Greek expression in research so far, it has been used since the Middle Ages to refer to objects that 'go beyond nature', that is, exceed the realm of empirically, sensually graspable things. Often, what is 'behind nature' has been and is sought as the (final or first) cause of things and thus ultimately the 'actual' or 'true' reality. Since this 'final' cause is only accessible to thought as 'actual reality', metaphysical concepts and theories are not empirically verifiable.

Res extensa und res cogitans: (from lat. *res*, thing, substance; *extensa*, extended; *cogitans*, thinking). René Descartes (1596–1650) has in his Meditationes de prima philosophia (Meditations on First Philosophy) as a starting point for a new foundation of sciences the self-certainty of a thinking consciousness determined in doubt. This results in a dualism of 'thinking substance' (*res cogitans*) and 'extended (three-dimensional) substance' (*res extensa*), so that the dualism of mind and nature (or material), soul and body, which is still valid today and shapes the sciences. Even in the juxtaposition of human geography and physical geography, this dualism is abolished.

3.1 Philosophical Basics

First of all, the place of the philosophy of science in a wide context of philosophical and other "science of sciences" (Stegmüller, 1973) will be identified and described below, and some epistemological questions and distinctions relevant to the philosophy of science will be introduced (Sect. 3.1.1). We will then address the question of which criteria can be given for "scientificity" (Sect. 3.1.2). Since science in its self-conception and in the expectations placed on it is aimed at truth, we will introduce some widely held theories of truth (Sect. 3.1.3).

3.1.1 Theory of Science and Other Sciences of Science

In addition to scientific theory, there are some other sciences that deal with "science" itself in a scientific way. This "meta" perspective on "science" branches out into considerations of specific aspects, just as any other object of possible scientific investigation can be questioned from different points of view. Philosophy has been concerned with the question of what knowledge and understanding consist of and what distinguishes science from non-science since the days of ancient Greece. Examples from ancient Greece include Plato's dialogue *Theaitetos* and Aristotle's *Second Analytics*, from the modern era Descartes' *Discours de la méthode*, Fichte's *Wissenschaftslehre*, Kant's *Kritik der reinen Vernunft* and Hegel's *Enzyklopädie der Wissenschaften*. The concept of truth (gr. *aletheia*; lat. *verum*) has been at the center of all these discussions from the very beginning.

However, philosophy has not only been concerned with the question of truth since its beginnings, but also with the questions of the "good" and the "beauty" (cf. Berr, 2020; Kurz, 2015). This "triad" of "true, good, beautiful" (Kurz, 2015) was differentiated in modern times into the autonomous cultural areas of the sciences, morality (and law) and art (Habermas, 1994). In philosophy, this differentiation is reflected in the philosophical disciplines (Pieper, 1998) of epistemology and philosophy of science (science: truth), ethics (morality and law: the good) and aesthetics (art: beauty). Accordingly, in addition to the theory of science, the corresponding disciplines of the science ethics, and the science aesthetics have also been established within the framework of philosophical considerations of "science" (cf. Kornmesser & Büttemeyer, 2020).

Science ethics asks, for example, generally about the responsibility of scientists for their research, research results and research applications. Discussions about animal testing, cloning of humans or research on weapon technologies are widely known. *Science aesthetics* examines, for example, the connection between aesthetically effective geometric symmetries and 'art forms of nature' (Ernst Haeckel), between the aesthetic elegance of mathematical formulas and their scientific relevance or the parallel between scientific and artistic creativity, each of which cannot be brought to rules (see Sect. 4.5).

In addition to the philosophy of science, science ethics and science aesthetics, there could be other disciplines mentioned (see Poser, 2009), which, however, are not yet established like "science policy" (Poser, 2009, p. 15), which investigates the (controllable?) connection between science as an institution and society (see Sects. 1.2 and 5.4)—or which does not yet exist as an established discipline (Poser, 2009), like the *psychology of science*, which could examine the motives of individual scientists in their research work (see but Sects. 5.2, 4.2, 4.5). For an understanding of the development of science, two other disciplines are significant: *history of science and sociology of science*. In the research literature, there is widespread agreement that the theory of science "operates in the spectrum of sociology of science and history of science" and that these three approaches "complement" and "contribute together to a better understanding of science" (Carrier, 2017, S. 10). These three specific 'sciences of science' will also determine the structure of this book.

The *history of science* takes the factor *time*, that is, the historicity of science as a historical context into account (Chap. 4). Here, the history of science is used to examine how sciences, theories, methodologies, or methods have changed and developed, regardless of whether they were or are correct or incorrect, useful or useless according to the current state of science. *Science sociology* conceives of science as social action (Poser, 2009) and as a subsystem of society (Sect. 5.3). Here, sociology is used to examine how the interaction and communication of scientists in research communities influence the development of sciences, theories, methodologies, or methods, again regardless of whether they were or are correct or incorrect, useful or useless according to the current state of science. The social embeddedness of scientific *knowledge* and the reciprocal relationship between 'knowledge' and 'social' is addressed by *knowledge sociology* (Sect. 5.1), which is also taken into account here. History of science, science sociology, and knowledge sociology together take the factor of the *social embeddedness* of scientific knowledge and of science into account (Chap. 5).

Theory of science deals with scientific knowledge and the question of how scientific knowledge can be obtained. 'Knowledge' is traditionally associated with truth, insofar as an 'assertion that has been shown to be true' is understood as 'knowledge' (Poser, 2009, p. 16). This puts it in the vicinity of *epistemology*. In contrast to the theory of science, which deals with scientific knowledge, the scope of epistemology is more comprehensive, insofar as it is to be determined as the 'investigation of the conditions, possibilities and limits of human knowledge' (Gabriel, 1993, p. 10). The leading distinction for epistemology is the relationship between a subject that knows and an object that is known. The attempt to transform epistemology into the theory of science (cf. Gabriel, 1993) would be misleading because then knowledge would be limited or reduced to scientific knowledge. This would devalue all the knowledge that people gain in everyday life or through art (literature, music, film, visual art) compared to scientific knowledge. The theory of science is therefore a subfield of epistemology (Gabriel, 1993; Poser, 2009) that is directed exclusively at *scientific* knowledge (Fig. 3.1).

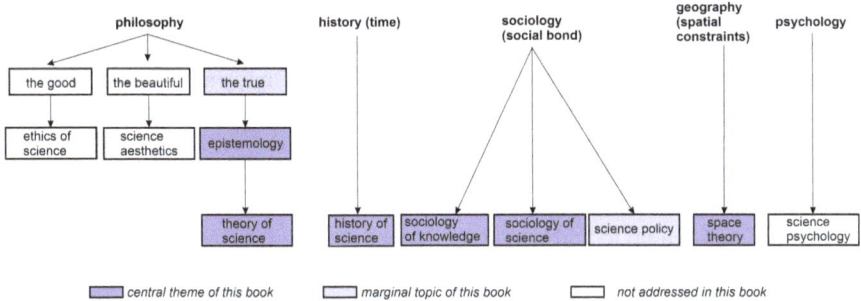

Fig. 3.1 Overview of different sciences and their thematization in this book (own representation)

For the development of philosophical theories on the subject of "science" and "scientificity", which are now dealt with under the title "theory of science", traditional *epistemological questions* were also determinative in antiquity, the Middle Ages and the modern era, which we would like to briefly sketch with Gabriel (1993).

1. The question of the *origin* of knowledge. It is asked about the sources of knowledge, whether they are to be found in reason (as in rationalism) or in experience (as in empiricism). This question will play an important role in Sects. 3.1.2 and 3.3.
2. The question of *reality*. It is asked whether there is an external world ("reality") independent of the recognizing subject (observer) (as claimed by realism and positivism) or not (as claimed by idealism). This question will play an important role in Sects. 3.1.2, 3.2, 3.4 and 6.10.
3. The question of the *nature* of the subject of knowledge and "world" as object of knowledge. It is asked whether the "world" is uniformly composed of *one* fundamental "substance" (monism: in materialism from matter, in spiritualism from spirit), from *two* (dualism: matter *and* spirit) or from *three* (Poppers "Three-Worlds-Theory"). This question will be particularly guiding for the argumentation logic of Chap. 6.
4. The question of the mode of being of *universals* (general concepts). This question properly belongs to metaphysics, but is often confused with question three (Gabriel, 1993). The question is whether universals really exist (realism or Platonism), are merely conventional 'names' or designations of things (nominalism), or conceptual concepts (conceptualism). It is only mentioned in Sects. 3.2 and 3.4.4.

Another important distinction that stems from epistemology is the one between "genesis" (from Greek *genesis*, birth, origin) and "validity" (cf. Gabriel, 2012; Schildknecht et al., 2008), that is, between the origin or historical development of something and its need for justification. Therefore, within the framework of epistemology and, consequently, also of the theory of science, "the genetic conditions of the acquisition of knowledge are always to be distinguished from the reasons for its validity" (Gabriel, 1998, p. 57). This epistemological distinction was taken up by Hans Reichenbach (1938) and introduced into the debate of the theory of science as a distinction between the context of discovery and the context of justification. This distinction affects in an eminent way the entire debate between the theorists of science up to and including Popper and the historians and sociologists of science following Popper and decisively determines Chaps. 4 and 5.

Finally, another question concerning the theory of science *in general* is to be addressed: the question of whether the theory of science is to be understood and pursued as a descriptive or normative project (cf. Kornmesser & Büttemeyer, 2020; Pfister, 2016; Schurz, 2014; Stegmüller, 1973). The question is: Has the theory of science been limited to the *description* of the sciences in their development as well as the factual structure of the *practice* of scientific work and the choice of methods, without thus making evaluative or prescriptive statements? This *descriptive* understanding therefore has to say what science *is*. It determines the approaches of the history of science, science- and knowledge sociology, science policy, science psychology. Science ethics and science aesthetics can—like ethics and aesthetics as philosophical basic disciplines—be conceived both descriptively and normatively. In addition to this descriptive understanding, an *analytical* understanding must also be taken into account. This is about the task of understanding how science works in which contexts, that is, understanding the function of science. The *normative* understanding wants to say what science *should* be (Chilla et al., 2016). Then it must be able to determine "what characterizes a good scientific method, explanation or theory" (Pfister, 2016, p. 12) and "in what scientific rationality consists and on the basis of which criteria a scientific hypothesis can be rationally justified" (Schurz, 2014, p. 21). This position is, for example, represented by the logical empiricists or positivists and by Karl Popper and critical rationalism.

As we will see, representatives of both positions have been and still are partly irreconcilable. A mediation seems to be possible by both approaches acting as each other's "corrective" (Schurz, 2014; Stegmüller, 1973). This means, simplifying strongly, that "of course, scientific theory needs a as accurate as possible image of how research is actually conducted"—only then "one arrives at a reasonable judgment about what a good scientific method, explanation or theory is" (Pfister, 2016, p. 13). This *descriptive* corrective "contains examples of successful scientific findings as well as counter-examples of refuted scientific hypotheses. There is rational consensus about these examples (counter-) which scientific theory may at least provisionally rely on" (Schurz, 2014, p. 23). But science also needs rules, so that there is also an *normative* corrective. According to Schurz (2014), this consists of an "epistemological goal", which is concretized by a "minimal common epistemological model of the sciences" and, for example, includes demands for a "minimal realism and empiricism", for the recognition of fallibility, as well as the "pursuit of objectivity" and logical clarity.

The previous explanations refer at different points and in different conceptual contexts to "theory" or "theoretical". It is therefore necessary to give some explanations for the concept of "theory" (Textbox 9).

Textbox 9: "Theory" Levels

The term 'theory' is very vague and is used differently in different sciences. The attempts at definition are hardly manageable and, in their contradictions, not very encouraging (cf. Egner, 2010; Zima, 2004). In a very general and prominent statement, 'theory' can be defined as follows:

"Theory is the net we throw to catch 'the world'—to rationalize, explain and control it" (Popper, 1989, p. 31).

In addition, the term 'theory' can be used at different levels, as the terms 'epistemology' and 'theory of science' already show. It is therefore useful to represent the hierarchy of 'theoretical' and 'metatheoretical' relationships with each other at the different levels and metalevels of the scientific process of cognition. Some explanations:

Epistemology as a subdiscipline of philosophy is the science of structures, conditions, origins, types and limits of human knowledge *in general*. The theory of science is a special case of epistemology in that it deals with structures, conditions, origins, types and limits of *scientific* knowledge. Theories are the 'framework' that is delineated, or the 'net' that is 'thrown', to catch a particular object- 'world' within a *discipline* (such as geography) that is specific to a research question and research area. Here a distinction is often made between 'theories of middle range' (Dahrendorf, 1972; Merton, 1957 [1949]) and theories of 'high complexity' or 'grand theories'. While the former relate to clearly social or spatial areas of the world (for example, the conflict theory of Ralf Dahrendorf, 1972, 1992 or the social constructivist landscape theory; see Sect. 6.5), provide further comprehensive explanations, for example for the development of societies, like the systems theory by Niklas Luhmann (see Sect. 6.3; in more detail by: Richter, 2016). Methodologies are a special or application case of the theoretical framework, in that they deal with the question of which methods can be used to "capture" and explain this object area. Within such theories and methodologies, different *methods* can be used as conceivable ways of research and data collection and evaluation in view of specific object areas and research perspectives (based on Lamnek (2005) (Fig. 3.2.)) ◄

3.1.2 Criteria of Scientificity

A decisive normative question of epistemology is: What distinguishes science, for example, from superstition or charlatanism, what distinguishes scientific knowledge from mere opinion? This question cannot simply be answered by describing the factual practice and self-conception of scientific work and the scientists acting

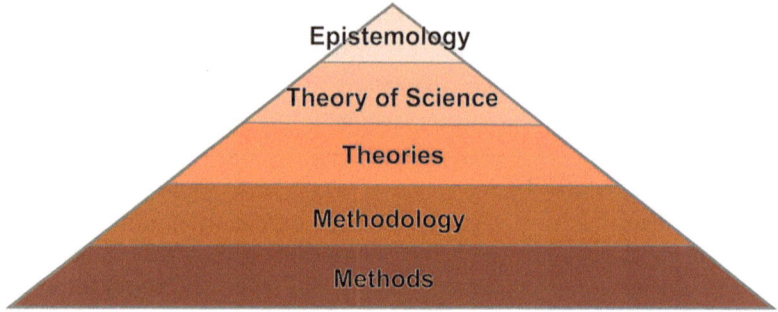

Fig. 3.2 Theoretical levels in their hierarchical dependence (own representation)

therein and deriving from this a general determination of what science is. Instead, a prior understanding of science and scientificity is already required in order to be able to describe a given context *as* science and corresponding action *as* scientific at all on the basis of this prior understanding. At least some "formal" conditions of scientificity must be met (Stegmüller, 1973). Consequently, a criterion is required that *prescribes* according to which standards something can be determined to be scientific or scientific.

But which criteria for scientificity can be given and how can these be won and justified? Initially, the answer to this question can be facilitated by first indicating what is *not* scientific in the narrower sense. The question is then: What is nonscientific, prescientific or extrascientific in contrast to "scientific" (cf. Schnädelbach, 1980 for the following distinction)? *Extra* -scientific is everything that has nothing whatsoever to do with science or scientificity before any criterion for science or scientificity has been established. For geography, for example, a river basin to be mapped is something extrascientific. *Pre* -scientific is everything that does not yet correspond to the criteria of scientificity, but already extends into the field of science. This includes, for example, the practically proven knowledge, such as in the early phase of cartography. This knowledge would only be scientific if it could at least prove this fact by means of convincing statistics, ideally also explain it. "nonscientific" denotes what clearly violates the criteria of scientificity recognized in a science. It would be nonscientific, for example, to conclude from the fact that creativity has not yet been explained scientifically in a clear-cut manner (cf. Sect. 4.5) that here a supernatural mechanism, a divine or other inspiration, is at work. The (still) unexplained cannot be explained with nonscientific assertions. Or, expressed differently: The unexplained cannot be explained with the even more unexplained.

In free association with the '*Logical Square*' used in logic, the opposite relationships of these concepts can be represented as follows (Fig. 3.3):

The ratio of scientific to nonscientific is a *contrary* opposition, in which the two corresponding statements (see Sect. 2.2) can not both be true but both be false. The ratio of scientific to extra-scientific (and that of nonscientific to prescientific) is a *contradictory* opposition, in which neither statement can *both* be true *and* both be

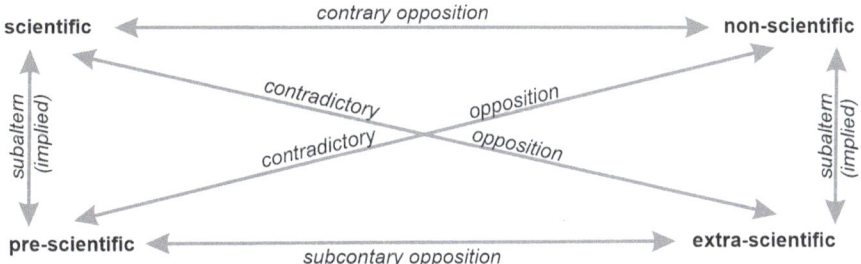

Fig. 3.3 The 'Logical Square' (own representation)

false. The ratio of prescientific to extra-scientific is a *subcontrary* opposition, in which the two corresponding statements can both not be false, but *both* be true. The ratio of scientific to prescientific and of nonscientific to extra-scientific is *subaltern*, the former statement *implies* the latter statement (see Sect. 2.2 for implication). That is: If something is scientific, then it can be concluded that something is prescientific (e.g. from scientific cartography to prescientific mapping); if something is nonscientific, then it can be concluded that something is extra-scientific (e.g. from the belief in supernatural mechanisms of creativity to non-scientific creativity techniques).

The question still stands to be answered: What makes knowledge *scientific* knowledge? Which *criteria* can or must be given here? If we take a cursory look at traditional scientificity criteria, philosophically we can distinguish between a fundamentalist and a fallibilist 'epistemological program' (Schurz, 2014) or 'certist' (lat. *certus*, reliable, certain) and 'justification-oriented' versus fallibilist philosophy concepts (Schroeder, 2004). Both programs agree that people are basically fallible in their cognition as well as in their actions. The fundamentalist-certist program draws the consequence from this that it wants to lay bare an unquestionable foundation for cognition, on which the sciences can find safe footing and gain certain knowledge. So the phenomenon of justification "has essentially to do with the fundamental *condition humaine* for the facts that (1) human achievements have a structure that allows for something like *errors*, and (2) that people are finite, weak, *fallible*, i.e. constantly threatened by the risk of error" (Kuhlmann, 2011, p. 320). For the fundamentalist epistemological program, 'true knowledge' is therefore only possible "if it rests on a foundation of certain and necessary principles, which are not gained through uncertain experience, but through rational intuition" (Schurz, 2014, p. 12). Fallibilism, on the other hand, transfers the insight into the fundamental fallibility of human cognition efforts also to the fundamentalist justification program itself and declares it to be unworkable.

According to the framework assumptions of the fundamentalist justification program, Aristotle (385-323 BC) already determines four characteristics of scientificity: 1) Scientific knowledge is "knowledge of the general" (Aristotle, 1991 [348-345 BC], 981a) as well as 2) justifiable by the specification of causes and principles which not only know the "that"—the concrete individual case in its facticity—but also the "why" and the cause "of the recognized" (Aristotle, 1991 [348-345 BC.], 981a) and in addition "have the concept" (Aristotle, 1991 [348-345 BC.], 981b), that is, conceptually secured knowledge. Since scientific knowledge "excludes the possibility of otherness", it has 3) "the character of necessity" and 4) "can be passed on to others and learned" (Aristotle, 2001, Book VI, 3). Aristotle determines scientificity *ontologically*, that is, from the *object* of knowledge. This means that

> "Knowledge must be knowledge of *something*, and this something must *be* and can not *not* be; knowledge of nothing would be nothing. But knowledge of the being (*to ón*) is ontological knowledge. It is brought about by the being itself." (Schnädelbach, 1991, p. 48)

Accordingly, science in the context of an "ontological paradigm" (Schnädelbach, 1991) is looking for the "essence" or the "substance" (lat. substantia) as the "underlying" (cf. Krieger, 2011) of the to be recognized and is therefore "essentialist" (cf. Kühne, 2019; Schneider, 2019; Schwemmer, 2004b) oriented. For example, Aristotle, as a main representative of the ontological paradigm in antiquity (next to Plato), expresses himself in his metaphysics and doctrine of categories explicitly to the "problem of substance" (gr. ousía) of things (cf. Rapp, 2010). "True" in this paradigm are such scientific statements or sentences that correspond to "reality". This understanding of truth is traditionally referred to as the "correspondence theory of truth" (cf. Sect. 3.1.3). The statement "The Black Forest is a low mountain range in the southwest of Baden-Württemberg" is true if and only if this statement can be factually confirmed. We will come back to essentialism (see Sect. 3.2) and theories of truth (see Sect. 3.1.3) in more detail.

In the 18th century, Immanuel Kant (1724–1804) carried out the epistemological turn from the *objects* of knowledge to the subject of knowledge and science. It is not the objects that determine independently of specific epistemic achievements of the recognizing human being, but the recognizing human being as the subject of knowledge determines the scientific quality of the achieved knowledge through general human epistemic achievements. Scientificity thus becomes a property not of objects but of *knowledge*:

> "Any doctrine, if it is to be a system, i.e. an ordered whole of knowledge, is called science [...]. *Proper* science can only be called that whose certainty is apodictic; knowledge that can only contain empirical certainty is only improperly called *knowledge*." (Kant, 1968, p. 468)

In addition, science must, in contrast to "mere groping", take a "sure step" (Kant, 1959 [1781], B VII), i.e. proceed *methodically*. Scientificity is thus determined by Kant's character as a 'doctrine', by systematic and methodological procedure and by apodictic certainty or necessity.

Even before Kant, René Descartes (1596–1650) had also made demands on *knowledge* that were to guarantee scientificity. In view of the inadequacies of the medieval scientific system, Descartes sought a new foundation for the re-foundation of modern science that is unquestionable and absolutely certain. As part of a *methodical* doubt, he rejected all known sources of knowledge (such as sensory perception) and all previously known knowledge (empirical, conceptual and mathematical) whose truth and correctness can be doubted, in order to, after the 'dismantling' of the inherited structure of knowledge and its supposedly secure sources of knowledge, rebuild the sciences on a new foundation. This *'fundamentum inconcussum'* (unshakeable foundation) is the 'I' (lat. *ego*). For in the act of doubting as a form of thinking (lat. *cogitare*) one can no longer be doubted: the doubt itself. But doubting is a mental act that refers to a thinking 'I'. The undeniable doubting as thinking therefore requires the existence of an 'I'. Descartes summarizes the result of the methodical doubt in a well-known formula as follows: *'ego cogito, ergo sum'* ('I think, therefore I am') (Descartes, 2008 [1637]. From this foundation, Descartes rebuilds the sciences anew. The New consists ultimately in the new justification or foundation in consciousness. In this way, the foundations of science and possible criteria of scientificity are sought and found in an analysis of consciousness (lat. *mens*), which is why Descartes can also be considered a protagonist of the 'mentalist paradigm' (Schnädelbach, 1991) and thus of consciousness philosophy. In this context, he sets out four rules in the *'Discours de la méthode'* ('Discourse on Method') which "precede all object knowledge" (Schnädelbach, 1991, p. 63) and are to serve as criteria of scientificity: (1) apodictic certainty as conceptual clarity and distinctness, the analytic-recompositive method as (2) analysis and (3) methodologically regulated synthesis of what has been analyzed before as well as (4) the system requirement as a requirement of completeness (Descartes, 1990 [1637]). The systematic thus appears here in two traditional versions (cf. Le Rond d'Alembert, 2011 [1751]) as a *sprit systématique*, i.e. as 'systematic mind' (requirement for a methodologically regulated procedure of knowledge acquisition) and (2) as *ésprit de système*, i.e. as 'mind striving for the system' (requirement for completeness of knowledge). The requirement for completeness is now controversial, it appears to many epistemologists hopeless and presumptuous to try to achieve a complete system of knowledge. The demand for the *esprit systématique* on the other hand is still current and hardly ever questioned. Scientists, of whatever discipline, who cannot show and justify the expediency and scientific foundation of the methods they claim to use, are considered to be unprofessional (summary: Tab. 3.1).

The brief overview of the considerations of Aristotle, Kant and Descartes on criteria of scientificity suggests that philosophers and scientists orient themselves to an "idea of science" (Detel, 2018; Tetens, 2013) which is factually more or less realizable as an "ideal" and serves as a criterion for the assessment of science. With regard to the research literature, some criteria can be named which are widely shared by scientists and partly coincide with the traditional philosophical criteria. Accordingly, science is to produce knowledge by means of research and to systematize it in the form of theories (2010, p. 3; Endruweit, 2015, p. 15; Karmasin & Ribing, 2017, p. 83) and to justify it (2018; Poser, 2009; Tetens, 2013). The difference between everyday knowledge and scientific knowledge is also relevant in this context: With Voss (2017, p. 34), everyday knowledge is based on life-prac-

Tab. 3.1 Traditional criteria of science in Aristotle, Descartes, Kant (own presentation)

	Aristoteles	Descartes	Kant
Nature of the determination of scientificity	*Ontological* Determination	*Methodical* Determination	*Epistemological* Determination
Criterion 1	Universality	System requirement	System requirement
Criterion 2	Justifiability	Analytical-recompositive method	Principles-based methodology
Criterion 3	Necessity	Apodictic certainty	Apodictic certainty
Criterion 4	Teachability and representability		Systematic science as teaching

tical and interest-related subjective experiences which are passed on colloquially in interaction and communication contexts—whereas scientific knowledge is intersubjectively verifiable and in principle comprehensible and independent of non-scientific interests for everyone, published in scientific language in discipline-specific media.

The philosopher of science Wolfgang Stegmüller (1923–1991) has named three formal conditions of scientificity which are to be "together as *rational search for truth*" (Stegmüller, 1973, p. 5):

1) the effort for linguistic clarity, because the "raising of questions of understanding and the willingness to answer them [constitutes] one of the external characteristics of scientific discussion and rational conversation in general. [...] if there is a predominance of a Babylonian confusion of languages, science is not possible" (Stegmüller, 1973, pp. 5–6).
2) Scientific statements must be basically controllable or verifiable in order to be intersubjectively valid.
3) Scientific assertions must be rationally justifiable: "Where there are simply assertions against assertions, there is no scientific discussion. To questions of the form: 'where do you know that from?' the person being questioned must be *able to give a justification*. The appeal to an authority or to divine inspiration is just as little a rational justification as the subjective assurance of being completely convinced of the truth of the assertion" (Stegmüller, 1973, p. 6).

With Tetens (2013) the 'idea of science' can be divided into five 'ideals' which abstractly summarize the mentioned and other criteria:

1) The 'ideal of truth' obliges scientists to arm their theories against errors and deception and to find out whether something is actually the case or not.
2) The 'ideal of justification' requires that argumentatively justifiable and verifiable reasons be brought forward for statements and theories.
3) In view of a complex 'reality', the 'ideal of explanation and understanding' requires "scientists to look for patterns, rules, structures, how the facts are related to each other in the world" (Tetens, 2013, p. 21).

Tab. 3.2 Criteria of scientificity according to Stegmüller and Tetens

	Stegmüller	Tetens
Determination of the scientificity	*Rational search for truth*	*Ideals of science*
Criterion 1	Linguistic clarity	Self-reflection
Criterion 2	Controllability; Intersubjective verifiability	Intersubjectivity
Criterion 3	Justifiability	Justification
Criterion 4	–	Truth
Criterion 5	–	Explanation and understanding

4) The 'ideal of intersubjectivity' requires that scientific results can be followed up and checked by other competent scientists.
5) The 'ideal of self-reflexivity' requires the willingness to recognize one's own errors and mistakes and to critically question the methods used and the results achieved (Tab. 3.2).

3.1.3 Truth Theories

The previous explanations show that science and therefore also scientific theory aim at truth. But then the question has to be answered what truth 'is' or what can be understood under truth. Whoever studies the philosophical research literature on the topic of 'truth theories' comes across a wealth of different and sophisticated concepts of truth (u. v.: Enders & Szaif, 2006; Gloy, 2004; Janich, 1996; Skirbekk, 1977). Truth and knowledge are linked to each other, and vice versa: "Knowledge, wherever it occurs, occurs with the claim to truth" (Gloy, 2004, p. 67). But if "truth goes hand in hand with knowledge and knowledge consists in the relationship of the knowing subject to the known or to be known object" (Gloy, 2004, pp. 67–68), then it makes sense to use the formal schema of the relationship between a subject of knowledge and the possible object of knowledge in order to "bring a certain order into the wealth of concepts of truth by deriving basic types that correspond to certain types of knowledge" (Gloy, 2004, p. 67).

The first basic type results from the position of the object, which exists as observer-independent and autonomous with respect to subjective knowledge achievements, the subject is dependent on the objects in the process of knowledge. Truth refers to the 'being' or the factual content of the objects, and can therefore be understood as 'ontological truth' (gr. *on*, being). This understanding corresponds to the position of the *common sense* (of the pre-scientific everyday understanding of the 'common sense' of the 'healthy mind') and philosophical variants of a Platonic realism.

The second basic type results from the position of the subject, which is independent of the objects and is significantly and actively involved in the constitution of the object, the object is subject to the subject in the process of cognition. Truth refers to the subject and its constitution and can therefore be understood as "subjective" or "logical truth", insofar as "it comprises the systematic connection of thinking. Truth in this sense means coherence, occasionally also consensus, if the agreement of the speakers is meant" (Gloy, 2004, p. 72). This understanding corresponds to variants of so-called "idealist truth theories" such as coherence theories, consensus theories and pragmatism.

The third basic type results from the mediation, "alignment" or "correspondence" of subject and object. Truth refers to this agreement between subject and object and can therefore be understood as the "correctness" of this agreement. This understanding corresponds to variants of so-called "mirror theories", for example.

In the following we will introduce some of the most important truth theories in more detail. We cannot go into all the details, the goal is a first orientation about basic approaches that are explicitly or implicitly brought into play again and again in the theory of science. For didactic reasons, we begin with the correspondence theory.

3.1.3.1 Correspondence Theory

The correspondence theory not only reflects the truth understanding of common sense, but it "is also based consciously or unconsciously on modern truth theories" (Gloy, 2004, p. 93). 'True' in this understanding are such scientific statements or sentences that express the equalization (lat. *adaequatio*) of the scientific understanding (lat. *intellectus*) to the recognized thing (lat. *res*) (cf. Aquin, 2013 [1256–1259]; Zoglauer, 1999): With Thomas Aquinas, therefore, the classic definition is '*adaequatio rei et intellectus*' (Aquin, 2013 [1256–1259]). This understanding of truth is traditionally referred to as the 'correspondence theory of truth' (cf. Gloy, 2004, pp. 92–167; Janich, 1996, pp. 30–39)—the truth is independent of opinions and contexts and exists solely in the "agreement of knowledge with its object" (Kant, 1959 [1781], B 82). With Hegel and Heidegger, this understanding of truth can also be referred to as 'correctness' (Hegel, 1970; Heidegger, 1992).

The following general objection can be raised against correspondence theories, in particular in the form of mapping theories: A correspondence between statements and objects or images and the mapped cannot be established at all. Only statements can be compared with statements, images with images. People cannot perceive objects unmediated, as they apparently seem to exist independently of the observer, but only filtered through the conditions of perception and understanding. In order to be able to compare both members of the relationship of knowledge, an objective ('quasi-divine') perspective outside of this relationship would be necessary, which people are fundamentally denied. People are only left with the internal perspective. In cognitive psychology and brain research, the attempt to justify an 'external appropriateness' (Merker et al., 1998) of perception and object is referred to as the 'fallacy of the homunculus', 'that is, the perceiving little man in the brain or consciousness, who looks at the image of the event [...] there, recognizes and evaluates it' (Janich, 1996, p. 33) and compares it with the perceived and makes a perceptual judgment. Such 'mapping theories are now considered to be refuted' (Janich, 1996, p. 36).

3.1.3.2 Coherence Theory

The strength of coherence theories lies in the weakness of correspondence and copy theories. If statements cannot be compared with "reality", but only with statements, then only statements are compared with each other, thus a consistent internal perspective is taken and only an "internal fit" (Merker et al., 1998) is sought. Statements within a sentence or theory system are compared with regard to coherence. Coherence means freedom from contradiction or compatibility or fit with other statements *within* this system. "Truth" is nothing more for coherence theorists than this freedom from contradiction of the statements. If a statement is contradictory, either this statement or the whole system is false. Truth is then always "context-dependent" and "depends on the beliefs of the members of a research community who accept a theory T as true. Truth is always relative to a theory T" (Zoglauer, 1999, p. 32). Coherence theorists reject the idea of given and secure truths. Instead, they assume that scientists work on theories that are always uncertain and in need of modification, without ever being able to enter a safe haven of comparison of the theory with a given "reality". The sociologist Otto Neurath (1882–1945) once compared this situation to the following image:

> "There is no *tabula rasa*. Like shipbuilders, we who have to rebuild our ship at sea without ever being able to dismantle it in a dock and rebuild it from the best parts." (Neurath, 1932, p. 206)

The coherence theory is subject to some objections, which we will briefly introduce.

a. **Isolation objection**: The objection "argues that the beliefs of a coherent system could be completely 'isolated' from the world. It could be that the beliefs only fit together coherently because they are all (or almost all) wrong"—an error would therefore "become invisible if it is assumed to be of sufficient size" (Flick et al., 2007, p. 92; Ernst, 2014, p. 92).
b. **Multiple-systems objection**: There can be several (partly contradictory) coherent systems side by side. The question of truth is shifted: *Which* system is then *right*? (Ernst, 2014, p. 92)

3.1.3.3 Consensus Theory

The consensus theory of truth, as developed in particular by Jürgen Habermas, ties 'truth' to statements that receive the consent and acceptance of the participants of a discourse as an regulated discussion context of equal discussants. 'Truth' consists in a consensus, an agreement: "We call sentences true whose claim to validity must be acknowledged by every reasonable person" (Habermas & Luhmann,

1972, p. 222). In the framework of his 'Theory of Communicative Action', Habermas distinguishes three claims to validity: 'truth', 'rightness', and 'truthfulness' (Habermas, 1995, p. 412). 'Truth' refers to an "objective world" of facts, 'rightness' to a "social world" of norms and values, 'truthfulness' to the "subjective world" of a person (Habermas, 1995, p. 439). A consensus on the truth claim of an assertion is reached if all discourse participants in an ideally and fairly organized 'non-dominant discourse' and after exchange of all reasonable arguments can agree to this claim. In the framework of this 'ideal speech situation', the discourse participants can ideally at some point no longer escape the 'unforced force of the better argument' (Habermas, 1983).

Two objections to the consensus theory of truth will be briefly mentioned:

1) A common objection to the consensus theory of truth criticizes the *lack of practicability* of the procedure. The 'ideal speech situation' requires the consent of all potential consenting parties. This ideal can lead in the practice of concrete discourse to the fact that they do not want to end, it comes to a 'dictatorship of the seat meat' (Weinrich, 1975) (compare Reese-Schäfer, 1991) and discourse often remains without result.
2) Another objection is directed against a widely held *consensus-seeking optimism* and a self-imposed consensus-seeking compulsion (cf. Berr et al., 2019). Critics point out that the "consensus expectation" associated with discourses is often too high, as the required "talking together" turns out to be "ambivalent" (Ottmann, 2012, p. 118). In the tradition of a human image that determines people as a '*zoon politikon*' (Aristotle, 2009), that is, as a political living being that is dependent on a life in and through community, communication can lead to 'community'. In the opposite tradition, which determines people as a '*homo homini lupus*' ('man is a wolf to man') (Hobbes, 2017 [1651]), human contact can also lead to ongoing dissent, conflict (Dahrendorf, 1972) and even hostility (cf. Grau, 2017; Mouffe, 2007, 2014; Schmitt, 1933, 2011 [1967]).

3.1.3.4 Pragmatic Truth Theory

Pragmatism (see Sect. 3.5.1) avoids the difficulties of correspondence, coherence, and consensus theories from the outset by not taking truth to be statements and thus theoretical, but rather binding it to the *practice* of acting people in everyday life and science. Pragmatism is oriented towards the *effects* of action, so that not internal or external agreement, but usefulness or success in concrete contexts become the criterion of 'truth' (Joas, 1988; Schubert et al., 2010; Steiner, 2014a, b). This confuses the 'true' as theoretical with the 'good' as ethical (see Sect. 3.1.1) and also reduces the 'good' to the 'useful', that is, to an evaluative ('suitable' vs. 'unsuitable') meaning (cf. Berr, 2019; Berr and Kühne, 2019) while ignoring the normative meaning ('good' vs. 'bad') (cf. Gloy, 2004).

As appealing as this approach may appear at first glance, it is also subject to a serious objection. The equation of truth and success has to face the fact that successful theories can be false (as in the Ptolemaic world view), while initially unsuccessful theories can be successful and established at a later point in time (Gloy, 2004; Zoglauer, 1999). Moreover, the question of usefulness is already *assumed* before the question of what is meant by 'good'. But with Bertrand Russell (1872–1970) we have to *know* what 'good' is, "before we can recognize that anything is 'true'; because only after we have decided that the effects of a belief are good, can we rightly call it 'true'. This makes the whole thing incredibly complicated" (quoted from Gloy, 2004, p. 224).

The truth criterion of 'usefulness' or 'success' was specified by some pragmatists. William James (1842–1910) developed a 'genetic truth theory' (Diaz-Bone & Schubert, 1996), which assumes that new 'ideas', new knowledge or new 'convictions' can be combined with the 'old' ideas, knowledge and convictions. Only in this way would learning and progress in knowledge be possible, 'truth' would be subject to constant change in an evolutionary development process (James, 1977). John Dewey (1859–1952) specifies the pragmatic understanding of truth by introducing his concept of 'warranted assertibility' (Dewey, 2016; cf. Neubert, 2004). This concept locates truth temporally between already established scientific methods and results on the one hand and methods and results lying in the future on the other hand, thus between revisions of previous truths and justification of new truths (cf. Neubert, 2004). In this context, the concept of "viability" (Glasersfeld, 1988, 1995) can also be located, which is advocated by representatives of radical constructivism (see further Sect. 6.3). "Viability" means usefulness or validity of constructions of reality, as long as they are useful in practical action or prove themselves in scientific practice. Viability moves away from the idea of a general "truth" here. In this radical constructivist context, the question is not whether a statement is "true", but whether it is goal-oriented or connectable in the chosen context. For example, the statement "Bielefeld is a city in North Rhine-Westphalia" is a viable statement regardless of the assumption of the "Bielefeld conspiracy" (which has the purpose that Bielefeld does not exist), because it simply appears to be useful to assume that Bielefeld exists, for example, because it would be too embarrassing if VfL Bochum had lost 0:2 against Arminia Bielefeld on 20.1.20 against a non-existent opponent.

As we have seen, Habermas distinguishes between the claims to validity "truth" and "truthfulness", which can be understood as part of the linguistic philosophy of the traditional triad of "true, beautiful and good". However, this distinction often leads to misunderstandings, which is why some comments may be helpful for a better understanding (Textbox 10)

Textbox 10: Truth and Truthfulness

The "true" or "truth" in the context of the "correspondence theory of truth" is understood as "agreement of knowledge with its object" (Kant, 1959 [1781], B 82). The statement "The Black Forest is a mountain range in the southwest of Baden-Württemberg" is true if this statement can be confirmed factually. But "truth" can also be understood as "agreement of a content with itself" (Hegel,

1970, p. 85), i.e. as agreement of an object with its *concept*. This understanding can already be found "in everyday language usage. For example, one speaks of a *true* friend and understands this to mean a friend whose behavior is in accordance with the concept of friendship" (Hegel, 1970, p. 85). This understanding of truth is not only found in everyday understanding, but also among landscape experts in planning, science and administration, so that in each case an idealized image of a "true landscape" (cf. Kühne, 2018, p. 50) is used as a starting point and the actual landscape is measured and evaluated against this idealized image.

Such measuring of reality against conceptual semantics is not problematic per se, but only if the measuring scale is understood as an unchangeable "essential" combination of characteristics, thus "essentialist" (see Sect. 3.2). What is overlooked is that concepts are subject to constant change—this change of concepts is dealt with, for example, by "Historical Semantics" (cf. Koselleck, 1979; Müller & Schmieder, 2016; Riecke, 2011).

The aforementioned understanding of truth moves the 'true' idea of something closer to the semantic concept of 'authenticity' in the sense of 'genuineness' or 'truthfulness'. This similarity is based on a traditional understanding of 'beauty', which can not only be the 'pleasing' and the 'harmonious', but also the 'authentic' or 'truthful'. In everyday life, for example, we find people who radiate a certain authenticity or truthfulness in the sense of naturalness or sincerity, 'beautiful'. This specification is similar in meaning to Habermas, who understands it as a concordance of external appearance and internal motive (Habermas, 1995, p. 411).

We should therefore only attribute statements and judgments to 'truth', 'truthfulness' to people in their personal appearance. ◀

The following figure 3.4 provides an overview of the presented theories of truth and arranges the already explained and the still to be processed scientific theories Positions and protagonists of influential philosophical approaches to these truth theories. Some positions (logical empiricism and pragmatism) or protagonists (Ludwig Wittgenstein and Charles Sanders Peirce) appear twice or more at different places; this is because these positions or these protagonists have undergone developments that led to different positioning and theory models. The figure also makes it clear that certain theories (here in particular essentialism) are received to different degrees of intensity (a loss of meaning in essentialism), while other theories are able to frame matters in a more 'viable' way. It is also clear that the development of truth theories—so it is to be assumed—will not be completed, after all society, environment, the relationship between the two, as well as—and this is particularly relevant for the theory of science—the scientific concern with these changes, will continue to develop. But: It can also turn out differently, because the future is—and here we turn to the philosophy of science and society by Karl Popper—after all open (Popper, 1959, 1984, 2003 [1945]).

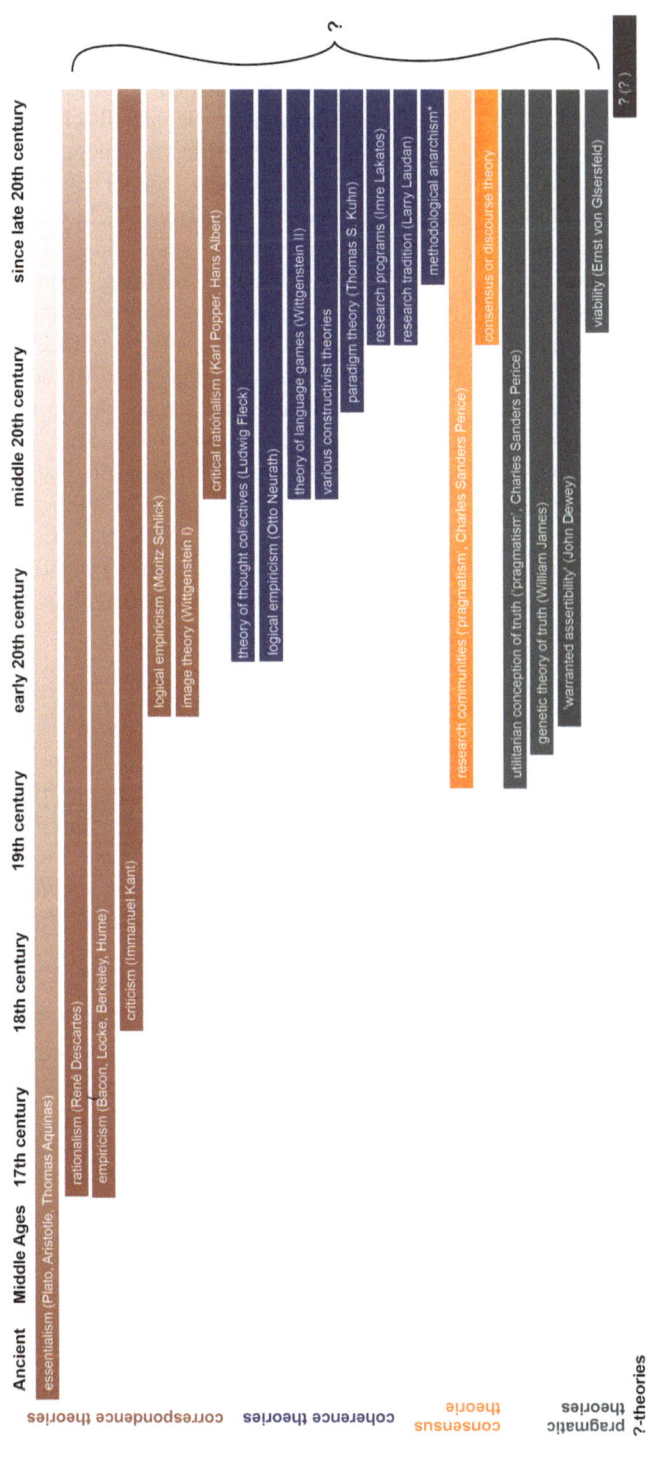

Fig. 3.4 Truth theories and scientific positions with protagonists over time. The intensity of the shading conveys an impression of the reception at different times. It is clear that the development of truth theories and scientific positions—most likely—is not yet completed (own representation)

3.2   Essentialism

Essentialism is a position in the history of philosophy that can be associated with the "ontological paradigm" mentioned at the beginning (Schnädelbach, 1991). In this context, attention is focused on the objects of scientific knowledge, according to which research is carried out on the basis of the viewer-independent "essence" or "essentials" of the object, or on its "substance" or *Essentia*. In exactly this sense, such a position can be referred to as "essentialist" (cf. Kühne, 2019; Schneider, 2019; Schwemmer, 2004b). With the "essence" of things also appears the "nature" of things (Schwemmer, 2004a). Essentialist positions assume in this sense that a) the essence is the basis or the "underlying" of the "existence", facticity or "being" (lat. *existentia*) of the individual thing or object and b) the "things have essential properties that make up their essence and are not due to the attribution of a viewer, but must be sought and justified in the object itself" (Lautensach, 1973, p. 24). These essential (necessary) properties are distinguished from accidental (contingent) properties, with the essential properties of a thing making it "what it is, while the accidental properties are not relevant to the existence of the thing" (Albert, 2005, p. 44). Already in ancient Greece, this way of thinking developed in two typical variants, which founded a Platonic-Aristotelian tradition of essentialism.

3.2.1   The Platonic-Aristotelian Tradition of Essentialism

Plato claimed in the context of his theory of ideas (see, for example, Bormann, 1987; Martin, 1973), the essence of things or of 'reality' is an 'idea' (gr. *idéa*) as a form (gr. *eidos*) of things, which is both the reason for existence and the 'nature' of things or of 'reality'. The 'essence', for example, of justice (gr. *dikaiosyne*) is the *'idea of justice'*, which first of all makes just actions possible in their existence and makes these just actions distinguishable and identifiable as just actions from other unjust actions. Plato's ideas are perfect, eternal, unchangeable and really existing archetypes (gr. *paradeigmata*) of all imperfect, transitory, subject to change individual things, people, artifacts, etc. For example, in the dialogue 'Parmenides' they are for him '*homoiômata*' ('images') (Burnet, 1962) of the archetypal idea, of which they can only participate (gr. *methexis*). Ideas are therefore on the one hand ordered before the things (lat. *universale ante rem*), on the other hand separated from them spatially (gr. *chorísmos*), since they are located in an 'extra-sensory place' (gr. *topos hyperouranios*) outside the sensually perceivable 'world'. This conception leads to numerous inconsistencies, which Plato himself reflected on in his late work (see Martin, 1973).

Aristotle wants to avoid these inconsistencies and difficulties of an doctrine of ideas. For him, the *'substance'* as essence or being (lat. *essentia, quidditas*) of things is not separated from the existence or being (lat. *existentia*) of things: everything that is, "is a unity of matter and form, which can only be divided mentally, but not really" (Hauskeller, 2005, p. 15). The essence of things is *in* the things (lat. *universale in re*) and not spatially separated from them. Therefore, the forms or the substance or the essence (gr. *ousía*) of things do not exist outside of the world, but *in* it (cf. Hirschberger, 1976). In the Categories (from gr. *kategoria*, statement) Aristotle distinguishes the individual (individual people, things, artifacts) as 'first substance' (gr. *próte ousía*) and bearer of properties from the kind or genus or the general as 'second substance' (gr. *deutéra ousía*), which, as a kind of belonging, decides "*what* a thing is and what its essential features are" (Rapp, 2010, p. 46). Concepts and their definitions are obtained by abstraction, the essential as the general can only be separated mentally from the individual and then 'abstracted'). This abstraction arises, "if, on the basis of many observations of experience, a general understanding of similar situations develops" (Aristotle, 1991 [348-345 BC], 981a). The general, the essence is therefore only really at things and only through and at them recognizable; but the things are also only the *what* they are, through the general.

Essentialism in this Platonic-Aristotelian tradition therefore claims first of all an ontological primacy of the "essence" or the "essentia" over the "existence" or the "existentia". Secondly, it assumes that the "essence" is ontologically and epistemically independent of subjective attributions: Scientific definitions aim to capture and describe the essential, the essence of things conceptually as they are also independent of a possible observer. The nominalist counter-position to both Platonism of ideas and Aristotelian substance-realism, according to which general concepts and their definitions do not mirror language- and knowledge-independent entities, but only serve as "names of things" (lat. nomina rerum), was first developed in the medieval "universal dispute" (cf. Gethmann, 2004; Stegmüller, 1978b). The "universals" (general concepts) are not *pre* -ordered (lat. ante rem) and not *in* them (lat. in re) anchored, but the things by means of naming *post* -ordered (lat. post rem).

Despite the nominalist countercurrent, essentialism has been attributed to Western philosophy up to Hegel (cf. Schwemmer, 2004b). Popper concretizes this general criticism and in particular characterizes the essentialism standing in the Platonic tradition as *methodological* essentialism. Characteristic of this essentialism is the view that "the goal of science is to reveal and describe essences with the help of definitions" (Popper, 2003 [1945], p. 40). Popper sets this against a methodological nominalism, which *not* has the goal of finding out and defining the "true nature of a thing", but "describing the behavior of a thing under different circumstances and in particular to indicate whether this behavior shows any regularities" (Popper, 2003 [1945], p. 40).

3.2.2 Normative Consequences of Essentialism

Since the essence of things is at the same time their "nature" (cf. Schwemmer, 2004a), essentialist positions are almost always normative as well, insofar as the "natural" is claimed to be the "normal" and thus the "normative" in one. The "natural" is opposed to the "unnatural" as valueless and contrary to the norm (cf. Birnbacher, 2006, pp. 17–64). This understanding is not only a theoretical problem, but also has concrete practical consequences. For example, transferred to "landscape", this understanding leads to the understanding of "landscape" as an objectively given and observer-independent entity (Kühne, 2013; Kühne et al., 2018; Lautensach, 1973) with specific "characteristics" and "peculiarities". "Landscape" or "cultural landscape" (cf. Weber & Kühne, 2019) is considered in this perspective as a "quasi-organic unity with special features" with its own "immovable value and identity" (Gailing & Leibenath, 2012, p. 97). This view can be found, for example, in the Federal Nature Conservation Act, according to which the "diversity, peculiarity and beauty as well as the recreational value of nature and landscape" are protected in § 1 (1) (BNatSchG, 2009 [1976]). "Nature" is therefore not only a consequential, but also a prerequisite-rich linguistic construct (Textbox 11).

Textbox 11: Nature

The term "nature" (from Latin *natura*, based on *nasci*, to be born, to arise; Greek *physis*) can be characterized as "a shimmering concept with blurred contours" (Falkenburg, 2017, p. 96). In first approximation, different basic concepts of nature can be distinguished (Birnbacher, 2006; Falkenburg, 2017; Gloy, 2005 [1995]; Schäfer, 1991).

1) Extensional (scope of the concept) as the 'physical world' from the immediate environment to the entire universe or as a '*area*' (the 'inclusion of phenomena'), i.e. as an area of eternal laws, invariant order, unbreakable necessity, as a cosmos, as a universe or as a 'divine artifact' (as a creation).
2) As a 'mode of being', i.e. intensional (conceptual content) as a certain 'property' of something (entities, beings): b with a c with b d with c
   a) as a mode of emergence (*physis*, *natura*: self-generated, growth, being alive) and as not made by humans or
   b) essentialistically as "the essence of something" or
   c) as a mode of action (targeted due to an internal program: entelechy).
3) as a '*norm*' (i.e. as the 'natural'). In this understanding, nature is both the original and the normal; and what is natural has priority over what is contrary to it: as abnormal, pathological, deviant, perverse. So what is natural moves into the rank of the normative and exemplary.

Furthermore, the concept of "nature" only "receives its determinacy through oppositions" (Schäfer, 1991, p. 480). Typical "contrasts to the concept of nature" (cf. Schäfer, 1991, pp. 480–482; cf. also Falkenburg, 2017, pp. 100–101) are, for example, the oppositions that arose in ancient Greece between

1) *physis* versus *techne*, that is, the distinction between things that have arisen through nature versus those that have arisen through human artistry.
2) *physei* versus *thesei*, that is, the distinction between legal relationships that exist by nature ('natural law' tradition) versus those that exist through human enactment ('positive law') as well as
3) the well-known oppositions of nature vs. art, nature vs. culture, nature vs. history, nature vs. freedom, nature vs. mind.

All these distinctions and concepts of understanding run through the history of the concept of nature in philosophy and sciences (cf. u. v. Falkenburg, 2017, S. 96–100; Schäfer, 1991, S. 482–501). This 'shimmering' of the concept often makes it difficult to have a rational and factual discussion about this concept and its suitability for scientific research and practical oriented discussions (for example in nature conservation and climate policy). ◀

3.3   Empiricism, Inductivism and the Principle of Causality

Empiricism is an epistemological position within modern philosophy. It is also a response to rationalism, which was founded in the 17th century by René Descartes (1596–1650) and further developed by Baruch de Spinoza (1632–1677), Gottfried Wilhelm Leibniz (1646–1716) and Christian Wolff (1679–1754). Both positions are historically attributable to the 'consciousness-philosophical' or 'mentalistic paradigm' (Schnädelbach, 1991). The reason for the switch from the 'ontological' to the 'mentalistic' paradigm (see Sect. 3.1.2) is a fundamental doubt about the possibility of knowledge of the 'true being' (gr. *óntos ón*) as the 'essence' (lat. *essentia*) of things and reality. The 'Adaequationstheorie der Wahrheit' (see Sect. 3.1.3), which is decisive for the 'ontological paradigm' and according to which knowledge and truth consist in the equalization (lat. *adaequatio*) of the scientific mind (lat. *intellectus*) with the thing recognized (lat. *res*) (lat. *Veritas est adaequatio rei et intellectus*; cf. Aquin, 2013 [1256–1259]), is increasingly being called into question. Especially the possibility of recognizing the '*res*': "Ontological philosophizing becomes impossible if it is first fundamentally doubted that knowledge of true being can succeed" (Schnädelbach, 1991, p. 59). The way to knowledge and truth is no longer paved in the objects of knowledge, but initially only the '*intellectus*' (mind, understanding) remains, from which this way must be newly paved. This leads to fundamental investigations of the functioning and performance of human consciousness. The ancient and medieval question 'What is …?' (e.g. 'What *is* space?'; 'What *is* landscape?'), which aimed at the 'What-Being' (lat. *quidditas*) of things, is now replaced by the question 'What can we *recognize* in which way and under which conditions and thus *know*?'.

Both positions have a common starting point in the framework of a fundamentalist-certist epistemological program (cf. Sect. 3.1.2): Starting from a first, immediate, unquestionable and secure intuitive insight into absolutely certain and necessary epistemological principles (cf. Schurz, 2014), a system of secure knowledge should be built strictly methodologically from this foundation and each step should be strictly necessary from the previous one. Rationalism and empiricism are "sister concepts" (Beckmann, 1997, p. 162) in this sense, significant differences only become apparent in different areas of application (e.g. mathematics or natural sciences) (cf. Engfer, 1996). The difference lies mainly in the different weighting of the two *sources* 'reason' and 'experience' for the acquisition of certain knowledge. The basic assumptions of rationalism in the form first developed by René Descartes consist first of all in "the clearer emphasis on the subject compared to the thinking of antiquity, secondly in the assumption of a consistently rational structure of reality and thirdly in the claim that human reason because of its special equipment is able to grasp the rationality of reality with conceptual clarity" (Beckmann, 1997, p. 164). However, it must be avoided that rationalism only recognizes *reason*, just as empiricism only recognizes *experience* as a certain source of knowledge. Thus, the *rationalism* grants reason a "priority" over experience or sensory knowledge (Gabriel, 1993, p. 29). From first indubitable principles of reason ('innate ideas' like 'God', '*res extensa*', '*res cogitans*') knowledge is derived deductively, mathematics serves as a scientific guideline and model (cf. Beckmann, 1997; Gawlick, 1995). The *empiricism* is based on experience: "Without experience there is no knowledge of reality at all, all certainty is based on sensory certainty; knowledge is built from elements that are given immediately and thus indubitable; the inductive methods of natural science are exemplary" (Gawlick, 1995, p. 11). This means "that all human knowledge [...] is primarily based on experience, not primarily on thinking" (Bauberger, 2016, p. 13) or on reason.

Rationalism and empiricism have been irreconcilable with each other from the beginning, despite some similarities. Only Immanuel Kant was able to make a synthesis of both positions in the Critique of Pure Reason (1959 [1781]) (Textbox 12).

Textbox 12: Kant's Synthesis of Rationalism and Empiricism

On the one hand, Kant exposes the false claims of each position, on the other hand, he at the same time mediates the respective right of both approaches. Kant ascribes a contradiction to empiricism to its self-conception as an empirical theory: The exclusive reference to experience as a source of knowledge is an assertion for which the field of experience must be left in order to speak about experience as the sole source of knowledge. In addition, he can show how and why experience-related conditions of the possibility of experience are necessary in order to come to empirical experiences: the forms of intuition of space and time and the categories. Kant reproaches rationalism for overlooking the autonomy of sensory perception as a source of knowledge. Just as experience cannot take place without a priori concepts and forms of intuition, so too are these sensory data indispensable. These arise from sensuality as an autonomous ability. Since Kant thus makes the limits of reason and sensuality recognizable in each case, his position is also referred to as 'criticism' (gr. *krinein*, to draw, to distinguish, to separate). Thus, according to Kant, sensuality and reason can be mediated by showing both to be indispensable for knowledge in their own way. Therefore, every factual knowledge necessarily draws from two sources of knowledge (Tab. 3.3). Kant's famous formulation is therefore:

"Thoughts without content are empty, intuitions without concepts are blind". (Kant, 1959 [1781], B 75)

Another famous formulation of Kant's is therefore:

"It seems that only so much is needed for the introduction or pre-memory, that there are two branches of human knowledge, which may perhaps spring from a common, but unknown root, namely *sensibility* and *mind*, through the former of which objects are given to us, but through the latter are thought." (Kant, 1959 [1781], B 29) ◄

3.3.1 Preparation of Empiricism by Criticism of Prejudice and Induction: Francis Bacon

Francis Bacon (1561–1626) is considered the "pioneer" (Schülein & Reitze, 2012, p. 67) of empiricism, with him "the movement of British empiricism begins" (Bauberger, 2016, p. 13). Corresponding to the role model function of the natural sciences for empiricism, he can also be considered the "first philosopher of modern natural science" (Carrier, 2017, p. 16). A "kind of founding document of the theory of science" and modernization of Aristotle's '*Organon*' (from Greek *organon*, tool; as a designation for his logical-scientific-theoretical writings) is Bacon's main work '*Novum organon scientiarum*' (Bacon, 1990 [1620]), "in which Bacon imposes the systematic consideration of experience on the emerging natural sciences" (Carrier, 2017, p. 16). His doctrine of methods in the sense of a 'new' methodological 'tool' (Latin *novum organon*) is based on three fundamental steps (cf. Carrier, 2017).

Tab. 3.3 Rationalism and Empiricism in Comparison (own representation)

	Rationalism	Empiricism
Protagonists	Descartes, Spinoza, Leibniz	Bacon, Locke, Berkeley, Hume
Method	Deduction	Induction
Source of knowledge	Reason	Experience
Kant's accusation	Thoughts without content are empty	Intuitions without concepts are blind
Kant's Note	Sensibility is an independent source of knowledge	Reason and mind are necessary sources of knowledge

First, the reliability of experience qua sensory perception depends on scientists freeing themselves from illusions (Latin *idola*), that is, the individual and socially mediated prejudices that could reduce the quality of the observations and "contribute to the imperfection of human knowledge" (Carrier, 2017, p. 18) and are therefore absolutely to be avoided. Bacon addresses (Bacon, 1990 [1620]):

1) stereotypical 'natural' illusions of the human species (lat. *idolae tribus*), for example the insistence on preconceived opinions and existing theories which are 'immunized' against refutations (compare also Popper's theory of falsification, Sect. 3.6);
2) illusions of the cave (lat. *idolae specus*), which are due to individual prejudices, when scientists, as it were, take refuge in the cave of individual views and convictions;
3) illusions of the market (lat. *idolae fori*), when—anticipating Wittgenstein's criticism of language (compare Sect. 3.4.2)—language, through imprecise formulations and attributions of meaning, impedes the unbiassed access to natural phenomena;
4) the illusions of the theater (lat. *idolae theatri*), that is the errors of the philosophical schools (today they would be the communities of scientists who are trapped in their paradigms), which "invent" concepts and theories to a certain extent (cf. also Carnap's criticism of metaphysics, Sect. 3.4.4). Positively, Bacon places the demand for experimentation as the "factual basis of science" next to the criticism of prejudice; "Bacon is the first to explicitly emphasize the importance of experimentation as a means of knowledge in natural science" (Carrier, 2017, p. 19).

Secondly, Bacon works as a method to achieve true knowledge, the *true induction* (lat. *inductio vera*; Bacon, 1990 [1620]). In order to avoid the hasty conclusions of observations on generalizations within the traditional 'anticipation of the mind' (lat. *Anticipatio mentis*), he relies on a gradual and slow approach of a 'interpretation of nature' (lat. *Interpretatio naturae*) (Bacon, 1990 [1620]). He demands the "specification of precautionary rules" (Carrier, 2017, p. 22) by means of creation, comparison and evaluation of tables (lat. *Tabulae*), in order to succeed in an 'exclusion analysis' (from lat. *excludere*, exclude, reject), which segregates 'essentialist' unessential from essential properties, to classification of the observed phenomena and successively to generalizing explanatory principles, generalizations and ultimately to general forms of nature. In Bacon's own example of the research of the phenomenon 'heat', he relies on the evaluation of the correlation tables on a weak analogy (cf. Sect. 2.4.3), insofar as he uses the "assumption of an analogy between the mechanisms in the large and in the small" (Carrier, 2017, p. 24) as a basis for the claimed identity of heat and microscopic movement.

Thirdly, the "deductive descent to observational consequences" follows from the inductive ascent from observations to generalizations (Carrier, 2017, p. 25). The so-called '*Instantiae crucis*' (Bacon, 1990 [1620], Part 2, Aphorism 36), which, in the sense of the indirect proof known in mathematics, set up two mutually contradictory hypotheses and submit them to further experience in an experiment, are known. If one is refuted by experience, the opposite is considered to be correct and proven. Popper later described this procedure as a criterion of scientific theories; his method of falsification therefore goes in a similar direction (see Sect. 4.1). With Bacon, therefore, "a interplay of empirical experience and the ordering activity of the mind" (Schülein & Reitze, 2012, p. 71) can be observed, which is also characteristic of the following protagonist of empiricism.

3.3.2   Empiricism's Justification Through a Return to the Experiential Origin of All Knowledge: John Locke

John Locke (1632–1704) is generally considered the "founder" of empiricism, "because he was the first to attempt to show the experiential origin of all our concepts in detail" (Gawlick, 1995, p. 74) and thus to formulate and epistemologically justify Bacon's empiricist program for the first time (Schülein & Reitze, 2012). In his epistemological main work *Essay Concerning Human Understanding* (Locke, 1981 [1689]), Locke first turns against the rationalist assumption that there are innate principles in the human mind (such as the 'law of the excluded middle'), or—as in the case of Descartes—in addition to 'acquired' ideas (of sensory perception such as 'yellow') and 'self-made' ideas (by composition such as 'golden mountain'), also the '*innate* ideas' 'God', 'res extensa' and 'res cogitans' (Descartes, 2008 [1637], Meditatio III), with which data can be ordered and from which deductively sound knowledge can be derived. Instead, the individual human mind of a person at their birth is a '*tabula rasa*'(blank or un inscribed table), that is, an "white paper, void of all characters, without any ideas" (Locke, 1981 [1689], p. 107, vol. 1). The 'ideas' all originate from

> "*Experience*. In that all our knowledge is founded; and from that it ultimately derives itself. Our observation employed either, about external sensible objects, or about the in ternal operations of our minds perceived and reflected on by ourselves, is that which supplies our understandings with all the materials of thinking. These two are the fountains of knowledge, from whence all the ideas we have, or can naturally have, do spring." (Locke, 1981 [1689], p. 108, vol. 1)

Accordingly, he distinguishes between the '*sensation*'as external sensory perception and the '*reflection*'as internal self-perception. Both modes of perception generate ideas, namely simple and complex ideas. Simple ideas are caused by passively received stimuli that emanate from the objects and are perceived by only one sense (colors, sounds), multiple senses (spaces, movements), by 'reflection' (thinking, willing) and finally by 'sensation' *and* reflection (joy, pain).

Locke also differentiates between primary and secondary qualities (Locke, 1981 [1689], vol. 1) in sensory perception. The primary qualities are consciousness-independent material qualities such as size, shape, position, movement, and they are ultimately quantitatively measurable. The secondary qualities such as colors, taste, smell, etc. are ontologically dependent *and* consciousness-dependent, because they can only appear through the mediation of the human mind. Locke thus represents an epistemological realism (Gabriel, 1993; see Sect. 3.1.1), insofar as he distinguishes between a 'world in itself' of primary qualities and a 'world as representation' of secondary qualities:

> "If our senses actually provide our mind with an idea, we can be sure that *at this moment* there really is a thing outside of us that affects our senses, makes itself known to our perception through their mediation and actually generates that idea which we then perceive." (Locke, 1981 [1689], p. 2, vol. 2)

From simple ideas, the human mind can generate complex ideas by means of comparison, separation, connection, and abstraction: substances (individual things or species), modes (ideas that occur in substances, such as distance and extension in space), relations (relationships such as cause and effect, spatial up and down). Knowledge or understanding is "the perception of the connexion and agreement, or disagreement and repugnancy of any of our Ideas" (Locke, 1981 [1689], p. 167, vol. 2; fig. 3.5).

3.3.3 Exaggeration of Empiricism in the Form of Immaterialism: George Berkeley

The Irish Bishop of Cloyne, George Berkeley (1685–1753), criticizes Locke's distinction between primary (perception-independent) and secondary (perception-dependent) qualities as untenable. For this distinction first of all opens an unbridgeable gap between the recognizing subject and the recognized object, between knowledge and world (dualism, see Sect. 3.1.1). Secondly, the question is inescapable where we can know of a material external world if our knowledge consists only in the comparison of ideas (see Sect. 3.1.3). How can material objects then be compared to ideas as non-ideas with the ideas they cause (Gabriel, 1993)? And thirdly, this is Berkeley's decisive objection, there can be no ideas of material primary qualities outside of consciousness (such as size, shape and movement), without at the same time imagining secondary qualities in consciousness (such as colors, sounds, heat, cold). There are therefore no general abstract ideas (of primary qualities). So there are also no *material substances* as carriers of primary qualities that exist independently of our perception. What exists is only what can be perceived: *"esse est percipi"* (Being is perceiving). Gabriel (1993, p. 104) summarizes the structure of the argument as follows: "The secondary qualities are conceded to be dependent on the recognizing consciousness, the primary qualities cannot be thought (i.e. imagined) independently of secondary qualities; therefore the primary qualities cannot be thought (imagined) independently of the recognizing consciousness". Berkeley does not deny the reality of a phenomenal external world at all, but he denies that this phenomenal external world is based on an *material world in itself*. The 'idealism' often attributed to Berkeley (see Sect. 3.1.1) is therefore, according to Berkeley's own statements, an 'immaterialism' or

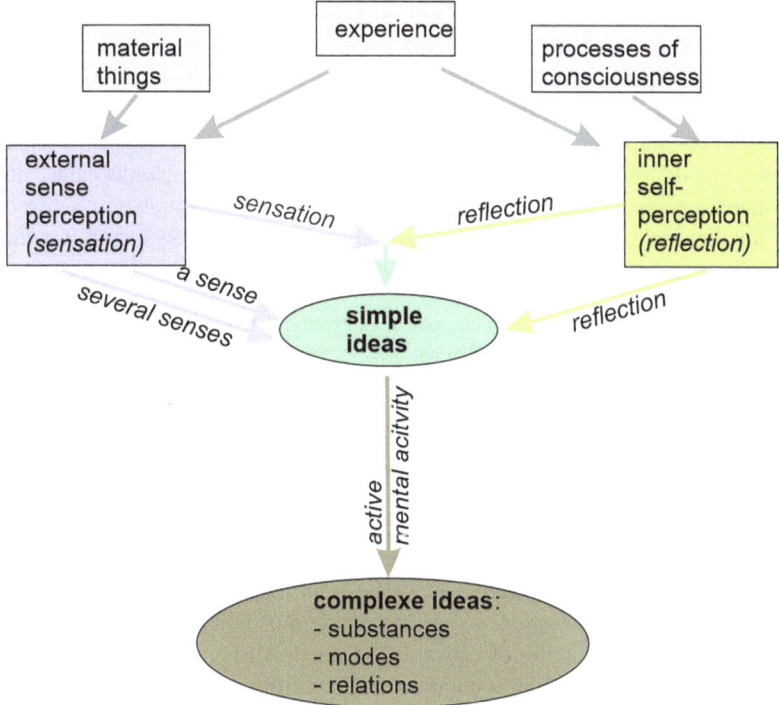

Fig. 3.5 John Locke's theory of ideas (own representation based on: Kunzmann et al., 1993, p. 118)

a 'spiritualism' (from Latin *spiritus*, breath, spirit). As counterintuitive as this position may appear at first glance—there is a connection via Fichte's idealistic theory of a self-positing 'I' (consciousness), which also sets itself a 'non-I' (external world) in front of itself, up to radical constructivism (see Sect. 6.3).

3.3.4 The Limits of Empiricism, the Principle of Causality and the Induction: David Hume

David Hume (1711–1776) avoids the ontological question of the reality of things, which is guaranteed by Locke's primary qualities and led to Berkeley's sharp criticism, by modifying Locke's terminology. In place of the superordinate concept of "idea" or "representation" he sets the concept of "perception" (from Latin *percipere*, to grasp, to understand), which is divided into impressions (from Latin *impressio*, impression) and ideas or representations. Impressions are immediate understandings of inner self-perception and external sensory perception. The ideas or representations are *images* of the *impressions* and not of the things: "All our representations or weaker understandings are images of our impressions or more vivid understandings" (Hume, 1961 [1748], p. 19). Impressions and ideas differ in the degree of intensity ('*more vivid* understandings'), insofar as a previously seen color is experienced more faintly in the present, and a pain suffered is experienced more weakly in the memory. From the impressions, simple representations arise, and complex ideas arise through the association of representations. With his recourse to impressions as the origin of all representations, Hume gains, on the one hand, a scientific criterion in the sense of empiricism, but also a "sense criterion for linguistic expressions of representations" (Gabriel, 1993, p. 57), which is directed in particular against metaphysics:

> "If we therefore suspect that a philosophical expression is used without any meaning or representation, which is only too often the case, we need only inquire from which impression this alleged representation originates?" (Hume, 1961 [1748], p. 22)

The 'logical empiricism' of the Vienna Circle applied this criterion of sense, which is based on ideas according to Hume, also to scientific *propositions* at the beginning of the 20th century and used it as a criterion of demarcation against 'meaningless' assertions of metaphysics (cf. Sect. 3.4.4).

Hume's criticism of induction and the principle of causality is of great importance, as both topics are interrelated: "The importance of the topic *causality* is transferred to the topic *induction*. For the problem of induction is an integral part of the question of the nature of causality" (Kienzle, 2010, p. 356). Since possible knowledge is associated with a claim to truth and only *statements* can be true or false (see Sect. 2.2), Hume distinguishes between statements that concern "relations of ideas" and statements that concern "matters of fact" (Hume, 1961 [1748], p. 35). Relations of ideas, such as mathematical theorems (Hume's example: Pythagoras' theorem), are independent (*a priori*) of experience true, as their opposite would be self-contradictory:

> In the case of facts, the opposite is always "possible, because it can never contain a contradiction and is represented by the mind with the same ease and clarity as if it were in agreement with reality. *That the sun will not rise tomorrow* is no less comprehensible a sentence and no more contradictory than the assertion that *it will rise*. We would therefore vainly try to demonstrate its falsity." (Hume, 1961 [1748], p. 35)

If the principle of contradiction cannot serve as a principle of justification for facts, then the question is which principle can take over this function of justification. The question is: "But with what right do we transfer past experience to the present case? [...] All of our experience-based conclusions, as Hume points out, contain an unspoken assumption, namely that the future will be similar to the past" (Gawlick, 1995, p. 142). Hume's answer is: the *causal principle* as a relationship between cause and effect. And this conclusion "from experience to the underlying causes proceeds in an inductive direction and is therefore an inductive procedure" (Carrier, 2017, p. 27). Typical for this conclusion is the "temporal succession and constant connection of events" (Gawlick, 1995, p. 143). But then the follow-up question arises: How do we even come to assume a *necessary* connection between the preceding and the following event as a connection between cause and effect?

In contrast to Locke, who derives the concepts of cause and effect from *experience*, Hume's still astonishing, 'scandalous' (Broad, 1952; cf. Wiltsche, 2013) and provoking answer within the framework of his "*regularity theory* of causality" (Carrier, 2017, p. 27) is different: Through repeated perception of the regularity of such a succession of different events, a *habituation effect* and an *expectation* in the subject arise, which almost compulsively associates the one event as cause of the following event as effect. The necessity of the causal connection is thus a psychological effect in the subject of knowledge, it is by no means an objective, ontological necessity inherent in things themselves. The observation of the two events can "never find the effect in the assumed cause, even with the most careful examination and testing. For the effect is quite different from the cause and can therefore never be discovered in this" (Hume, 1961 [1748], p. 39). Ultimately, the causal relationship as a necessary connection is based on faith and instincts, that is, on non-rational, if not irrational, factors:

> "This belief is the necessary result if the mind gets into such circumstances. It is a mental process that is as inevitable in this situation as the affect of love when we receive benefits or the hatred when someone does us harm. All these processes are a kind of natural instinct, which neither produces nor prevents any rational or intellectual procedure" (Hume, 1961 [1748], p. 59)

This result also touches on the problem of induction, that is, the question of whether and why we may legitimately conclude from an observed sequence of events to a preceding event "that this will also be the case in the future or always" (Kienzle, 2010, p. 363).

Hume's considerations lead consistently to a form of epistemological "moderate skepticism" which at first "distrusts all scientific results, is therefore tolerant and undogmatic, remains open to correction through experience, and deliberately confines itself to the investigation of such objects which lie within the realm of possible experience" (Gawlick, 1995, p. 145). Since induction and causality are nothing more than psychological principles of order which have no basis in reality, an objective material reality is ultimately unrecognizable, and theories which claim such knowledge are speculative. These two basic ideas are later taken up again by Popper, who turns them into a positive fallibilist and falsificationist methodology (cf. Sect. 3.6), that is, a methodology which takes into account the fallibility and error-proneness of scientific thought and action by choosing the falsification (rather than confirmation) of hypotheses as the criterion of scientific validity.

Causality (from Latin *causa*, cause; Neo-Latin *causalitas*, causality) is, despite the obvious impossibility of providing a rational justification for it, something which cannot be dispensed with in either everyday thought or in the sciences. This principle of causality is highly controversial in the theory of science even today, as the following chapters will show. Nevertheless, causality is still routinely used as an explanatory category and fundamental research assumption within the sciences (Textbox 13).

Textbox 13: The Causal Principle

British philosopher Sir Peter Frederick Strawson (1919–2006) assumed that this fundamental principle belongs to concepts or categories such as 'truth', 'space' or 'time', without which everyday life and science is not possible (Strawson, 2003). They are inescapable and cannot be reduced to other, more fundamental concepts or facts, for example data from experience or scientific concepts. They are therefore neither empirically verifiable (see Sect. 3.4.3) nor falsifiable (see Sect. 3.6).

A very cursory glance at the history of philosophy shows the importance of the principle of causality. Already Aristotle showed in his *Metaphysics* in the doctrine of the four causes (lat. *causae*; gr. *aitiai*), that something natural as well as something man-made can only be understood if four questions can be answered (Aristotle, 1991 [348-345 BC], 1013a-1014a) (1) What is it made of? (*'causa materialis'*: Material cause), (2) What is it for? (*'causa finalis'*: Final cause), (3) Who or what made it? (*'Causa efficiens'*: Efficient cause), (4) In what structure or form does it present itself? (*'causa formalis'*: Formal cause). In the current sciences, only the efficient cause and the final cause have been able to hold on. The efficient cause establishes a causal connection between past and present: In physical geography, physical efficient causes are sought for the physically given, in human geography human efficient causes are sought for what is caused by humans and used as an explanatory basis. The final cause establishes a causal connection between present and future: For example, in climatology and demography, future (predicted, see Sect. 2.3.2) events are used to draw conclusions about current data, their interpretation and even about derived recommendations for action.

The Inescapability and non-reducibility of the causal principle becomes particularly apparent in the famous *law of sufficient reason* (lat. *principium rationis sufficientis*), which was formulated by Gottfried Wilhelm Leibniz (1646–1716). In addition to the principle of contradiction, so Leibniz, "rational inferences" are also based on the principle

"[…] of the *sufficient reason*, according to which we consider that no fact can be regarded as true or existing and no statement as correct, without there being a sufficient reason for it to be so and not otherwise, although we may not be aware of these reasons". (Leibniz, 2019 [1714], p. 27)

In other words: Nothing happens without a reason (lat. *Nihil fit sine causa*) or: Nothing comes from nothing (lat. *Ex nihilo nihil fit*).

In his *Critique of Pure Reason*, Immanuel Kant no longer reformulates these variants of the principle of causality, formulated ontologically by Aristotle and Leibniz, as principles of being, but as principles of understanding. In the *Critique of Pure Reason*, Kant determines causality as one of 12 categories or "basic concepts of the mind" (Kant, 1959 [1781], B 105–109). Kant solves the problem of causality left by Hume by not binding this concept to reality, but by understanding it as an a priori, unshakeable form of knowledge given by the human cognitive apparatus. People cannot help but see things in cause-and-effect relationships, just as they cannot always already perceive things in space and time (see "space" in Sect. 6.2). It is not the things that people observe that cause and effect each other, but people always already see and try to see things in this way. It is this view that also guides the interpretation mentioned by Strawson.

Ultimately, the ontological and the mentalist interpretation of the principle of causality refer to the two opposed epistemological positions of realism and constructivism or "anti-realism" (see Egner, 2010, pp. 26–43 for an overview). In a figurative sense, scientists come to a point of decision or belief when trying to justify or explain why one or the other position is followed as a "rule" of scientific orientation and action, at which point no further foundations can be dug deeper:

"If I have exhausted the justifications, I have now arrived at the hard rock, and my spade bends back. I am then inclined to say: 'That's how I act.'" (Wittgenstein, 1995 [1953], p. 350)

Every scientist is therefore referred to their own '*decisionistic*' decision (from Latin *decisio*, agreement, decision) with regard to this question, which cannot be fully rationally justified. Therefore, an *informed* decision on the basis of fundamental epistemological and scientific theoretical knowledge is all the more important, since scientists are not relieved of at least explaining *how* they came to their decision. With Strawson, causality can therefore be understood as a concept of 'descriptive metaphysics'; with Tetens (2013), such a decision can be made with regard to 'aprioristic experience frameworks' (see Textbox 14).
◀

3.4   Positivism, Ideal- and Normalsprache and Logischer Empirismus

The empiricist positions presented so far stand in a tradition of consciousness philosophy that ties experience to the conditions of consciousness and thus largely does not yet take into account the linguistic conditions of this experience. The 'linguistic' turn thematizes the linguistic conditions of experience and knowledge and influences the 'Logischer Empirismus' as a further intensification of the original empiricist program. The empiricist *knowledge* criticism is modified to a *sense* criticism as part of a language analysis. These intensifications ultimately lead to an exaggeration of empiricism in the form of the coupling of the sense criterion with supposed 'pseudo-problems' and the attempt at justification and construction of a physicalist ('Einheitswissenschaft') based on physics.

3.4.1 Positivism

Positivism is a "variety of empiricism" (Bauberger, 2016, p. 13) and originally a philosophical direction introduced by the mathematician and philosopher Auguste Comte (1798–1857). It can be *historically* characterized first "by the emphasis on the *factual* (the positive facts) and the rejection of metaphysical speculation" as well as second "by the commitment to the idea of *progress*" (Gansland & Carrier, 2004, p. 302). *Systematically* it serves the "characterization of such arguments and positions which see the rational foundation of scientific theories and institutional orientations solely in *factual* assertions, the so-called 'positive'" (Kambartel, 2004, p. 303).

In the six-volume 'Treatise on Positive Philosophy' (1830–1845) Comte addresses the question of the structure, function and development of knowledge in society. For this purpose, he refers back to the 'law of the three stages' already stemming from the Enlightenment, which individual human beings as well as humanity as a whole go through. In the 'theological state', mythical and religious interpretation offers with fictitious supernatural beings are predominant; in the 'metaphysical state', supernatural beings are replaced by abstract beings; in the 'scientific' or 'positive state', science turns away from fictitious or abstract beings to observable facts. These facts are explained by the fact that they are related to each other and derived from them laws. This 'logical positivism' leads to a "strengthening of empiricism by denying the possibility in principle that there can be knowledge about unobservable objects" (Bauberger, 2016, pp. 13–14).

John Stuart Mill (1806–1873) radicalizes this 'logical positivism' in his 'System of Deductive and Inductive Logic' (1968 [1843]) to a unified methodology for *all* sciences on the basis of an '*inductive logic*'. Not only the natural sciences, but also the deductive sciences (mathematics and formal logic) and the humanities are based on inductive generalizations, by inferring general laws from individual events that regularly recur. He starts from a 'uniformity in the course of nature', that is, a "causal order of reality" (Schülein & Reitze, 2012, p. 112). For Mill, science is therefore a "method of active induction by observation and experiment. Its results are objective knowledge of established causalities that reality contains. Any knowledge of the 'true nature' of the world is not connected with it" (Schülein & Reitze, 2012, p. 113). This approach "frees itself from the blinkers of a dogmatic metaphysics, but runs the risk of hardening into a dogma itself" (Höffe, 2001, p. 233)—namely, into a methodological dogmatism, since all areas of knowledge are now to be accessed by a single method—a problem that will reappear under the term 'unified science' in 'logical empiricism' (cf. Sect. 3.4.5).

3.4.2   Ideal and Normal Language

The older positivism of Comte and Mill is still under the spell of an optimism that quantitative methods guaranteed by observation and experiment a reliable object reference and thus objectivity and truth. The reference to the 'positive', the 'facts', is made within the mentalist paradigm, the object reference, objectivity and truth to the experience conditions of human *consciousness* binds. The classical formulation of this mentalist object reference is found in Kant's Critique of Pure Reason:

> "The conditions of the *possibility of experience* in general are at the same time conditions of the *possibility of the objects of experience*, and therefore have objective validity." (Kant, 1959 [1781], B 197)

The binding of the experience and validity of knowledge to methods and an inductive logic is a binding to the conditions of consciousness. In this mentalist framework, the role and significance of *language* for all kinds of human cultural achievements (including science) in their comprehensive scope have not yet come into view. A criticism of language is required, which in the course of the so-called 'linguistic turn' (cf. Leerhoff et al., 2010; Newen, 2018; Rorty, 1967) leads to a change from the 'mentalist' to a 'linguistic paradigm' (Schnädelbach, 1991). The empiricist skepticism as epistemological criticism is transformed into a criticism of the sense of linguistic constructs. The question of consciousness philosophy "What can I know?" Is replaced by the question of philosophy of language "What can I understand?" (Schnädelbach, 1991, p. 69)—this is the question of the meaning and the *analysis* of linguistic expressions. In contrast to the ontological paradigm, whose representatives begin philosophizing with the attitude of wonder at the beauty and order of the cosmos (gr. *kosmos*, beautiful order) and of 'being', and to the mentalist paradigm, whose representatives begin philosophizing with the attitude of doubt about existing knowledge, Wittgenstein's attitude as the protagonist of the linguistic paradigm is confusion:

> "A philosophical problem has the form: 'I do not know my way around.'" (Wittgenstein, 1995 [1953], p. 302)

The American philosopher Richard Rorty summarizes the concern of the philosophy of language and thus of the "linguistic turn" as follows:

> "I shall mean by 'linguistic philosophy' the view that philosophical problems are problems which may be solved (or dissolved) either by reforming language, or by understanding more about the language we presently use". (Rorty, 1967, p. 3; translation by author)

The different paradigms of the history of philosophy and their characteristics are illustrated in the following Table. 3.4.

Currently, philosopher Dirk Hartmann has written a seven-volume "New System of Philosophical Sciences in Outline" and introduced the history of philosophy in the first volume already published on "Theory of Knowledge"—only

Tab. 3.4 Paradigms of the history of philosophy (according to Schnädelbach, 1991, p. 69, supplemented)

Paradigm	Epoch	Area	Initial Attitude	Initial Question	Protagonists
ontological	Ancient/Middle Ages	Being	Astonishment	What is?	Socrates, Plato, Aristotle
mentalistic	Modern period	consciousness	doubt	What can I know?	Descartes, Locke, Kant
linguistically	20th Century	language	confusion	What can I understand?	Wittgenstein, Habermas

Hartmann does not speak of "paradigms", but of "phases" or "epochs" (Hartmann, 2020, pp. 1–36).

Occasionally, it is objected to such right-woodcut-like divisions and distinctions that they could not do justice to the variety of historical phenomena and the complexity of systematic relationships and would therefore be inadequate. Often, this criticism is based on the belief that the variety and complexity of research fields can be grasped and examined directly. Against this belief in immediacy, it can be objected with Hartmann that "the drawing of such provisional distinctions is not only useful, but also unavoidable [...], because while one can experience the inadequacy of distinctions, one sees and learns nothing without any distinctions" (Hartmann, 2020, p. 1).

Within so-called 'analytical philosophy' (cf. Leerhoff et al., 2010; Newen, 2018; Schönwälder-Kuntze, 2020), two contrary approaches therefore developed in the sense of Rorty's initial statement: an ideal-language and a normal-language approach. The ideal-language strategy wants to develop an ideal formal language that can avoid the imprecisions and equivocations (ambiguities; see Sect. 2.4.1) of everyday language (normal language) and meet the requirements of logic. Representatives of this approach are Gottlob Frege (1848–1925), Bertrand Russell (1872–1970), Rudolf Carnap (1891–1970) as well as in particular Ludwig Wittgenstein (1889–1951) with his 'Tractatus-logico-philosophicus' (Wittgenstein, 1995 [1953]), but also Wilhelm Kamlah (1905–1976) and Paul Lorenzen (1915–1994) as founders of the ortho-linguistic Erlangen constructivism (Kamlah & Lorenzen, 1967). This idea of an ideal language can already be found in Gottfried Wilhelm Leibniz (1646–1716) as the 'characteristica universalis', a universal language modeled after mathematics. In this sense, Frege spoke of a 'formal language of pure thought', which is 'modeled' on arithmetic (Frege, 1879). Wittgenstein and, in particular, Russell advocated a 'logical atomism', which assumes that "the world consists of smallest, indivisible elements, the *logical atoms*"—this approach is intended to "ensure that the language corresponds to this finest structure of the world: that the rules of the language, i.e. the grammar, the possibilities of the world, the possibilities of combining the logical atoms, reflect" (Leerhoff et al., 2010, p. 40). The grammar (syntax) of the ideal language is to be "designed so finely, the rules to be made so precise, that every sentence that corresponds to them is guaranteed to be meaningful and thus also has a truth value" (Leerhoff et al., 2010, p. 40). The rules of syntax also determine the meaningfulness of semantics.

Wittgenstein is not only considered the main founder of the *'linguistic turn'*, but he is also a philosopher who first developed an ideal language, later a normal language program. The latter approach is skeptical towards a formal universal language and relies on the everyday language proven in everyday communication and cooperation (the *'ordinary language philosophy'*). The main representatives of this approach are, for example, George Edward Moore (1873–1958), Gilbert Ryle (1900–1976) and again Ludwig Wittgenstein.

Wittgenstein justified the ideal-language program of analytic philosophy and linguistic criticism as *sense criticism* with his *'Tractatus-logico-philosophicus'*. Consciousness does not work in the medium of thought without language, but thinking is essentially linguistic. In place of a demarcation (criticism: Greek *krinein*, to distinguish, to draw a line) between the knowable and the unknowable (Kant's program of a Critique of Pure Reason) or between the observable and the unobservable (the programs of Comte and Mill), Wittgenstein therefore sets a demarcation in the *'Tractatus'* between the thinkable and the unthinkable and thus between the sayable and the unsayable. The clarification of thoughts is thus identified with a clarification of sentences:

> "The book therefore wants to draw a boundary for thinking, or rather—not for thinking, but for the expression of thoughts [...] The boundary will therefore only be drawn in language and what lies beyond the boundary will simply be nonsense." (Wittgenstein, 1995 [1953], p. 9)

Wittgenstein held the thesis that many philosophical and scientific problems are based on misunderstandings of the logic of human language, ultimately 'bewitch' the mind and lead to pseudo-problems—in the *'Philosophical Investigations'* it says about this:

> "These problems are solved, not by teaching new experience, but by assembling the long-known. Philosophy is a fight against the bewitching of our minds by the means of language." (Wittgenstein, 1995 [1953], p. 299)

In the "Tractatus" it says: "The thought is the meaningful sentence" (Wittgenstein, 1995 [1953], p. 25). And this leads to Wittgenstein's almost identical scientific criterion of meaning:

> "What can be said at all can be said clearly; and whereof one cannot speak, thereof one must be silent" (Wittgenstein, 1995 [1953], p. 9)

With an almost identical sentence, Wittgenstein also concludes the "Tractatus". We will return to this linguistic critique of meaning in the next Sect. (3.4.4).

Wittgenstein believed that with the "Tractatus" he had "essentially solved the problems" (Wittgenstein, 1995 [1953], p. 10), that is, developed an ideal language that is conceived as the coincidence of syntax (logical grammar) and semantics (meaning, meaningfulness). Against the traditional ontology of things or substance metaphysics, for which the "world" consists of things like a large container (cf. Sect. 6.1), Wittgenstein determines the "world" as "the totality of facts, not of things" (Wittgenstein, 1995 [1953], p. 11). Facts are determined as "the existence of states of affairs" and these as a "connection of objects" (Wittgenstein, 1995 [1953], p. 11). It remains unclear to Wittgenstein what the objects are exactly (Leerhoff et al., 2010). The 'objects' can vary depending on the "logical [form]" (Wittgenstein, 1995 [1953], p. 33) many and 'countless' combinations with other objects arise: the 'facts'. "These ontological ideas are now transferred to language", by giving them each a name (for example, 'Tübingen': T, 'is a city': S, 'Tübingen is a city': TS)—and this "logical form of the objects is passed on to the ideal language, reflected in it and forms the grammar of the language" (Leerhoff et al., 2010, p. 42). The simplest sentences are the elementary sentences and they claim the "existence of a fact" and they "consist of names", which are 'linked' (Wittgenstein, 1995 [1953], p. 38). The specification of all elementary sentences "describes the world completely" and they can be true or false (Wittgenstein, 1995 [1953], p. 39); the corresponding truth values can be calculated in truth tables and combined into sentences. In this way, every sentence can be represented purely logically as a truth function of its elementary sentences: "The sentence is a truth function of the elementary sentences" (Wittgenstein, 1995 [1953], p. 45). Unfortunately, Wittgenstein himself does not give an example of an elementary sentence, because "he did not consider it the task of the logician to answer this question. At the time of the composition of the Tractatus, he thought that this question was of an empirical nature" (Gabriel, 1993, p. 158). Some of the 'logical empiricists' have subsequently tried to answer this question logically *and* empirically and thus burdened themselves with the difficulties of the 'protocol sentence discussion' (cf. Popper, 1989) (cf. Sect. 3.4.3).

Wittgenstein later retracted the concept of a unified language as an ideal language himself. The decisive point is the *lack of pragmatics*. This means that the "use of language" (Wittgenstein, 1995 [1953], p. 238) is not taken into account in communication and interaction contexts. The meaning (semantics) of the words and sentences of a language is therefore dependent on concrete situational and contextually bound language-use practices with speakers and listeners who pursue intentions (from Latin *animo intendere*, to intend) and must be able to understand each other. Wittgenstein calls these language-use practices '*language games*' (Wittgenstein, 1995 [1953], p. 241), which in turn only receive their meaning relatively to concrete '*forms of life*' (Wittgenstein, 1995 [1953], p. 146): "The meaning of a word is its use in the language" (Wittgenstein, 1995 [1953], p. 262). This thesis is often referred to as the 'use theory of meaning' (cf. Leerhoff et al., 2010, pp. 50–52) (cf. Leerhoff et al., 2010, pp. 50–52). Against a possible essentialist understanding of concepts that postulates a clearly defined and unchangeable meaning of linguistic expressions, Wittgenstein sets the concept of '*family resemblance*' (Wittgenstein, 1995 [1953], pp. 277–299), according to which the meanings of a linguistic expression can only have *similarities* like the members of a family. Ultimately, Wittgenstein himself has to admit that "there is no uniform use of language; each language game has its own rules. This makes it impossible to have a clear understanding of language" (Schülein & Reitze, 2012, p. 149).

This circumstance also affects the "language of science accordingly. It is practice-oriented and has its place in this practice—but only there" (Schülein & Reitze, 2012, p. 149). This practice-orientedness of science also had to be specifically addressed and thus brought to public attention only in the second half of the 20th century (we will return to this in Sects. 3.5.1 and 4.2).

3.4.3   Neopositivism or Logical Empiricism of the "Vienna Circle": Verification Problem and Induction Problem

From 1923 to the strengthening of Austrofascism in the mid-1930s, a group of scientists and philosophers met in Vienna: the so-called "Vienna Circle". This circle included, for example, the physicist Moritz Schlick (1882–1936) and the philosopher Rudolf Carnap (1891–1979) as the *spiritus rector*" (leading and guiding spirit), the sociologist Otto Neurath (1882–1945) and occasional guests such as Carl Gustav Hempel (1905–1997), Kurt Gödel (1906–1978), Ludwig Wittgenstein (1889–1951) and Hans Reichenbach (1891–1953). In contrast to older positivism and in connection with, in particular, Frege, Russell and Wittgenstein, the members of the "Vienna Circle" gave positivism a "linguistic turn" on the one hand (Bauberger, 2016, p. 14); on the other hand, a connection was made to the empiricist tradition.

The connection to these two traditions led to an understanding of science as a logically-methodically regulated systematic context of statements. The fundamental distinction is between analytic and synthetic sentences, that is, "the a priori statements of logic and empirical statements about observable phenomena. In view of this, the neopositivism also called itself logical empiricism" (Ströker, 1998, p. 439). The sentences of mathematics and logic are analytic and a priori, they are always true regardless of experience, but they do not say anything factual (real) about the world; the sentences of the "real sciences": "natural sciences, humanities and psychology" (Carnap, 1931, p. 432) are synthetic and a posteriori, they can be wrong because they depend on phenomena in the world, and they extend knowledge about the world. With the 'observable' phenomena, the question of the empirical basis of the sciences is addressed. The question is how empirical statements can support the sciences in what way. This question therefore arises as a double legacy of empiricism in the double form of the question of genesis (origin, acquisition) as well as the validity (justification) of knowledge. This is connected with the verification problem (question of validity) and (again) the induction problem (question of genesis).

The distinction between genesis and validity (see Sect. 3.1.1) ultimately goes back to Gottfried Wilhelm Leibniz (1646–1716), who sought to counter John Locke's criticism of the "innate ideas" by claiming that the genetic impetus for the formation of ideas "does not provide the basis for the validity of the corresponding truths" (Leibniz, 1961), (Gabriel, 1993, p. 47). For whether "a knowledge is justified can be determined independently of how it came about. And to explain how it came about, one does not need to appeal to justification. Therefore, the validity of knowledge is to be judged independently of its origin" (Pfister, 2019, pp. 111–112). Gabriel (1998) illustrates this distinction using the arithmetic equation '7 + 5 = 12'. We first have to learn to count, for example with fingers, blocks or something else, so we have to have experiences with counting in order to be able to see whether an equation is valid or not. But an equation is not valid or not valid for empirical reasons, but a priori (independent of experience). Even though the ability to count must therefore be ontogenetically (with regard to individual development history) first experienced inductively, the validity of an equation is not to be justified empirically-inductively. For example, if someone miscounts, they will not question the validity of the equation, but assume that they (empirically) miscounted. In the light of this distinction between genesis and validity, the attempt at an empirical justification of knowledge can be "saved" on a *case-by-case* basis, if it is taken into account that "one can justify a priori knowledge *independently* of experience, but not that it *must* be justified in this way. [...] It is not impossible to justify a priori knowledge on an individual basis with experience" (Pfister, 2019, p. 112)—in the example, for example, by (re)counting or calculating with a calculator.

Gabriel's example "can be generalized. The genetic conditions for the acquisition of knowledge are always to be distinguished from the reasons for its validity" (Gabriel, 1998, p. 57). In the theory of science, this epistemological basis has led to the distinction between the context of discovery and the context of justification introduced by Hans Reichenbach (1938) (*context of discovery—context of justification*) has become common. In current discussions, the social context of utilization (cf. Schurz, 2014) and the associated 'utilization pressure' (Carrier, 2005) are also taken into account. In the context of discovery, it is about the innovative function (Schnädelbach, 1980) of science, about the origin or discovery of new knowledge. In the context of justification, it is about the evaluative function (Schnädelbach, 1980) of science, about the empirical verifiability or validity of knowledge or knowledge claims. Within the framework of the distinction between descriptive and normative philosophy of science (cf. Sect. 3.1), this distinction has been criticized in view of the findings of the history of science (cf. Chap. 4). The claim of the 'normativists' that the "factual history of origin" of a hypothesis, that is, the "reasons for its discovery or invention", are "completely irrelevant" for its justification (Schurz, 2014, p. 21) is based on the omission of 'knowledge-internal moments' relevant for knowledge, such as scientific methods, data extrapolations or evaluation procedures, which Scientists actually use and which play "a decisive role" in the empirical justification (Schurz, 2014, p. 22). That a scientist has the inspiration for a new scientific idea during an epistemic *external* walk in the forest or while playing with a model railway may be irrelevant for the validity of the idea; relevant, however, can be a specific epistemic *internal* social empirical method of data collection (such as ero-epic conversations), which leads to new knowledge. This addresses the pragmatic context in which scientists are always embedded, and the epistemic relevance of which had to be specifically worked out by protagonists of the history of science.

The question of *validity* is the question of the acquisition of first immediate data in the form of "protocol sentences" or "observation sentences". In this "protocol sentence discussion" (see Popper, 1989; Stegmüller, 1978a) it is ultimately "about the interface of language and world" (Leerhoff et al., 2010, p. 96) From observations of allegedly "infallible" (see Schurz, 2014, p. 14) facts or from their documentation (Carnap, 1931) generalizations in the form of laws and theories were to be derived inductively. Within such a "reductionist program", "finally, all statements of a scientific theory should simply be representable as truth functions of observation sentences" (Ströker, 1998, pp. 439–440; see Sect. 3.4.2). The legacy of empiricism therefore consists in the conviction that all scientific statements can only be justified by *appeal* to experience. In this way, logical empiricism burdens itself with the "verification problem", that is, the question of how scientific statements can be verified, or justified. This is the question of how, through individual indications (e.g. in the experiment), the truth content of general statements (laws, theory statements) can be justified. This is already the question of how Hume (see Sect. 3.3.4) discussed question of the transition from the individual to the general, thus the problem of induction, which also plays the decisive role in the genesis of knowledge. Popper will solve these problems with his theory of "falsification" and the "hypothetical-deductive method" of a preliminary solution (see Sect. 3.6). Later, logical empiricism also had to recognize the fallibility of observational sentences and the view that "theoretical concepts such as 'force', 'electric field' or 'human character' are not defined by observational concepts, but go far beyond what is directly observable" (Schurz, 2014, p. 14 and 'theory-laden' (this is a foretaste of Textbox 16).

The question of the *genesis* of knowledge in logical empiricism takes over the legacy of empiricism with the conviction that not only *allknowledge*, but also the *means* of systematizing experience, come from experience. The induction, whose conclusion, as already explained (see Sect. 3.3.4), cannot be justified logically, because the conclusion (the general) has a higher information content than is contained in the/the premise (s) (the individual) (Fig. 3.6). Even the inductive conclusion schema itself, if it were to come from experience, would have to be won as a general again by induction from the individual. This inevitably leads to an infinite regress or a logical circle (Ströker, 1998, p. 440). The development led to the fact that, although not giving up induction, it was to be unproblematicized to the extent that the claim to absolute validity of scientific statements is to be given up in favor of specific probability (Ströker, 1998, p. 440). Popper dispenses with the empiricist induction principle, for him induction is merely "to be understood as a heuristic procedure for generating new hypotheses" (Poser, 2009, p. 112).

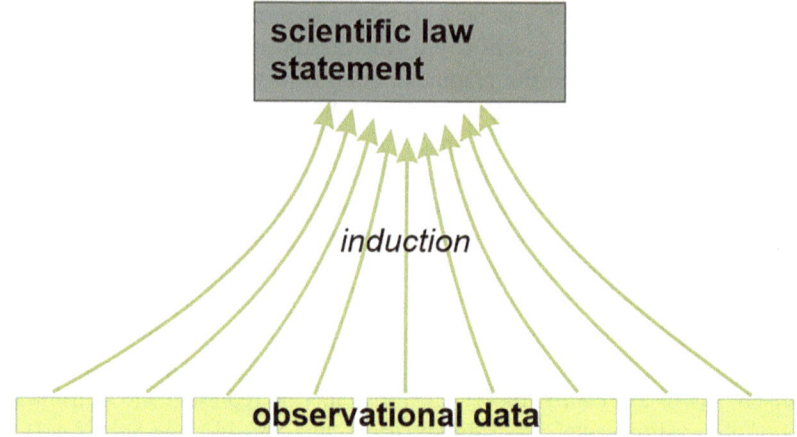

Fig. 3.6 Induction in logical empiricism (own representation)

3.4.4   Criterion of Meaning and Pseudo-Problems

David Hume had mobilized the *experience* as a criterion of scientificity and a "criterion of meaning for linguistic expressions of representations" (Gabriel, 1993, p. 57) "below the level of sentence or statement" (Gabriel, 1993, p. 145) against metaphysics. Rudolf Carnap as an outstanding protagonist of logical empiricism had applied the criterion of meaning related to *representations* ('ideas') at Hume also to scientific *sentences* and had used it also as a criterion of demarcation against 'meaningless' assertions of metaphysics, but also of philosophy in general and of all sciences which violate this criterion (Carnap, 1931).

Statements that violate this criterion are not only unscientific, but they also create "pseudo-problems" (Carnap, 1966 [1928]). Scientific statements that cannot be reduced to last empirical observation or protocol sentences and thus are basically verifiable and therefore factual are disqualified as meaningless and unscientific: "conceivable verifiability defines factualness, and factualness is the criterion of meaningful statements" (Gabriel, 1993, p. 145). Carnap calls this "view that the protocols form the basis for the entire structure of science, [...] "methodological positivism"" (Carnap, 1931, p. 461). Carnap's "logical analysis" of metaphysical sentences, but also of many sentences of the 'humanities, cultural and social sciences' also comes to the conclusion that they are

> "Pseudo-statements, because they are in no derivation relation (neither a positive nor a negative) to the sentences of the protocol language. They contain either words that are not reducible to words of the protocol language, or they are composed of reducible words in a syntactically incorrect way." (Carnap, 1931, p. 452)

This also applies to concepts such as "objective mind" or "meaning of history" (Carnap, 1931, p. 451). For example, Carnap applies the meaning criterion to the philosophical discussion of realism and idealism, that is, to the epistemological question of the "reality" of the external world (see Sect. 3.1.1), which he regards as a "pseudo-problem" (Carnap, 1966 [1928]). Using the example of two geographers—one is an idealist and does not believe in the reality of an external world, the other is a realist and believes in it—it can be seen that both can agree on all *empirical* questions with regard to a specific mountain in Africa. Only the *philosophical* question of whether the mountain "exists" leads to a split. Carnap now argues that both statements—the mountain "exists", the mountain "does not exist"—are meaningless and constitute a pseudo-problem, since neither can be proven (verified) nor refuted (falsified).

The only meaningful task of philosophy, according to Carnap, is "to clarify the concepts and sentences of science" (Carnap, 1931, p. 433). It therefore has to be related to the "empirical sciences" as a "pure science of science, namely a reflection on the form and methods of empirical science. It serves the empirical sciences in a similar way to logic and mathematics" (Bauberger, 2016, p. 15).

As correct as the non-decidability of the philosophical question of "existence" may be, Carnap underestimates the importance of a fundamental distinction (see Textbox 14).

Textbox 14: A Priori Experience Frames

Carnap underestimates the importance and consequences of the distinction between 'aprioric frame of experience' and interpretations of empirical data within such a frame, and thus the "meta-fact of the aspectual character of reality" (cf. Tetens, 2013, 83–93, 90). If we followed Carnap, for example, the discussion we described about the scientific positions of essentialism, positivism and constructivism would be disqualified as a pseudo-problem. These positions can, however, be characterized as an 'aprioric frame of experience' by Tetens, which precedes all concrete experiences and provides principles for each case, with which scientists can make certain experiences (perceptions, observations, etc.) of or in the 'world' at all and interpret them correctly within the respective framework. The respective principles cannot be proven or refuted, but they determine in the frame of experience which data are to be interpreted as 'correct' and thus can function as an instance of refutation of generalizing statements.

It is also not a matter of arbitrariness or subjectivism or 'methodological solipsism' (Carnap, 1931, p. 461), but a

"*objective* property of the world that we make certain experiences E_1 in the aprioric frame of experience R_1 and different experiences E_2 in the frame of experience R_2. The world has a *perspectival* or *aspectual* character." (Tetens, 2013, p. 86)

With reference to Nelson Goodman (1978), this means that,

"*we should finally stop talking about one world and acknowledge that there are many worlds* in which we are all *constructors*, although we remain dependent on *inputs* in all of them" (Stegmüller, 1984, p. 29).

However, this "objective fact" of the inescapable relatedness of all scientific (and all non-scientific) approaches to the world to a priori frameworks of experience is not located on the object or theory level, but "on the so-called meta-level [...], because in this sentence we do not simply refer to the world, but we refer to how we relate to the world and how we experience" (Tetens, 2013, p. 116). In other words: "Not in the theory, but only in the metatheory does reality appear" (Hübner, 1978, p. 71). This circumstance could be called "a 'hard' meta-fact" (Tetens, 2013, p. 86). Last but not least, it is also

"an objective property of the world that not any principles are suitable as a priori framework of experience and that in any case only a limited number of fundamentally different a priori frameworks of experience have been established in the history of culture" (Tetens, 2013, p. 86)—above all the fundamental one of a *scientific* world view.

It is evident from these considerations that Carnap himself also claims an a priori framework of experience with his "methodologically positivist" sense criterion, which would also be subject to the verdict of senselessness when applied to itself (cf. Gabriel, 1993). The perspective of world constitution is occasionally accused of relativism of truth—we will return to this in Sect. 4.2. ◄

3.4.5 Physics as a Unified Science

Carnap not only formulated a criterion of sense for scientific statements, he also propagated that the language of physics should be the unified language of the sciences. The occasion is the fact of the diversity of different sciences (cf. Berr, 2018), which exists not only for "practical reasons of division of labor", but also because

"they fundamentally differ in terms of their objects, their sources of knowledge, their methods. In contrast, it is to be argued here that science forms a unity: all sentences can be expressed in one language, all facts are of one kind, knowable by one method." (Carnap, 1931, p. 432)

The aim is a "unified science" (Carnap, 1931, p. 435) on the guideline of the 'physical language' as both 'intersubjective' and 'universal language' (Carnap, 1931):

"Not only the languages of the various scientific disciplines, but also the protocol languages of the various subjects are only partial languages of the physical language; *all sentences, both the protocols and the scientific system, which is constructed in the form of a hypothesis system in connection with the protocols, are translatable into the physical language; this is a universal language and, since no other such language is known, the language of science* " (Carnap, 1931, pp. 460–461)

Accordingly, for Carnap all areas of science are "*parts of the unified science, physics*" (Carnap, 1931, p. 465). This program of a unified science was later to be implemented on the model of a *Scientia generalis* by Leibniz in the *Encyclopedia of Unified Science* (Neurath et al., 1970). The "individual volumes followed a structure that Neurath had compared to an onion [...] Layer by layer, the empirically confirmed and sufficient scientific areas were to be layered around this core like onion skins" (Poser, 2009, pp. 142–143).

Such a "unity *in* diversity" (Gethmann, 1991, p. 352) also wants to explain a specific science (in Carnap: physics) or a specific scientific paradigm (in Carnap: 'methodological positivism') as the exemplary model of science in general or even claim that all theories can be reduced to a single theory or incorporated into a comprehensive theory. Such ambitions lead to "unity instead of diversity" (Gethmann, 1991, p. 352) and would result in the dissolution of the diversity of the individual sciences in such a 'unity science'—either epistemologically in the sense of an encyclopedically compiled 'unity science' or a "unity of knowledge" (Nicolescu, 2008, p. 2) or in the sense of an ontologically conceived 'unity of the world', which in each case want to level all differentiations between concepts, theories or paradigms (see, for example, Balsiger, 2005; Eisel, 1992; Gethmann, 1991, 2005; List, 2004; Mittelstraß, 1998).

A *reduction* strategy (Latin *reductio*, reduction) pursues the program demanded by Carnap to reduce all sciences and theories to physics or translate them into physical language (see Lorenz, 2004; Tetens, 1999). An example of an *incorporation* strategy (Latin *incorporatio*, incorporation) is the "unification program of physical theories" pursued by Werner Heisenberg (exemplarily: Heisenberg, 1942) and subsequent scientists under the title "unity of nature", which is based on the "idea of the one nature", "the one unified theory correspond to which involve all natural laws" (Gethmann, 1991, pp. 356–357). A look at the actual structure of current sciences in their hardly overviewable diversity of different sciences, disciplines and subjects shows (cf. Sect. 3.1.1), that a 'unity' of the sciences is not given. They "are and were a reality-less idea, which is not confirmed by the theory and practice of current scientific disciplines. This is neither the case in nor between the sciences" (Gamm, 2009, p. 159; cf. Berr, 2018). It is "rather a scientific-theoretical norm than a claim in accordance with reality" (Gamm, 2009, p. 159), with Gethmann it is not to be understood as "cognitive unification" (1991, p. 350), but as "operative task" or as "pragmatic project" (1991, p. 351) to be striven for. Gethmann (Gethmann, 1991, p. 350) suspects that such yearnings for unity are based on an ideological "scientism" (lat. scientia, knowledge, science), as it was represented in a classical way by Comte (see Sect. 3.4.1) and logical empiricism (see Sect. 3.4.3). The following text box 15 explains what "scientism" is about.

Textbox 15: Szientismus

'Szientismus' refers to the "view that only the sciences, especially the natural sciences, can formulate appropriate explanations and descriptions of the world" (Demmerling, 2019, p. 872; Mittelstraß, 2004). Examples in history include, in addition to physics, mathematics in particular (Galileo Galilei), more recently biology, prominently represented by Ulrich Kutschera:

"In the natural sciences there is [...] an observable progress in knowledge, based on objective data or facts and [...] theory formation. In contrast, many 'humanities scholars' only give subjective speculation, often without the factual basis." (Kutschera, 2011, pp. 241–242)

This dictum would also affect geography as a 'diffuse' or 'hybrid discipline', its non-scientific components—such as the term 'stereotypical landscape'—would be nothing more than 'subjective speculation' in this reading. ◄

3.5   Pragmatism and Hermeneutics

Pragmatism and hermeneutics occupy a certain intermediary position between positions which, without expressly taking into account the contexts of scientific thought, cognition, knowledge, and action, focus rather on the epistemological, logical, and methodological conditions of the sciences in both the discovery and the construction contexts, and positions which also take these contexts into account, such as historical and social contexts. Pragmatism emphasizes *action* and the primacy of practical criteria over theory, even in efforts at cognition and knowledge. Hermeneutics addresses the conditions and functioning of human understanding, as well as the *linguistic* context of these efforts at understanding.

3.5.1   Pragmatism

American pragmatism was developed by Charles Sanders Peirce (1839–1914), William James (1842–1910), George Herbert Mead (1863–1931), and John Dewey (1859–1952) at the end of the 19th and beginning of the 20th centuries and spread beyond America. It plays an important role in the philosophy, history, and theory of science in that, parallel to the "linguistic turn" of philosophy of language (see Sect. 3.4.2), it promoted a "pragmatic turn" (Gamm, 2009, p. 232) in the sense of a "pragmatic tendency" from 'consciousness to action' (Gethmann, 1987) and thus contributed to the "pragmatic turn of the philosophy of science" (Gamm, 2009, p. 232) in the wake of Thomas S. Kuhn and his conception of the history of science (cf. Sect. 4.2). Pragmatism represents the primacy of practical criteria over theory. It reflects on the relationship between actions and relevant expected effects for the subject of action in its imagination, which are reflected, for example, in concepts. Peirce's famous and much-cited "pragmatic maxim" as an "instruction" for concept analysis reads:

"Consider what effects, which might conceivably have practical bearings, we conceive the object of our conception to have. Then, our conception of these effects is the whole of our conception of the object." (Peirce, 2018 [1878], p. 63)

This anticipation of sequences of actions constitutes the respective way of viewing the world as well as the understanding of truth and theory. The term "hard", for example (Peirce, 2018 [1878]), can be clarified by mentally imagining the resistance of materials by cracks (e.g. with a stone), blows (e.g. with a hammer) or pressure (e.g. with a thumb) and experimentally testing and, if necessary, correcting them. *Convictions* determine behavioral habits; from these, the sense of the convictions determining them can be clarified by a thought experiment by conceptually deriving their practical consequences.

By upgrading the act and binding theory, knowledge and truth back to *practice* also *pragmatics* comes into focus of scientific interest. In the consciousness-philosophical or mentalist "tradition of Descartes" and Kant knowledge and truth are still largely thought and justified context- and person-independent. Peirce developed a theory of signs as a forerunner of modern semiotics, which assumes that thinking and cognition are basically embedded in a triadic sign relation: Every sign is "first [...] a sign in relation *to* a thought that interprets it; second, it is a sign *for* an object for which it [...] stands, third, it is a sign *in* a respect [...] which it brings into connection with its object" (quoted after Gamm, 2009, pp. 236–237). Shorter: *"In the sign, something is interpreted as something "* (Poser, 2009, p. 227). Following Peirce, signs "can only relate representatively to objects through the mediation of other signs—their interpretants. This just means that thinking can only be related to an external reality of consciousness mediated." (Gamm, 2009, p. 237). Scientific findings must be proven intersubjectively in a communicative process of all scientists, whose "truth" arises as agreement in an "infinite community of researchers":

> "The opinion that everyone must ultimately agree, who researches, is what we understand as truth, and the object represented in this opinion is the real. So I would explain reality." (Peirce, 2018 [1878], pp. 87–89)

In contrast to later currents of pragmatism, which link 'truth' to success (see Sect. 3.1.3), this truth concept can be assigned to consensus theories of truth, insofar as 'truth' is linked to the consent of the scientific community. Peirce himself also conceptually distinguished the two truth concepts from each other, by saying, "who, in his own development, passed from one to the other, reserved the term 'pragmatism' for the utility interpretation of truth, while he applied the term 'pragmaticism' to the pragmatic consensus theories" (Gloy, 2004, p. 222). Peirce's consensus theory of truth can also be related to Popper's understanding of truth, insofar as Popper assumes that science approaches truth in an unending process (see Sect. 4.2). The *difference*, however, lies in the fact that Popper does not represent a consensus theory, but a correspondence theory of truth. In connection with the basic idea that science works according to the principle of 'trial and error' (engl. *trial and error*), which in turn is shared by pragmatism and Popper's approach, a bridge can be built to '*neopragmatic*' approaches (see Sect. 6.11).

3.5.2 Hermeneutics

Already at the end of the 19th century, with the rise of the natural sciences and their scientific criteria, a criticism of positivism took place by those sciences to which scientificity was denied by these criteria and which were then summarized under the title "humanities" and have since been further differentiated as humanities, cultural and social sciences (Diemer, 2019). Wilhelm Dilthey builds on an emerging use of the term "humanities" and promotes this development through the publication of his influential "Introduction to the Humanities" (2017 [1883]) as a "science of the spiritual world" (Diemer, 2019), which he further developed in his late phase (Dilthey, 1970 [1910]). He propagates the division of the sciences into natural and human sciences. The division principle is not ontological, but psychological in the sense of different "attention directions": attention is directed outward to nature and justifies the natural sciences, inward to "spiritual" and justifies the humanities. Dilthey determines the "going inward" in reception of the hermeneutic tradition as "understanding" (Schnädelbach, 1993, p. 75).

Wilhelm Windelband puts in the place of the distinction 'natural science/humanities' a "purely methodological [...] division of the empirical sciences" (Windelband, 1924, p. 141) in 'nomothetic sciences of law' (gr. *nomos*, law; gr. *thesis*, setting: setting of laws) and 'idiographic sciences of events' (gr. *idios*, particularly, characteristic; gr. *graphein*, (to) describe, to draw: description of the particular) (cf. Schnädelbach, 1993, pp. 75–76). This distinction is not extensional, but intensional, thus concerning the *'method of treatment'* of objects, not the objects themselves. An object (e.g. the Black Forest) can therefore be examined both nomothetically within the framework of general natural scientific questions (e.g. physical-geographical) and idiographically as an individual or singular phenomenon within the framework of humanities questions (e.g. human-geographical social constructivist).

For further, especially also present development, influential was also the replacement of the distinction 'natural science vs. humanities' by the distinction between 'natural science vs. cultural science' by Heinrich Rickert (Rickert, 1986 [1902]). Building on Kant, Rickert defines 'nature' as 'the existence of things under laws', 'culture' as 'the existence of things under values'. This means that Rickert defines 'culture' as "the totality of real objects to which generally accepted values or sense-data constituted by them adhere" (1986 [1902], p. 46). Accordingly, cultural sciences "deal with the objects related to the general cultural values and therefore understandable as meaningful, and as historical sciences they represent their unique development in their particularity and individuality" (1986 [1902], p. 125).

In summary, the distinction between "natural and social sciences" can be determined as follows: natural sciences relate "to the general, to laws, etc. Their approach is nomothetic; their task is to explain. In contrast, the social sciences as historical sciences relate to the (human) individual, the unique and irreplaceable as the "historical", which they "understand"" (Diemer, 2019, p. 214). In this way, the

still effective antithesis and the unresolved dispute about "explaining vs. understanding" are set in the scientific world.

When discussing the HO-schema (see Sect. 2.3.3), it was already pointed out
that *every* (whether natural or social) explanation can never be final and conclusive, but must always be based on a assumed explanation or interpretation background or on scientific "pre-understandings" (cf. Poser, 2009), p. 212). This
"connecting horizon of world view" (Poser, 2009, p. 212) is not accessible with
the HO-schema itself, but requires another method, namely the *hermeneutic*
method. Within this framework, pragmatics (see Sect. 2.1), that is, "the engagement with the respective speech situation, which should be precisely left out of
the explanation concept of Hempel and Oppenheim for the sake of scientificity"
(Poser, 2009, p. 213), is the indispensable starting point. Speaking or communication takes place—as Peirce pointed out—by means of signs.

The meaning of the signs is not, however, fixed a priori and context-free. Such
an understanding of language, its signs and concepts, would stand in the Platonic-
Aristotelian tradition of the "doctrine of the generality, invariance and incorpo-
reality of meaning" of language, the "doctrine of the language-free or general
nature of thinking" (Borsche, 1996, p. 10) as well as the doctrine that speaking
or language is "the expression of general thoughts" and that thinking is only the
"grasping of natural phenomena [...], which it is truly represented by words or
even better by numbers and symbols" (Borsche, 1996, p. 11). Language is understood in this understanding as a means of transport of object-given immaterial and
invariable meanings in a 'world', which is to be grasped truly by a 'language-free
reason' (Borsche, 1996, p. 11) and to be mediated for an appropriate world knowledge. With the '*linguistic*' and the '*pragmatic turn*', however, an understanding of
linguistic meanings prevailed, according to which they arise and pass away in and
through communication and can solidify or stabilize their content—for example,
"by convention or institution [...]; concepts live, they live in the mutual recognition and community of those who use them in their words" (Borsche, 1996, p. 12).

The meaning of signs is thus only accessible in a language or action context in
which they are used and conveyed. This understanding of the meaning of the signs
takes place only in the back and forth between the conditions (the prior understanding) and the result (the understood) of the understanding process, which was
modeled in tradition as the relationship between the "parts" and the "whole" of the
sign-mediated meaning. That is, with a view to a vague but assumed prior understanding of what is to be understood, for example, individual letters, phonemes,
words, sentences, signs, gestures, facial expressions, gestures, the forms of a work
of art, etc. are combined into larger units of meaning until the "parts" increasingly
give rise to a "whole" of meaning. And from this achieved understanding of the
"whole", the synthesized "parts" can in turn be seen anew and often differently, re-
grouped, re-arranged and re-understood as a "whole". This movement from "part"
to "whole" and vice versa is referred to as the "hermeneutic circle".

It is always about understanding "something external as a sign and inferring
something spiritual behind it" (Poser, 2009, p. 214). In the words of Wilhelm

Dilthey: "We call the process in which we recognize an inner from signs that are given to us externally by the senses, understanding" (Dilthey, 1924, p. 318). This understanding can be directed at linguistic signs (spoken, texts), at mimicry, or gestures as signs of psychological processes in other people, at pictorial/visualized of all kinds and at works of art as expression signs, at actions and behavior as external signs of inner motives, drives or intentions and even at historical events as signs of historical developments (cf. Poser, 2009). In principle, this understanding is possible based on Giambattista Vico (1668–1744) and the basic idea of his *Principii di Scienza Nuova* (1725): We only recognize and understand what we have produced ourselves, nature as the 'not made' remains forever incomprehensible to us (Vico, 1966 [1725]).

Deriving from the tradition of rhetoric and, in particular, from Protestant biblical exegesis, hermeneutics also poses the question of how an interpreter can understand the text as the product of a *foreign* author. Friedrich Schleiermacher (1768–1834) answered this question by saying that understanding is a 'putting oneself in the place of' another person. *"Feeling and empathy* [...] Understanding thus becomes a *reconstruction process* of the foreign individuality" (Poser, 2009). The question is how to 'correctly' understand. Historicism made it clear that each epoch—including that of the later interpreter and his or her interpretive assumptions—is irreplaceably unique and can only be understood from within itself: How is understanding from today's perspective then possible? This question applies to all understanding, of whatever objects. All attempts so far to overcome the relative validity of such acts of understanding have not been convincingly successful (cf. Poser, 2009). There is also no unified methodology of understanding in sight or even possible:

> "Any methodology starts from a prior understanding without which there can be no understanding in principle; and even if this prior understanding is revisable, no methodology can do more than secure a relative validity." (Poser, 2009, pp. 219–220)

Even the most recent attempt by Hans Albert to subsume the understanding under the HO-schema has to be considered a failure (Poser, 2009). According to Albert, in understanding, interpretation *hypotheses* are formulated which, like all other hypotheses, are to be explained by the HO-schema. However, the counter-criticism is that

> "the general demand to replace hermeneutics with the concept of explanation cannot hold, because Albert himself has to speak of the '*interpretation*' of phenomena. But in this interpretation, it is not about the verification of law-like statements—and with that, any possibility of subsumption under the HO-schema is lacking. The concept of explanation of analytic philosophy is therefore not able to solve the problem of hermeneutics: explaining is not thinkable without understanding." (Poser, 2009, p. 232; emph. by the author)

This means that every attempt at explanation already presupposes a lifeworldly or scientific pre-understanding of what is to be explained. This also means that the sciences are just as dependent on hermeneutic methods as is scientific theory. In Sects. 4.2 and 5.2 we will see that and how the so-called 'objectivity' of the individual sciences is dependent on the discipline-specific foundation and mediation of a shared pre-understanding in the sense of a 'pre-judgmental structure' (Gadamer, 1975) of the subject matter, the methodology, the methods, the standards of scientificity, etc. of the discipline. Hermeneutics can clarify these conditions of the practice of understanding and explanation of scientific disciplines.

3.6   Falsificationism: Karl Popper

A key intermediate position from logical empiricism as part of analytic philosophy in the transition to the various positions of a history of science (see Chap. 4) is occupied by the falsificationism of Karl Popper. As the previous considerations have shown, the principle of induction is fraught with insoluble difficulties; the principle of "verification" as a principle of generalization of data obtained to "true" natural laws is also problematic. Popper drew two radical consequences from this: The principle of induction is "logically inadmissible", there is therefore "no induction"; and hypotheses or theories are "never empirically verifiable" (Popper, 1989, p. 14). The principle of induction is *logically* not justified, because the conclusion from even so many individual observations to a general law is invalid, since the higher information or factual content of the conclusion (law) is not covered by the lower information or factual content of the premises (observational data). Also, the inductive conclusion schema itself cannot be derived from experience, because then it would have to be justified as experience by induction. The *verification principle* is inadmissible because, in both the justification and the discovery context, the *truth* content of the general law statements or theorems derived from observational data cannot be justified by even so many individual indications (for example, by experiments). Popper therefore replaces the verification principle with the falsification principle.

Instead of trying to verify hypotheses or theories, the task now is to keep them open to falsification (disproof). For such a falsification, a counterexample is sufficient to disprove a law hypothesis or theory as false. Popper's famous example is the all-statement 'All swans are white', logically equivalent to 'There is no non-white swan'. The attempt to verify this hypothesis as a law claim is impossible, as this would require observing all swans in the past, present and future in all corners of the world. Falsification is simpler and logically unproblematic. A counterexample is sufficient for the refutation, which according to the valid conclusion form of the '*modus tollens*' (Popper, 1989, pp. 44–46) works: 'a → b; ¬b; ¬a'. Let's set the variable 'S' for 'All swans are white', the variable 'W' for 'This swan is white', we get the following valid conclusion: 'S → W; ¬W; ¬S' (see Tab. 3.5).

The validity of the *modus tollens* is shown by the fact that, with a distribution of true premises, the conclusion is also true. A corresponding transformation into statement forms would show that the *modus tollens* is always a true tautological statement form (see Sect. 2.3.2).

Tab. 3.5 Truth table of the modus tollens (from Latin tollere, to cancel: 'cancelling conclusion')

Partial statement	Partial statement	Premise 1	Premise 2	Conclusion
S	W	S → W	¬W	¬S
w	w	w	f	f
w	f	f	w	f
f	w	w	f	w
f	f	w	w	w

However, the method of falsification necessarily includes the falsifiability of the hypotheses. This means that it must be possible to indicate when a hypothesis can be considered failed and thereby ensure that such a refutation is possible. The possibility of falsification thus serves at the same time as an *criterion of demarcation* between empirical and non-empirical, metaphysical or even unscientific theories or sentences: "A *scientific empirical system must be able to fail in experience* " (Popper, 1989, p. 15). Popper thus builds on Kant's criticism of such theories that "no touchstone of experience is recognized anymore" (Kant, 1959 [1781], A VII), and is therefore expressly directed against 'immunization strategies' with which scientists try to make their theories unassailable against criticism and refutation. Such 'ad hoc hypotheses', could, for example, when a black swan appears, doubt that it is a swan, that there is a mistake here. Or bad lighting conditions are criticized during the observation or the color is attributed to external factors such as illness or injury. Whoever absolutely wants to protect a theory from refutation will always find such apparent reasons.

Bacon (1990 [1620]) had already demanded that one proceed against the illusions of the human species (*idolae tribus*), for example the insistence on preconceived opinions and existing theories, which are thus immunized against refutations (see Sect. 3.3.1). Popper expressly points out that this criterion of demarcation is non-empirical and therefore also not empirically refutable. It is a "useful '*suggestion*'" for the sciences, a "philosophical thesis: a thesis of metascience" (Popper, 2010, p. XXXIII), which is "set" as a normative criterion in the sense of "rules of the game 'empirical science'" (Popper, 1989, p. 25) and can be accepted or not. This is a "decision for science", which "is and remains linked to the decision for openness to criticism, because only through this openness can dogmatism be overcome" (Poser, 2009, pp. 133–134).

For the connection between hypotheses and the empirical basis (observational data), Popper's approach leads to a reversal of the methodological function of observation: Instead of deriving and justifying laws from observational data by means of inductive generalization, within the context of a scientific problem, research questions are developed and corresponding hypotheses are formulated which, first and foremost, guide the observations—for example, in an experiment or a research setting—and direct them towards the corresponding "objects" of the investigation. Scientific observations do not arise randomly, they are "made" deliberately and provide corresponding "answers" of the object of investigation (Popper, 1989)—with Kant: Natural scientists (like Galileo) "realized that reason only sees what it itself produces according to its design, that it must proceed with principles of its judgments according to constant laws and force nature to answer its questions, but not let itself be led by it as if by a leash." (Kant, 1959 [1781], B XIII)

This means that scientific reason designs hypotheses about nature and its possible laws, brings them to nature and checks the correctness or falsity of the hypotheses by the reaction of nature, that is, by experience and corresponding observational data (Popper, 1989). Observations and observation sentences—Popper calls them 'basic sentences' (Popper, 1989)—are, according to Popper, the 'control instance' (Ströker, 1998) and *basis of validity* for hypotheses, but not for their confirmation, but for their possible falsification. Within this *"deductive methodology of verification"* (Popper, 1989, p. 5), observations therefore serve the deductive verification of hypotheses—namely, *ex negativo* (Fig. 3.7). Positive confirmation of law statements and theories is in principle not possible; what remains is the openness to refutations. The falsifiability is at the same time a quality criterion of the seriousness and scientificity of scientific theories.

Popper further develops the falsificationism in later work on the "fallibility", that is, an approach to science that clearly takes into account the fundamental and inescapable susceptibility to error and mistake of all theory and all scientific thinking and knowledge (Popper, 1963). Nobody can claim scientifically certain knowledge about which concepts, theories or "natural laws" with regard to relevant phenomena, pressing problems or specific research questions offer the "undoubtedly" "right" explanations or solutions:

"Secure knowledge is denied to us. *Our knowledge is a critical guessing; a net of hypotheses; a fabric of conjecture"* (Popper, 1989, p. XXV)

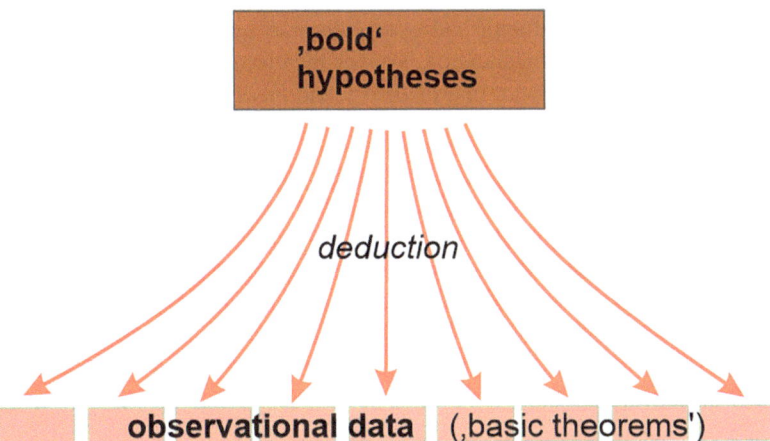

Fig. 3.7 The deductive-hypothetical method at Popper. 'Bold' hypotheses are brought to nature in order to then be checked on the basis of observational data ('basic sentences'). These represent the basis of validity of the falsifiability of the hypotheses (own representation)

This sobering diagnosis can be turned into a positive by Popper by precisely viewing this fallibility as the motor of scientific dynamics and progress. The dynamics of science does not depend on an incessant "accumulation of discoveries and truths"—also referred to as the "accumulation theory of the history of science" (Carrier, 2017, p. 142)—but on the compulsion of the sciences to revise their theories and develop new theories in the face of constant falsifications. For this purpose, new and "bold" hypotheses must be conceived, formulated and deductively checked:

> "It is only the idea, the unfounded anticipation, the bold thought that we try to catch nature with, setting it at stake again and again: Whoever does not expose his thoughts to refutation does not play along in the game of science" (Popper, 1989, p. 224)

According to Popper, it is irrelevant how these hypotheses *psychologically* come about (Popper, 1989), the "discovery context [...] contributes nothing to the justification of the hypothesis" (Poser, 2009, p. 123). The progress of science therefore consists "in the formulation of ever sharper hypotheses, that is, hypotheses for which there are more possibilities of falsification than for the previous version" (Poser, 2009, p. 122). Popper adapts a basic idea of Darwin's theory of evolution, because the "preferred assumptions are the result of selection, the struggle for the survival of hypotheses under the pressure of criticism, which represents an artificially reinforced selection pressure"—he explicitly calls this the "Darwinian process of selection of beliefs and actions" (Popper, 1997, p. 98).

In view of some weaknesses of his falsification approach, which cannot be discussed in detail here for reasons of space (cf. Bauberger, 2016; Poser, 2009), Popper had to make changes to this approach. For example, the falsification principle proved to be "naive" to some critics (cf. Lakatos, 1974), because it is only applicable to simple generalizations (the existence of *one* black swan falsifies the all-statement "All swans are white"), but not to complex theories; it also makes theory formation more difficult and even calls into question natural laws due to the possibility of simple falsification. Popper therefore replaced the juxtaposition of theory and *empiricism* with that of theory and *theory*. Instead of testing theory directly against empiricism, the task is now to compare several theories with each other with regard to the question of which theory has proven itself to be the best in view of the empirical data. This success consists solely in "surviving attempts at refutation" (Ströker, 1998, p. 443). The advocacy of theory competition leads to the rejection of a monotheoretical and to the advocacy of a polytheoretical model of theory testing, thus to the advocacy of a theory pluralism. This pluralism together with theory competition is scientifically productive, because the open horizon of this pluralism requires scientists to intensify their efforts to justify and thus increases the chances of generating, to the best of human knowledge, the best possible theories and corresponding knowledge. Scientific objectivity is therefore not bound by *"factual adequacy"*, but by *"mutual control and criticism"* (Carrier, 2017, p. 43). It is therefore the "competition of contrasting approaches from which the increase in knowledge arises" (Carrier, 2017, p. 43).

Another "weakness" of falsification theory concerns the status of the "basic statements" called by Popper (Popper, 1989) which serve as a "control instance" (Ströker, 1998) and basis of validity for falsifiable "bold" hypotheses. On the one hand, the fallibility of these basic statements cannot be circumvented, because the general fallibility of all scientific activity also applies to them (cf. Bauberger, 2016, p. 47). On the other hand, the so-called "theory-ladenness" or "theory-dependence" of the observation is unavoidable (see Textbox 16), which also affects Popper's basic statements (cf. u. v. Bauberger, 2016; Carrier, 2017; Poser, 2009).

Another difficulty concerns Popper's rejection of the principle of induction once again. Critics have pointed out that in Popper's method of "verification" and the "trial-and-error principle" a form of "pragmatic induction" goes into "obvious" (Schurz, 2007). It is therefore obviously not possible to circumvent the principle of induction, even if it stubbornly resists a generally accepted justification. It is also not meaningful to deny the principle of induction. Because it is neither in everyday life nor in the sciences to eradicate, on the contrary: In both areas, this principle is always silently claimed. The English philosopher Thomas Reid (1710–1796), for example, argued that induction is part of the human nature and the Commons-sense understanding without which a human life and learning from experience would not be possible at all (Reid, 2000 [1785]). The British philosopher Bertrand Russell (1872–1970) also argued similarly: "If the principle is untenable, then we may not *from a reasonable reason* expect that the sun will rise tomorrow morning, that bread will always be more nutritious than stones, that we will fall if we jump from the roof of our house. […] Our entire behavior is based on associations that have worked in the past and that we assume will work in the future. The reliability of these assumptions depends on the validity of the principle of induction." (Russell, 1967 [1912], p. 61)

In the sciences, too, the generation of new knowledge would hardly be possible without the principle of induction. It can not be justified, but scientists work with it because it "works" so to speak. It is claimed just as much as other controversial concepts, such as "truth", "experience", "causality", "space and time" and many more.

Textbox 16: On the Theory-ladenness of Observation

After everything said so far, the basic idea of "theory-ladenness" is easily comprehensible: There are neither concepts nor observations that are not already "impregnated" by theories and therefore cannot guarantee "immediacy" of observation, which could legitimate the observation data as a "control instance" for falsifications. This is already evident in everyday experience. Supposedly theory-free words like "here", "now" or "next to" (the classical topos: Hegel, 1980), all the more concepts like "dog" or "cat" presuppose other concepts or entire theories, without which the intersubjective use of these expressions would not be possible at all. Whoever speaks, for example, of "before" and "after" presupposes the concept of time and thus theories about what "time" is. Even if these conditions are not thematic in everyday life, they act unthematic as necessary conditions. Correspondence theories of truth in the form of image or mirror theories (see Sect. 3.1.3) are just as implausible from the outset as positivism (see Sect. 3.4.1) with its claimed immediacy of the "positive". This is especially true for the sciences, for example in experiments. If Popper is right in forcing the "nature" or "reality" to "answer" scientific questions by means of hypotheses in the wake of Kant, then even the observation data obtained in this way are subject to this theory-ladenness and thus to their quality as a "control instance" for falsifications.

But the child must not be poured out with the bath. The merit of Popper, to have introduced the method of falsification regardless of possible weaknesses or inadequacies, can not be overestimated. At this point we agree with a corresponding assessment by Hans Poser:

"Nevertheless, Popper's methodological basic rule that there must be possibilities of refutation for a scientific statement, a methodological basic postulate, which is indispensable if there is to be a difference between a figment of the imagination and a science," (Poser, 2009, p. 131)

This assessment fits into the mentioned insight that scientific theory needs a normative corrective in addition to a descriptive one (see Sect. 3.1.1). Fallibility and falsifiability belong to such a normative corrective just as much as the concept of "truth" (see Sect. 3.1.3), which is supposed to protect us from nonsense or gross errors, or the concept of "experience" (see Sect. 3.3), which is supposed to keep us from unscientific speculation or the not infrequent occurrence of metaphysical "cloud-cuckoo-lands" as part of a "conceptual poetry". Even if these concepts—as often described—are always associated with theoretical difficulties or have to struggle with practicality problems, they are indispensable for the sciences as research-guiding regulations. ◄

Textbox 17: Interim Conclusion on Theory of Science—Philosophical Foundations and Positions
This chapter first introduced the philosophical foundations of the theory of science. It showed how the theory of science is related to other "science-sciences" how attempts have been made to better define science and scientificity and how difficult it is to define the decisive concept of "truth" for science. Subsequently, the decisive philosophical positions of the engagement with science in their partly temporally succeeding, partly argumentatively referring comprehensive philosophical discourse could be presented and reconstructed in their basic features, statements and difficulties. The two decisive epistemological positions of empiricism and rationalism as well as essentialism, positivism, pragmatism, hermeneutics, philosophy of language

and falsificationism were taken into account as well as the problems of induction, verification, the principle of causality and the theory-ladenness of observation. This passage makes it possible to understand the historical and thematic richness of current scientific-theoretical discussions, the knowledge of which is certainly not dispensable for an informed reception and engagement with these discussions. On the other hand, it became clear in many places of the reconstruction how an appropriate understanding of science in theory and practice depends on the consideration of its historical and social contexts. Therefore, the historical and social contextualization of the sciences is the topic of the next two chapters.

Further Reading

Leerhoff et al. (2010): A clear and well-readable introduction to analytical philosophy with corresponding chapters on the importance of language and on philosophy of language.

Kornmesser and Büttemeyer (2020): A well-founded and well-readable book that covers the most important milestones and main directions as well as systematic aspects of the theory of science and also deals with specific theories of science of different scientific disciplines.

Schülein and Reitze (2012): An introduction to the theory of science for beginners, which comprehensively deals with the history and the systematic problems of the theory of science and also goes into special theories of science of different scientific disciplines.

Original Literature

Pfister (2016): A text collection that unites classic texts of the theory of science, including some that were discussed and quoted in this introduction. The individual texts are each briefly and precisely classified and introduced in outline.

Popper (1997): In this reader, important writings by Popper are reproduced in text excerpts according to themes. This selection is a very good introduction to the almost unmanageable complete works and the most important theses of Popper for students.

Wittgenstein (1995 [1953]): In addition to the "Tractatus", this volume also contains the "Philosophical Investigations" as Wittgenstein's main work on the philosophy of "normal language".

References

Albert, G. (2005). *Hermeneutischer Positivismus und dialektischer Essentialismus Vilfredo Paretos*. VS Verlag.

von Aquin, T. (2013 [1256–1259]). *Über die Wahrheit. Quaestiones disputatae de veritate*. Marix (In der Übersetzung von Eith Stein).

Aristoteles. (1991 [348–345 v.u.Z.]). *Metaphysik. Schriften zur ersten Philosophie*. Reclam.

Aristoteles. (2001). *Die Nikomachische Ethik. Griechisch-deutsch* (Sammlung Tusculum). Artemis & Winkler (Übersetzt von Olof Gigon, neu heraugegeben von Rainer Nickel).

Aristoteles. (2009). *Politik. Neuausgabe.* Rowohlt (Nach der Übersetzung von Franz Susemihl mit Einleitung, Bibliographie und zusätzlichen Anmerkungen von Wolfgang Kullmann).

Bacon, F. (1990 [1620]). *Neues Organon. Herausgegeben und mit einer Einleitung von Woflgang Krohn* (Teilband 1 und Teilband 2). Wissenschaftliche Buchgesellschaft.

Balsiger, P. W. (2005). *Transdisziplinarität. Systematisch-vergleichende Untersuchung disziplinenübergreifender Wissenschaftspraxis.* Fink.

Balzert, H., Schäfer, C., Schröder, M., & Kern, U. (2010). *Wissenschaftliches Arbeiten. Wissenschaft, Quellen, Artefakte, Organisation, Präsentation* (Soft skills). Herdecke W3L.

Bauberger, S. (2016). *Wissenschaftstheorie. Eine Einführung.* Kohlhammer.

Beckmann, J. P. (1997). Rationalismus I. In G. Müller (Hrsg.), *Theologische Realenzyklopädie.* (Theologische Realenzyklopädie, Bd. 28, S. 161–170). de Gruyter.

Berr, K. (Hrsg.). (2018). *Transdisziplinäre Landschaftsforschung. Grundlagen und Perspektiven.* Springer VS.

Berr, K. (2019). Konflikt und Ethik. In K. Berr & C. Jenal (Hrsg.), *Landschaftskonflikte* (S. 109–129). Springer VS.

Berr, K. (2020). Vom Wahren, Schönen und Guten. Philosophische Zugänge zu Landschaftsprozessen. In R. Duttmann, O. Kühne & F. Weber (Hrsg.), *Landschaft als Prozess.* Springer VS.

Berr, K., & Kühne, O. (2019). Moral und Ethik von Landschaft. In O. Kühne, F. Weber, K. Berr, & C. Jenal (Hrsg.), *Handbuch Landschaft* (S. 351–365). Springer VS.

Berr, K., Jenal, C., Kühne, O., & Weber, F. (2019). *Landschaftsgovernance. Ein Überblick zu Theorie und Praxis.* Springer VS.

Birnbacher, D. (2006). *Natürlichkeit.* de Gruyter.

BNatSchG. (2009 [1976]). *Gesetz über Naturschutz und Landschaftspflege (Bundesnaturschutzgesetz).* https://www.gesetze-im-internet.de/bnatschg_2009. Zugegriffen: 17. Mai 2018.

Bormann, K. (1987). *Platon.* Alber.

Borsche, T. (1996). Einleitung. Sprachphilosophische Überlegungen zu einer Geschichte der Sprachphilosophie. In T. Borsche (Hrsg.), *Klassiker der Sprachphilosophie. Von Platon bis Noam Chomsky* (S. 7–13). Beck.

Broad, C. D. (1952). *Ethics and the history of philosophy. Selected essays.* Routledge.

Burnet, J. (1962). *Platonis Opera recognovit brevique adnotatione critica instruxit Ioannes Burnet. Tomus II.* Clarendon.

Carnap, R. (1931). Die physikalische Sprache als Universalsprache der Wissenschaft. *Erkenntnis, 2*(1), 432–465. https://doi.org/10.1007/BF02028172.

Carnap, R. (1966 [1928]). *Scheinprobleme in der Philosophie. Das Fremdpsychische und der Realismusstreit.* Suhrkamp.

Carrier, M. (2005). Verwertungsdruck und Erkenntnisgewinn: Philosophische Reflexion angewandter Forschung. *Information Philosophie, 3,* 7–19.

Carrier, M. (2017). *Wissenschaftstheorie zur Einführung* (Zur Einführung, 4., überarb. Aufl., Bd. 353). Junius.

Chilla, T., Kühne, O., & Neufeld, M. (2016). *Regionalentwicklung* (UTB, Bd. 4566). Ulmer.

Dahrendorf, R. (1972). *Konflikt und Freiheit. Auf dem Weg zur Dienstklassengesellschaft.* Piper.

Dahrendorf, R. (1992). *Der moderne soziale Konflikt. Essay zur Politik der Freiheit.* Deutsche Verlags-Anstalt.

Demmerling, C. (2019). Szientismus. In J. Ritter (Hrsg.), *Historisches Wörterbuch der Philosophie* (Völlig neubearbeitete Ausgabe des „Wörterbuchs der philosophischen Begriffe" von Rudolf Eisler, Sonderausgabe, *St – T,* Bd. 10, S. 872–876). WBG Academic.

Descartes, R. (1990 [1637]). *Discours de la méthode. Französisch – Deutsch* (Philosophische Bibliothek, Bd. 261, Unveränderter Nachdruck). Meiner (Von der Methode des richtigen Vernunftgebrauchs und der wissenschaftlichen Forschung. Übersetzt und herausgegeben von Lüder Gäbe).

Descartes, R. (2008 [1637]). *Meditationes de prima philosophia. Lateinisch – Deutsch* (Philosophische Bibliothek, Bd. 597). Meiner.

Detel, W. (2018). *Grundkurs Philosophie* (Erkenntnis- und Wissenschaftstheorie, Bd. 4, 3., vollst. durchgesehene u. erw.). Reclam.

Dewey, J. (2016). *Logik. Die Theorie der Forschung* (Suhrkamp Taschenbuch Wissenschaft, 2. Aufl.). Suhrkamp.

Diaz-Bone, R., & Schubert, K. (1996). *William James zur Einführung*. Junius.

Diemer, A. (2019). Geisteswissenschaften. In J. Ritter (Hrsg.), *Historisches Wörterbuch der Philosophie* (Völlig neubearbeitete Ausgabe des „Wörterbuchs der philosophischen Begriffe" von Rudolf Eisler, Sonderausgabe, G – H, Bd. 3, S. 211–215). WBG Academic.

Dilthey, W. (1924). Die Entstehung der Hermeneutik. In G. Misch (Hrsg.), *Gesammelte Schriften* (Die geistige Welt. Einleitung in die Philosophie des Lebens. Erste Hälfte. Abhandlungen zur Grundlegung der Geisteswissenschaften, Bd. V). B.G. Teubner.

Dilthey, W. (1970 [1910]). *Der Aufbau der geschichtlichen Welt in den Geisteswissenschaften* (Suhrkamp-Taschenbuch Wissenschaft, Bd. 354). Suhrkamp.

Dilthey, W. (2017 [1883]). *Einleitung in die Geisteswissenschaften: Versuch einer Grundlegung für das Studium der Gesellschaft und ihrer Geschichte. Neuausgabe mit einer Biographie des Autors*. Hofenberg.

Egner, H. (2010). *Theoretische Geographie*. WBG.

Eisel, U. (1992). Über den Umgang mit dem Unmöglichen. Ein Erfahrungsbericht über Interdisziplinarität im Studiengang Landschaftsplanung – Teil 2. *Gartenamt, 41*(10), 710–719.

Enders, M., & Szaif, J. (Hrsg.). (2006). *Die Geschichte des philosophischen Begriffs der Wahrheit*. de Gruyter.

Endruweit, G. (2015). *Empirische Sozialforschung. Wissenschaftstheoretische Grundlagen* (UTB Sozialwissenschaften, Bd. 4460). UVK Verlagsgesellschaft.

Engfer, H.-J. (1996). *Empirismus versus Rationalismus? Kritik eines philosophiegeschichtlichen Schemas*. Schöningh.

Ernst, G. (2014). *Einführung in die Erkenntnistheorie* (5., bibliograph. akt. Aufl.). WBG – Wissenschaftliche Buchgesellschaft.

Falkenburg, B. (2017). Natur. In T. Kirchhoff, N. C. Karafyllis, D. Evers, B. Falkenburg, M. Gerhard, & G. Hartung et al. (Hrsg.), *Naturphilosophie. Ein Lehr- und Studienbuch* (S. 96–102). Mohr Siebeck/UTB GmbH.

Flick, U., Kardorff, E. v., & Steinke, I. (Hrsg.). (2007). *Qualitative Forschung. Ein Handbuch*. Rowohlt.

Frege, G. (1879). *Begriffsschrift. Eine der arithmetischen nachgebildete Formelsprache des reinen Denkens*. Verlag von Louis Nebert.

Gabriel, G. (1993). *Grundprobleme der Erkenntnistheorie. Von Descartes zu Wittgenstein* (3., durchgesehene. Aufl.). Schöningh.

Gabriel, G. (1998). Erkenntnistheorie. In A. Pieper (Hrsg.), *Philosophische Disziplinen. Ein Handbuch* (Reclam-Bibliothek, Bd. 1643, S. 52–71). Reclam.

Gabriel, G. (2012). Geltung und Genese als Grundlagenproblem. *Erwägen Wissen Ethik, 23*(4), 475–486.

Gadamer, H.-G. (1975). *Wahrheit und Methode. Grundzüge einer philosophischen Hermeneutik*. Mohr.

Gailing, L., & Leibenath, M. (2012). Von der Schwierigkeit, „Landschaft" oder „Kulturlandschaft" allgemeingültig zu definieren. *Raumforschung und Raumordnung, 70*(2), 95–106. https://doi.org/10.1007/s13147-011-0129-8.

Gamm, G. (2009). *Philosophie im Zeitalter der Extreme. Eine Geschichte philosophischen Denkens im 20. Jahrhundert*. Primus.

Gansland, H. R., & Carrier, M. (2004). Positivismus (historisch). In J. Mittelstraß (Hrsg.), *Enzyklopädie Philosophie und Wissenschaftstheorie* (3, P – So, unveränderte Sonderausgabe, S. 301–303). Metzler.

Gawlick, G. (Hrsg.). (1995). *Geschichte der Philosophie in Text und Darstellung* (Empirismus, Bd. 4). Reclam.

Gethmann, C. F. (1987). Vom Bewusstsein zum Handeln. Pragmatische Tendenzen in der deutschen Philospohie der ersten Jahrzehnte des 20. Jahrhunderts. In H. Stachowiak (Hrsg.), *Pragmatik. Handbuch pragmatisches Denken* (2. Aufl., S. 202–232). Meiner.

Gethmann, C. F. (1991). Vielheit der Wissenschaften – Einheit der Lebenswelt. In Akademie der Wissenschaften zu Berlin (Hrsg.), *Einheit der Wissenschaften* (S. 349–371). de Gruyter.

Gethmann, C. F. (2004). Universalienstreit. In J. Mittelstraß (Hrsg.), *Enzyklopädie Philosophie und Wissenschaftstheorie* (4, Sp – Z, unveränderte Sonderausgabe, S. 411–412). Metzler.

Gethmann, C. F. (2005). Ist das Wahre das Ganze? Methodologische Probleme Integrierter Forschung. In G. Wolters & M. Carrier (Hrsg.), *Homo Sapiens und Homo Faber. Epistemische und technische Rationalität in Antike und Gegenwart. Festschrift für Jürgen Mittelstraß* (S. 391–404). de Gruyter.

von Glasersfeld, E. (1988). Einführung in den radikalen Konstruktivismus. In P. Watzlawick (Hrsg.), *Die erfundene Wirklichkeit. Wie wissen wir, was wir zu wissen glauben? Beiträge zum Konstruktivismus* (5. Aufl., S. 16–38). Piper.

von Glasersfeld, E. (1995). *Radical constructivism. A way of knowing and learning* (Studies in mathematics education series, Bd. 6). Falmer Press.

Gloy, K. (2004). *Wahrheitstheorien. Eine Einführung*. Francke.

Gloy, K. (2005 [1995]). *Die Geschichte des wissenschaftlichen Denkens. Das Verständnis der Natur*. Komet.

Goodman, N. (1978). *Ways of worldmaking* (Harvester studies in philosophy, Bd. 5). Harvester Press.

Grau, A. (2017). *Hypermoral. Die neue Lust an der Empörung* (2. Aufl.). Claudius.

Habermas, J. (Hrsg.). (1983). *Moralbewußtsein und kommunikatives Handeln*. Suhrkamp.

Habermas, J. (1994). Die Moderne – ein unvollendetes Projekt. In W. Welsch (Hrsg.), *Wege aus der Moderne. Schlüsseltexte der Postmoderne-Diskussion* (2., durchgesehene. Aufl., S. 177–192). Akademie.

Habermas, J. (1995). *Theorie des kommunikativen Handelns* (Handlungsrationalität und gesellschaftliche Rationalisierung, Bd. I). Suhrkamp.

Habermas, J., & Luhmann, N. (1972). *Theorie der Gesellschaft oder Sozialtechnologie. Was leistet die Systemforschung?* Suhrkamp.

Hartmann, D. (2020). *Neues System der philosophischen Wissenschaften im Grundriss* (Erkenntnistheorie, Bd. I). Mentis.

Hauskeller, M. (2005). *Was ist Kunst? Positionen der Ästhetik von Platon bis Danto* (Beck'sche Reihe, 8. Aufl.). Beck.

Hegel, G. W. F. (1970). *Enzyklopädie der philosophischen Wissenschaften im Grundrisse 1830. Erster Teil: Die Wissenschaft der Logik. Mit mündlichen Zusätzen* (8 Bände). Suhrkamp.

Hegel, G. W. F. (1980). *Phänomenologie des Geistes* (Gesammelte Werke, Bd. 9). Meiner.

Heidegger, M. (1992). *Der Ursprung des Kunstwerkes*. Reclam.

Heisenberg, W. (1942). *Die Einheit des naturwissenschaftlichen Weltbildes*. Barth.

Hirschberger, J. (1976). *Geschichte der Philosophie* (Altertum und Mittelalter, Bd. 1). Komet.

Hobbes, T. (2017 [1651]). *Leviathan oder Stoff, Form und Gewalt eines kirchlichen und bürgerlichen Staates. Herausgegeben und eingeleitet von Iring Fetscher. Übersetzt von Walter Euchner* (Suhrkamp-Taschenbuch Wissenschaft, Bd. 462, 16. Aufl.). Suhrkamp.

Höffe, O. (2001). *Kleine Geschichte der Philosophie*. Beck.

Hübner, K. (1978). *Kritik der wissenschaftlichen Vernunft*. Alber.

Hume, D. (1961 [1748]). *Eine Untersuchung über den menschlichen Verstand. Übersetzt von Raoul Richter* (Philosophische Bibliothek, Bd. 648). Meiner.

James, W. (1977). *Der Pragmatismus. Ein neuer Name für alte Denkmethoden* (Philosophische Bibliothek, Bd. 297). Meiner.

Janich, P. (1996). *Was ist Wahrheit? Eine philosophische Einführung* (Beck'sche Reihe C. H. Beck Wissen, Bd. 2052, Original-Ausgabe). Beck.

Joas, H. (1988). Symbolischer Interaktionismus. Von der Philosophie des Pragmatismus zu einer soziologischen Forschungstradition. *Kölner Zeitschrift für Soziologie und Sozialpsychologie, 40*, 417–446.

Kambartel, F. (2004). Positivismus (systematisch). In J. Mittelstraß (Hrsg.), *Enzyklopädie Philosophie und Wissenschaftstheorie* (3, P – So, unveränderte Sonderausgabe, S. 303–304). Stuttgart: J. B. Metzler.

Kamlah, W., & Lorenzen, P. (1967). *Logische Propädeutik oder Vorschule des vernünftigen Redens*. Bibliographisches Institut.

Kant, I. (1959 [1781]). *Kritik der reinen Vernunft*. Felix Meiner.

Kant, I. (1968). Metaphysische Anfangsgründe der Naturwissenschaften. In I. Kant (Hrsg.), *Kants Werke. Akdemische Textausgabe* (Bd. IV, S. 465–566). de Gruyter.

Karmasin, M., & Ribing, R. (2017). *Die Gestaltung wissenschaftlicher Arbeiten. Ein Leitfaden für Facharbeit/VWA, Seminararbeiten, Bachelor-, Master-, Magister- und Diplomarbeiten sowie Dissertationen* (utb Schlüsselkompetenzen, 9., überarb. u. akt. Aufl., Bd. 2774). Facultas Verlags- und Buchhandels AG.

Kienzle, B. (2010). David Hume – Kausalprinzip und Induktionsproblem. In A. Beckermann & D. Perler (Hrsg.), *Klassiker der Philosophie heute* (S. 352–372). Reclam.

Kornmesser, S., & Büttemeyer, W. (2020). *Wissenschaftstheorie. Eine Einführung*. Metzler.

Koselleck, R. (Hrsg.). (1979). *Historische Semantik und Begriffsgeschichte*. Klett-Cotta.

Krieger, G. (2011). Substanz. In P. Kolmer & A. G. Wildfeuer (Hrsg.), *Neues Handbuch philosophischer Grundbegriffe* (S. 2146–2158). Alber.

Kuhlmann, W. (2011). Begründung. In M. Düwell, C. Hübenthal & M. H. Werner (Hrsg.), *Handbuch Ethik* (3., akt. Aufl., S. 319–325). Metzler.

Kühne, O. (2013). *Landschaftstheorie und Landschaftspraxis. Eine Einführung aus sozialkonstruktivistischer Perspektive*. Springer VS.

Kühne, O. (2018). Die Landschaften 1, 2 und 3 und ihr Wandel. Perspektiven für die Landschaftsforschung in der Geographie – 50 Jahre nach Kiel. *Berichte. Geographie und Landeskunde, 92*(3–4), 217–231.

Kühne, O. (2019). Landschaftsverständnisse in ihrer historischen Gebundenheit – zwischen Gegenständlichkeit, Essenz und Konstruktion. In K. Berr & C. Jenal (Hrsg.), *Landschaftskonflikte* (S. 23–36). Springer VS.

Kühne, O., Weber, F., & Jenal, C. (2018). *Neue Landschaftsgeographie. Ein Überblick (Essentials)*. Springer VS.

Kunzmann, P., Burkard, F.-P., & Wiedmann, F. (1993). *dtv-Atlas zur Philosophie. Tafeln und Texte*. dtv.

Kurz, G. (2015). *Das Wahre, Schöne, Gute. Aufstieg, Fall und Fortbestehen einer Trias*. Fink.

Kutschera, U. (2011). *Darwiniana Nova. Verborgene Kunstformen der Natur*. LIT.

Lakatos, I. (1974). Falsifikation und die Methodologie wissenschaftlicher Forschungsprogramme. In I. Lakatos & A. Musgrave (Hrsg.), *Kritik und Erkenntnisfortschritt* (Abhandlungen des Internationalen Kolloquiums über die Philosophie der Wissenschaft, Bd. 4, S. 89–189). Vieweg.

Lamnek, S. (2005). *Qualitative Sozialforschung. Lehrbuch* (4., vollst. überarb. Aufl.). Beltz PVU.

Lautensach, H. (1973). Über die Erfassung und Abgrenzung von Landschaftsräumen [Erstveröffentlichung 1938]. In K. Paffen (Hrsg.), *Das Wesen der Landschaft* (Wege der Forschung, Bd. 39, S. 20–38). WBG.

Leerhoff, H., Rehkämper, K., & Wachtenhofer, T. (2010). *Einführung in die Analytische Philosophie*. Wissenschaftliche Buchgesellschaft.

Leibniz, G. W. (1961). *Neue Abhandlungen über den menschlichen Verstand*. Insel.

Leibniz, G. W. (2019 [1714]). *Monadologie. Französisch/Deutsch* (durchgesehene und bibliographisch ergänzte Ausgabe). Reclam.

List, E. (2004). Einleitung. Interdisziplinäre Kulturforschung auf der Suche nach theoretischer Orientierung. In E. List (Hrsg.), *Grundlagen der Kulturwissenschaften. Interdisziplinäre Kulturstudien* (S. 3–12). Francke.

Locke, J. (1981 [1689]). *Versuch über den menschlichen Verstand. 2 Bände* (Philosophische Bibliothek, Bd. 75, 4., durchgesehene Aufl.). Meiner.

Lorenz, K. (2004). Einheitswissenschaft. In J. Mittelstraß (Hrsg.), *Enzyklopädie Philosophie und Wissenschaftstheorie* (1, A – G, unveränderte Sonderausgabe, S. 530). Metzler.

Martin, G. (1973). *Platons Ideenlehre*. de Gruyter.

Merker, B., Mohr, G., & Siep, L. (Hrsg.). (1998). *Angemessenheit. Zur Rehabilitierung einer philosophischen Metapher*. Königshausen & Neumann.

Merton, R. K. (1957[1949]). *Social theory and social structure* (Revised and enlarged Edition). Free Press.

Mittelstraß, J. (1998). Interdisziplinarität oder Transdisziplinarität? In *Die Häuser des Wissens. Wissenschaftstheoretische Studien* (S. 29–48). Suhrkamp.

Mittelstraß, J. (2004). Szientismus. In J. Mittelstraß (Hrsg.), *Enzyklopädie Philosophie und Wissenschaftstheorie* (4, Sp – Z, unveränderte Sonderausgabe, S. 872–876). Metzler.

Mouffe, C. (2007). Pluralismus, Dissens und demokratische Staatsbürgerschaft. In M. Nonhoff (Hrsg.), *Diskurs – radikale Demokratie – Hegemonie. Zum politischen Denken von Ernesto Laclau und Chantal Mouffe* (S. 41–53). transcript.

Mouffe, C. (2014). *Agonistik. Die Welt politisch denken* (Bd. 2677). Suhrkamp.

Müller, E., & Schmieder, F. (2016). *Begriffsgeschichte und historische Semantik. Ein kritisches Kompendium*. Suhrkamp.

Neubert, S. (2004). Pragmatismus – thematische Vielfalt in Deweys Philosophie und in ihrer heutigen Rezeption. In L. A. Hickman, S. Neubert, & K. Reich (Hrsg.), *John Dewey. Zwischen Pragmatismus und Konstruktivismus* (Interaktionistischer Konstruktivismus, Bd. 1, S. 13–27). Waxmann.

Neurath, O. (1932). Protokollsätze. *Erkenntnis, 3*(1), 204–214. https://doi.org/10.1007/BF01886420.

Neurath, O., Carnap, R., & Morris, C. W. (Hrsg.). (1970). *Foundations of the unity of science. Towards an international encyclopedia of unified science*. University of Chicago Press.

Newen, A. (2018). *Analytische Philosophie zur Einführung* (3., unveränd. Aufl.). Junius.

Nicolescu, B. (Hrsg.). (2008). *Transdisciplinarity. Theory and practice*. Hampton Press.

Ottmann, H. (2012). *Geschichte des politischen Denkens* (Das 20. Jahrhundert, Bd. 4). Metzler (Teilband 2: Von der Kritischen Theorie bis zur Globalisierung).

Peirce, C. S. (2018 [1878]). *Über die Klarheit unserer Gedanken. How to Make Our Ideas Clear* (4., erw. Aufl.). Vittorio Klostermann GmbH.

Pfister, J. (Hrsg.). (2016). *Texte zur Wissenschaftstheorie*. Reclam.

Pfister, J. (2019). *Philosophie. Ein Lehrbuch*. Reclam.

Pieper, A. (Hrsg.). (1998). *Philosophische Disziplinen. Ein Handbuch* (Reclam-Bibliothek, Bd. 1643). Reclam.

Popper, K. R. (1959). *The logic of scientific discovery*. Harper & Row.

Popper, K. R. (1963). *Conjectures and refutations. The growth of scientific knowledge*. Routledge & Kegan Paul.

Popper, K. R. (1984). *Auf der Suche nach einer besseren Welt. Vorträge und Aufsätze aus dreißig Jahren*. Piper.

Popper, K. R. (1989). *Logik der Forschung*. Mohr Siebeck.

Popper, K. R. (1997). *Lesebuch. Ausgewählte Texte zur Erkenntnistheorie, Philosophie der Naturwissenschaften, Metaphysik, Sozialphilosophie* (2. Aufl.). UTB GmbH.

Popper, K. R. (2003 [1945]). *Die offene Gesellschaft und ihre Feinde* (Der Zauber Platons, Bd. 1, 8. Aufl.). Mohr Siebeck.

Popper, K. R. (2010). *Die beiden Grundprobleme der Erkenntnistheorie. Aufgrund von Manuskripten aus den Jahren 1930–1933* (Karl R. Popper – Gesammelte Werke, Bd. 2, 3. Aufl., durchgesehen und ergänzt). Mohr Siebeck (Herausgegeben von Troels Eggers Hansen).

Poser, H. (2009). *Wissenschaftstheorie. Eine philosophische Einführung*. Reclam.

Rapp, C. (2010). Aristoteles – Das Problem der Substanz. In A. Beckermann & D. Perler (Hrsg.), *Klassiker der Philosophie heute* (S. 38–58). Reclam.

Reese-Schäfer, W. (1991). *Jürgen Habermas*. Campus.

Regenbogen, A., & Meyer, U. (Hrsg.). (1998). *Wörterbuch der philosophischen Begriffe*. Meiner.

Reichenbach, H. (1938). *Experience and prediction. An analysis of the foundations and the structure of knowledge*. University of Chicago Press.

Reid, T. (2000 [1785]). *An inquiry into the human mind. On the principles of common sense*. Edinburgh University Press.

Richter, R. (2016). *Soziologische Paradigmen. Eine Einführung in klassische und moderne Konzepte* (2. Aufl.). UTB.

Rickert, H. (1986 [1902]). *Kulturwissenschaft und Naturwissenschaft*. Reclam.

Riecke, J. (Hrsg.). (2011). *Historische Semantik*. de Gruyter.

Rorty, R. M. (Hrsg.). (1967). *The linguistic turn. Essays in philosophical method*. University of Chicago Press (With Two Retrospective Essays).

Russell, B. (1967 [1912]). *Probleme der Philosophie* (2. Aufl.). edition Suhrkamp.

Schäfer, L. (1991). Natur. In E. Martens & H. Schnädelbach (Hrsg.), *Philosophie. Ein Grundkurs* (Bd. 2, S. 467–507). Rowohlt.

Schildknecht, C., Teichert, D., & van Zantwijk, T. (Hrsg.). (2008). *Genese und Geltung. Für Gottfried Gabriel*. Mentis.

Schmitt, C. (1933). *Der Begriff des Politischen*. Hanseatische Verlagsanstalt.

Schmitt, C. (2011 [1967]). *Die Tyrannei der Werte* (3., korr. Aufl.). Duncker & Humblot.

Schnädelbach, H. (1980). *Probleme der Wissenschaftstheorie. Studienbrief 3302 der FernUniversität Hagen*. Kurseinheit 01: Grundfragen philosophischer Wissenschaftstheorie, Hagen.

Schnädelbach, H. (1991). Philosophie. In E. Martens & H. Schnädelbach (Hrsg.), *Philosophie. Ein Grundkurs* (S. 37–76). Rowohlt.

Schnädelbach, H. (1993). *Probleme der Wissenschaftstheorie. Eine philosophische Einführung. Kurseinheit 1: Grundfragen philosophischer Wissenschaftstheorie*. Vorlesung, Fernuniversität Hagen.

Schneider, H. (2019). Essentialismus. In J. Ritter (Hrsg.), *Historisches Wörterbuch der Philosophie* (Völlig neubearbeitete Ausgabe des „Wörterbuchs der philosophischen Begriffe" von Rudolf Eisler, Sonderausgabe, D – F, Bd. 2, S. 751–753). WBG Academic.

Schönwälder-Kuntze, T. (2020). *Philosophische Methoden zur Einführung* (3., erw. Aufl.). Junius.

Schroeder, P. (2004). Certismus. In J. Mittelstraß (Hrsg.), *Enzyklopädie Philosophie und Wissenschaftstheorie* (1, A – G, unveränderte Sonderausgabe, S. 385–386). Metzler.

Schubert, H.-J., Joas, H., & Wenzel, H. (2010). *Pragmatismus zur Einführung. Kreativität, Handlung, Deduktion, Induktion, Abduktion, Chicago School, Sozialreform, symbolische Interaktion* (Zur Einführung, Bd. 382). Junius.

Schülein, J. A., & Reitze, S. (2012). *Wissenschaftstheorie für Einsteiger* (3., akt. u. erw. Aufl.). facultas wuv.

Schurz, G. (2007). Popper und das Problem der Induktion. In H. Keuth (Hrsg.), *Karl Popper: Logik der Forschung* (3., bearb. Aufl., S. 25–40). de Gruyter.

Schurz, G. (2014). *Einführung in die Wissenschaftstheorie* (4., überarb. Aufl.). Darmstadt: WBG.

Schwemmer, O. (2004a). essentia. In J. Mittelstraß (Hrsg.), *Enzyklopädie Philosophie und Wissenschaftstheorie* (1, A – G, unveränderte Sonderausgabe, S. 590–591). Metzler.

Schwemmer, O. (2004b). Essentialismus. In J. Mittelstraß (Hrsg.), *Enzyklopädie Philosophie und Wissenschaftstheorie* (1, A – G, unveränderte Sonderausgabe, S. 591–592). Metzler.

Skirbekk, G. (Hrsg.). (1977). *Wahrheitstheorien. Eine Auswahl aus den Diskussionen über Wahrheit im 20. Jahrhundert*. Suhrkamp.

Stegmüller, W. (1973). *Probleme und Resultate der Wissenschaftstheorie und Analytischen Philosophie. Personelle und statistische Wahrscheinlichkeit. Erster Halbband: Personelle Wahrscheinlichkeit und Rationale Entscheidung* (Bd. 4). Springer.

Stegmüller, W. (1978a). *Hauptströmungen der Gegenwartsphilosophie. Eine kritische Einführung* (Bd. 1, Nachdruck, 6. Aufl.). Kröner.

Stegmüller, W. (Hrsg.). (1978b). *Das Universalien-Problem* (Wege der Forschung, Bd. 83). Wissenschaftliche Buchgesellschaft.

Stegmüller, W. (1984). Evolutionäre Erkenntnistheorie, Realismus und Wissenschaftstheorie. In R. Spaemann, P. Koslowski & R. Löw (Hrsg.), *Evolutionstheorie und menschliches Selbstverständnis. Zur philosophischen Kritik eines Paradigmas moderner Wissenschaft; Referate und der Bericht über die Schlußdiskussion* (Civitas-Resultate, Bd. 6, S. 5–34). Acta Humaniora.

Steiner, C. (2014a). *Pragmatismus – Umwelt – Raum. Potenziale des Pragmatismus für eine transdisziplinäre Geographie der Mitwelt* (Erdkundliches Wissen, Bd. 155). Franz Steiner.

Steiner, C. (2014b). Von Interaktion zu Transaktion – Konsequenzen eines pragmatischen Mensch-Umwelt-Verständnisses für eine Geographie der Mitwelt. *Geographica Helvetica, 69*(3), 171–181.

Strawson, P. F. (2003). *Einzelding und logisches Subjekt. Ein Beitrag zur deskriptiven Metaphysik.* Reclam.

Ströker, E. (1998). Wissenschaftstheorie. In A. Pieper (Hrsg.), *Philosophische Disziplinen. Ein Handbuch* (Reclam-Bibliothek, Bd. 1643, S. 437–456). Reclam.

Tetens, H. (1999). Wissenschaft. In H. J. Sandkühler (Hrsg.), *Enzyklopädie Philosophie* (S. 1763–1773). Meiner.

Tetens, H. (2013). *Wissenschaftstheorie. Eine Einführung.* Beck.

Vico, G. (1966 [1725]). *Die neue Wissenschaft über die gemeinschaftliche Natur der Völker* (2. Aufl.). Rowohlt.

Voss, R. (2017). *Wissenschaftliches Arbeiten … leicht verständlich!* (utb Schlüsselkompetenzen, 5. Aufl.). UVK/Lucius.

Weber, F., & Kühne, O. (2019). Essentialistische Landschafts- und positivistische Raumforschung. In O. Kühne, F. Weber, K. Berr & C. Jenal (Hrsg.), *Handbuch Landschaft* (S. 57–68). Springer VS.

Weinrich, H. (1975). System, Diskurs, Didaktik und die Diktatur des Sitzfleisches. In F. Maciejewski (Hrsg.), *Theorie der Gesellschaft oder Sozialtechnologie. Theorie-Diskussion Supplement 1* (S. 145–161). Suhrkamp.

Wiltsche, H. A. (2013). *Einführung in die Wissenschaftstheorie.* Vandenhoeck & Ruprecht.

Windelband, W. (1924). Geschichte und Naturwissenschaft. In W. Windelband (Hrsg.), *Präludien* (Bd. 2). Mohr.

Wittgenstein, L. (1995 [1953]). *Tractatus logico-philosophicus. Tagebücher 1914–1916. Philosophische Untersuchungen* (Werkausgabe Bd. 1, 10. Aufl.). Suhrkamp.

Zima, P. V. (2004). *Was ist Theorie? Theoriebegriff und Dialogische Theorie in den Kultur- und Sozialwissenschaften.* Francke.

Zoglauer, T. (1999). *Einführung in die formale Logik für Philosophen.* Vandenhoeck & Ruprecht.

The Contextualization of Science I: Time

4

Poppers approach can be seen as the culmination of the theory of science up to those approaches which are characterized by the inclusion of the contexts of scientific knowledge and corresponding knowledge. This contextualization is taken up in Chaps. 4 and 5 but intertwine, the conceptions dealt with in these chapters go back to different theoretical traditions: In Chap. 4, positions are presented which have been developed from the history of science and therefore take into account the factor "time". In Chap. 5 we present positions which have arisen from the tradition of sociology and therefore focus on the factor of "social embeddedness". In Chap. 4, after the identification of the characteristics of the transition from Popper to a post-Poppersian theory of science, the historical-scientific conceptions of Thomas S. Kuhn, Imre Lakatos, Larry Laudan and Paul Feyerabend are dealt with in their argumentative reference to each other. Important terms for this chapter are introduced in Textbox 18.

> **Textbox 18: Important Terms for Chap. 4**
>
> **Accumulation**: (from Latin *accumulare*, to accumulate). In the theory of science, it means the accumulation of knowledge or theories, in economics the accumulation of capital (Karl Marx), in geology the accumulation or stratification of, for example, rock material.
>   **Anomaly**: (from Greek *anomalia*, unevenness, irregularity). Means generally the deviation from norms or rules, from laws or traditions, from customs or practices. In Thomas S. Kuhn's theory of paradigms, anomalies

refer to significant deviations of research results from the assumptions and standards of the currently established theory.

Theory of evolution: (from Latin *evolvere*, to develop). In particular, the view established by Charles Darwin (1809–1882) that all life is subject to a process of development, which leads to changes and further development of the forms of life through the pressure of adaptation to environments and the resulting environmental adaptation. This development is not purposeful ('teleological'), but contingent and only retrospectively *as* development *reconstructable*.

Heuristics (from Greek *heuriskein*, to find, to invent) is sometimes also referred to as the art of invention (Latin *ars inveniendi*), insofar as the art of finding or inventing something new is meant. In general, any art is meant to come to results or solutions by means of rules of thumb, assumptions, 'trial and error', exclusion methods and other search methods.

Incommensurability (from Latin *mensura*, measure) means the 'unmeasurability' of two sizes or objects that cannot be measured with a common measure. In Thomas S. Kuhn, incommensurability refers to the incomparability of different paradigms (see Sect. 4.2).

Rationality (from Latin *rationalitas*, reasonableness) means in the broadest sense the reasonableness of human thinking and acting. In turn, in the broadest sense, thinking or acting 'rational' or 'reasonable', if it can be justified or (re) constructed intersubjectively and can exceed individual or group-specific interests, values, convictions in favor of generalizable orientations.

Teleology, teleological: The term goes back to the Greek word parts *télos* (goal, purpose, completion) and *lógos* (doctrine). 'Teleology' is the view that processes are determined by certain purposes or ideal final states and move towards them.

4.1   Karl Popper as the High Point and Culmination of Classical Philosophy of Science

The following approaches to the philosophy of science are characterized by the fact that they bring to the fore aspects of scientific change, theory development and scientific rationality that have so far been neglected or only marginally considered. They can also be understood as indirect or direct responses to some of the deficiencies of the Popper approach. In content, these approaches connect the thematization and elaboration of the context-dependence, in particular the historical, but also the social dimension of all science and research. There are four main differences between Popper and the following approaches:

1. Popper had rejected a central aspect of the 'accumulation theory of the history of science' by reconstructing scientific progress not as the cumulative summation of allegedly secure knowledge, but as a succession of incompatible theories. Against the 'bucket theory' of knowledge of the main representatives of logical empiricism, according to which scientific reason collects knowledge like in a bucket (Popper, 1997), Popper considers "scientific progress as a continued trial-and-error chain" (Poser, 2009, p. 140). On the other hand, he formulates a 'correspondence principle' "of the theory dynamics, according to which the later theory contains the earlier one in approximation under suitable circumstances […]. Ultimately, the accumulation theory is thus rather limited than abandoned by Popper" (Carrier, 2017, p. 146; cf. Popper, 1984). In contrast, Kuhn will exclude any similarity relationship within the framework of his thesis of the 'paradigm change' and describe the change of theories as non-accumulative, fundamental and radical.

2. Popper considered the discovery and origin context to be irrelevant for the establishment and validity of scientific hypotheses and theories. In contrast, the following science theorists precisely address the importance of the historical, social and pragmatic backgrounds of theory formation, whereas they pay less attention to the justification context of hypothesis and theory justification.

3. The meaning of the "context-dependence of scientific knowledge generation" (Vilsmaier & Lang, 2014, p. 92) and research is therefore to be taken into account. This demands the already mentioned pragmatics, that is, the consideration of the actions of scientists in the social environment of the "scientific community" (Franck, 2007; Heidegger, 1963, pp. 69–104), which is always also connected with the individual and collective interests of scientists and research communities as well as with questions of power (cf. exemplary Bourdieu, 1992; Kühne, 2008). According to Poser (2009), the question of the role of pragmatics became virulent after the Second World War, when natural scientists

"have completely replaced the concept of natural law with what they alone can accomplish—with the concept of the model. Against this background, it becomes almost imperative to see the way into the history of science as a *way into pragmatics*. It was imperative to understand the model assumptions also as assumptions of one's own practical scientific action." (Poser, 2009, p. 153)

This also makes it imperative to take a sociological, historical and psychological look at the genesis of knowledge and theory. This also led to the fact that, unlike Popper, one no longer primarily reconstructs scientific theory logically and methodologically, but also investigates the concrete, context-related and interest-based work of scientists and researchers (cf. Bauberger, 2016; Poser, 2009; Ströker, 1998).

4. The rationality (reasonableness and transparency) of the scientists has also been increasingly questioned. The work of research groups, for example, is "by no means always rational […], but to a large extent, and in dramatic episodes of scientific change in particular, irrational" (Ströker, 1998, p. 447).

4.2   Paradigm Shift: Thomas S. Kuhn

Incorporating some of the basic ideas of the Polish physician Ludwik Fleck (see Sect. 5.2), the physicist and philosopher of science Thomas S. Kuhn (1922–1996) developed a concept of theory dynamics that triggered a "historiographical revolution in the study of science" (Kuhn, 1976, p. 17), thus a "reorientation of the theory of science" that brought the "characteristics of scientific change and the historical development of theories into the foreground" and, like Fleck, oriented the "methodological discussion more towards historically elaborated case studies" (Carrier, 2017, pp. 146–147). At the center of his approach (Kuhn, 1976) are the concepts of 'paradigm', 'paradigm shift' and 'scientific revolutions' as well as the thematization of context-dependence and the limits of scientific objectivity and rationality.

Kuhn's central thesis is that the development of science proceeds in four phases from a "pre-paradigmatic" to a "normal scientific" phase of routine research, through emerging "crises" to subsequent scientific "revolutions". He describes the "path to normal science" (Kuhn, 1976, pp. 25–36) as a "pre-paradigm time" (Kuhn, 1976, p. 32) and as "exceptionally difficult" (Kuhn, 1976, p. 30), because in this phase scientists are still "faced with a chaos" (Kuhn, 1976, p. 30). Still "without any order [they are] gathered together like the cabinets of curiosities of the Renaissance and the early Baroque, whatever seems 'remarkable', without being able to connect it with a specific why-question or even an explanatory approach, because there is still no basis for this in the form of a paradigm" (Poser, 2009, p. 145). Their theories had "at best family resemblance" (Kuhn, 1976, p. 29); as an example, Kuhn gives a closer look at the development of electricity theory in the 18th century (Kuhn, 1976, pp. 31–36).

Scientists of the 'normal' sciences are connected by a common basic view as a general and binding frame of reference for their scientific work, which Kuhn calls 'paradigm'. In the first edition of 1962 this term remained vague with Kuhn himself, which brought him considerable criticism; only in the 'Postscript' of 1967 does he go into more detail on this decisive concept (Kuhn, 1976, pp. 194–203). The 'expression paradigm' stands on the one hand

> "for the whole constellation of opinions, values, methods, etc. that are shared by the members of a given community. On the other hand, it denotes an element in this constellation, the concrete problem solutions, which, used as models or examples, can replace explicit rules as a basis for solving the remaining problems of 'normal science'." (Kuhn, 1976, p. 186)

A 'paradigm' therefore expresses "the commitments shared by a scientific community. These are content-related ideas of the nature of the relevant subject area, the recognition of certain problem solutions as exemplary (or 'paradigmatic'), and the assumption of certain quality standards for explanations" (Carrier, 2017, p. 147). In other words: "A paradigm first of all defines what there are objects in the world, what is researchable, and what can be used as an explanation for observations" (Bauberger, 2016, p. 73), that is, which methods and methodologies are recognized in the light of which target images. Examples of such paradigms are Newtonian mechanics, Copernican astronomy and the theory of evolution.

A paradigm cannot be justified or problematized in the phase of "normal science" itself, but rather provides, as an exemplary model of proven research, the criteria for a generally accepted justification *within* a field of research, for example for the comparison of different theory approaches (Bauberger, 2016; Carrier, 2017; Poser, 2009). Without such paradigms as orientation models, which open up areas of phenomena and prescribe the "rules of the game", successful science would not be possible at all. The main work in the normal science phase after the establishment of the paradigm is "puzzle solving", (Kuhn, 1976, pp. 49–56) within the framework of the paradigmatic rules of the game. Contrary to Popper's claim, it is not about falsifying hypotheses, initiating attempts to refute the paradigm or regarding anomalies as a refutation of the paradigm. Instead, hypotheses are "modified, restricted by additions and transformed" (Poser, 2009, p. 148), anomalies are explained by immunizations (Kuhn, 1976, pp. 90–95). Theory and facts are to be brought "into better agreement" and the inability to "find a solution in this matter only discredits the scientist and not the theory" (Kuhn, 1976, p. 93) or the paradigm. What does not fit the paradigm is simply "made to fit". Such "mature" sciences are, according to Kuhn, monoparadigmatic and, for the most part, concerned with deepening this paradigm and solving the problems and puzzles associated with it, rather than with scientific innovation (Kuhn, 1976, p. 38). Such research is more like "tidying up" (Kuhn, 1976, p. 38) and an "esoteric form of research", which is at the same time to be seen as "a sign of maturity in the development of any particular scientific discipline" (Kuhn, 1976, p. 26).

As the natural sciences become increasingly professional, precise and deep in their exploration of phenomena within the hitherto unassailable paradigm (Kuhn, 1976, p. 38), more and more anomalies (from Greek *anomalia*, unevenness, irregularity) inevitably arise that can hardly be immunized or interpreted away. The paradigm is increasingly eroded by anomalies and enters into a severe crisis of foundations (Kuhn, 1976, pp. 90–103). At some point, individual scientists or research groups emerge and propose a *new* paradigm that is *incommensurable* with the previous one. Kuhn distinguishes between 'epistemological' (from Greek *epistéme*, knowledge, and *lógos*, science, doctrine) and 'semantic incommensurability' (cf. Kornmesser & Büttemeyer, 2020). 'Epistemological incommensurability' addresses the change in the fundamental perception or view of the 'world' that accompanies a paradigm shift; 'semantic incommensurability' refers to the "significant shifts in meaning between the conceptual systems of two successive paradigms" (Kornmesser & Büttemeyer, 2020, p. 93).

As examples of paradigm shifts, Kuhn himself lists, inter alia, the displacement of classical Newtonian physics by Einstein's theory of relativity and the Copernican revolution from a geocentric to a heliocentric world view. A typical example within *geography* is the paradigm shift from the landscape paradigm to the space paradigm, which was initiated at the famous geography congress in Kiel in 1969 (cf. Kühne et al., 2018). If the new paradigm is actually accepted, a '*scientific revolution*' in the form of a '*paradigm shift*' has taken place. The 'scientific revolution' described by Kuhn is a "non-accumulative change in theory" in the form of a "rebuilding of the discipline concerned", insofar as "core components of the previously accepted theoretical framework are abandoned and new ideas are put in their place" (Carrier, 2017, p. 149):

> "The transition from a crisis-ridden paradigm to a new one from which a new tradition of normal science can emerge is far from being a cumulative process, as would be the case with an articulation or extension of the old paradigm. It is much rather the rebuilding of the field on new foundations, a rebuilding that changes some of the most elementary theoretical generalizations of the field as well as many of its paradigmatic methods and applications." (Kuhn, 1976, pp. 97–98)

The switch from one paradigm to another is neither rationally justifiable nor controllable, it takes place "more in the form of a generational change" (Bauberger, 2016, p. 75). There remains a 'rationality gap' (Poser, 2009), insofar as the choice between competing paradigms proves to be a choice between incompatible lifestyles of a community in the struggle for recognition (Eisel, 1997; Honneth, 1992) and the arguments of the proponents remain "necessarily circular", because each "group uses its own paradigm to defend that paradigm"—the arguments therefore have more the "status of an attempt at persuasion" (Kuhn, 1976, p. 106). The representatives of the respective research communities live in different worlds, insofar as a paradigm shift leads to a 'change of world view' as a 'change of the world of the scientist' (Kuhn, 1976, p. 123) (Fig. 4.1).

Paradigms cannot be compared to each other nor is it possible to decide if and to what extent a new paradigm is to be valued as 'progress' in comparison to an old paradigm. In comparison to Popper's theory of progress, a "relativism of scientific paradigms seems unavoidable" (Poser, 2009, p. 154); this relativism can only be circumvented if both Kuhn's approach and Popper's own are each understood as paradigms which are not only to be seen as antagonists, but which can complement each other: "For Popper, it was only about the *theory dynamics*, for Kuhn, however, it is about the *research dynamics*" (Poser, 2009, p. 155). Popper therefore primarily illuminates epistemological, logical and methodological questions, Kuhn sociological, historical and pragmatic questions.

The question of relativism raised by Kuhn's concept of paradigms is a serious problem for the sciences, since, as has been mentioned several times, they claim a

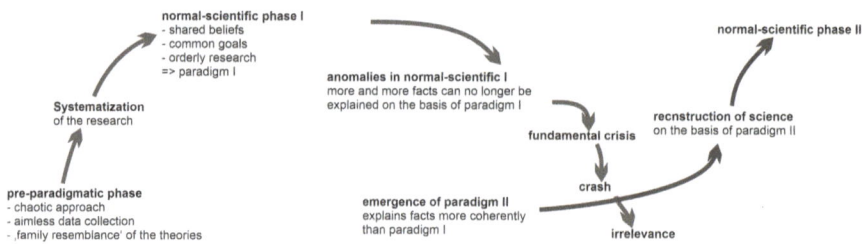

Fig. 4.1 Kuhn's phases model of the development of science (own representation)

unique selling point over other knowledge production by virtue of their reference to and orientation towards 'truth'. The question is therefore how to conceptualize pluralism (e.g. of paradigms) and the claim to truth together. Karl Popper explicitly addressed this question, which is why we insert his position on this issue at this point (Textbox 19).

Textbox 19: Popper: Critical Pluralism and Similarity of Truth

Popper himself addressed the problem of relativism associated with plurality in a lecture in 1981 (Popper, 1984, pp. 213–230) and propagated '*critical pluralism*' as a counterposition to relativism. He characterizes the 'idea of *truth*' as the decisive distinguishing criterion. Relativism is "the position that one can assert everything, or almost everything, and therefore nothing. Everything is true, or nothing. Truth is therefore meaningless" (Popper, 1984, p. 217). "Critical pluralism" on the other hand is "the position that *in the interest of truth-seeking* every theory—the more theories, the better—should be allowed to compete between the theories" (Popper, 1984, p. 217). It is therefore "about the truth of the competing theories […]: the theory that in the critical discussion of truth seems to come closer, is the better; and the better theory displaces the worse theories. It is therefore about the truth" (Popper, 1984, p. 217). Popper claims, with regard to the history of science and the comparison of theories, that while one cannot know "how close or how far away from the truth we are, that we but *always come closer and closer to the truth* can and do" (Popper, 1997, p. 177). Popper speaks of an "idea of (degrees of) *truthlikeness*" as well as of "*truthfulness*" (Popper, 1997, p. 179). This 'idea of truthlikeness' is "*not an epistemological or epistemic idea*", but it has "ideal or regulative character" (Popper, 1997, p. 181) in the sense of a 'regulative idea'. Progress then means that theories become increasingly 'truthlike', scientists orient themselves to this idea, without ever being able to be sure that they are reaching an 'absolute truth' with their theories. The sociologist Ralf Dahrendorf (1929–2009), who was strongly influenced by Popper, builds on this basic idea when he emphasizes that it is the task of the scientists, even in the awareness of truth, to "struggle for truth, knowing that they will not find it" (Dahrendorf, 2008, p. 61). He continues: "The lone warriors of truth […] do not proclaim truth, but set out in search of it. That they are drifting in a horizon of uncertainty is always taken into account" (Dahrendorf, 2008, p. 61). Tetens sharpens the considerations of both when, with regard to the limited number of a priori frames of experience (see Sect. 3.4.4), which have been able to establish themselves so far, he *never* wants to "give up" the concept of truth: "The talk of a truth relativism is and remains nonsense" (Tetens, 2013, p. 86; see 'truth theories' in Sect. 3.1.3). ◀

Regardless of this fundamental philosophical statement by Popper, the individual sciences have also addressed the question of the plurality of theories or paradigms. Within the framework of the individual disciplines, corresponding analyses of the respective basic constitution of the disciplines have been carried out and considerations have been made as to how to deal with this plurality—also in geography (see Textbox 20).

Textbox 20: Geography as a Polyparadigmatic Science and the Handling of Theory and Paradigm Plurality

The question of whether a plurality of theories and paradigms is *basically* advantageous or disadvantageous has been intensively and extensively discussed in different sciences. A fundamental approval of *plurality*, which can be observed in almost all areas of contemporary culture, can be found, for example, in the philosophy of science (Feyerabend, 1981; Stegmüller, 1985), in philosophy (Gabriel, 2005; James, 2005 [1909]; Tetens, 1999; Welsch, 1996, pp. 541–610), in the social sciences (Dahrendorf, 1980 (cf. Kühne, 2017, pp. 15–25); Kneer & Schroer, 2009), in ethics (Düwell, 2001; Gehlen, 2016) and in geography (cf. Kühne, 2018; Weichhart, 2006). Kneer and Schroer (2009, pp. 12–13) have reconstructed five different ways of dealing with plurality for sociology: the "convergence", the "integration", the "competition", the "complementarity", the "indifference" and—if this systematization in turn claims a theoretical perspective—the "unclear perspective". Weichhart lists five 'reaction types' for human geography: "ignorance, persistent dogmatism, rejection of paradigm diversity, evolutionary pragmatism, acceptance" (Weichhart, 2006, p. 196); as a systematic basis he gives the (sixth) position of an "unclear situation" (2006, p. 197). Weichhart associates the type of 'acceptance' with the position of the "complementarity idealist", who starts from an "essentially non-reducible complementarity structure of reality" (2006, p. 196). Descriptions of reality of a specific paradigm can at best capture '*aspects*' of reality, which are also "determined as specific projections by the perspective of the observation model" (2006, p. 196). In analogy to Schurz and Weingartner (1998) Weichhart has the advantage of a multiparadigmatic structure in that "one gets additional projections as it were from the perspective of competing approaches, which are complementary to each other and can only provide us with a more complete picture of reality through the comparison of the different and not reducible findings" (Weichhart, 2006, p. 196). Therefore, a complementary view of paradigm diversity seems to be the "only adequate and politically sensible response, which is also most in line with the 'postmodern spirit of the age'" (2006, p. 197). In the age of "postmodernity", he therefore pleads for "postmodern tolerance" (2006, p. 182) and, contrary to any form of scientific "messianic zeal", for understanding one's own scientific work as "intellectual play" and "taking into account that there are several serious and honorable ways to deal research-wise with reality" (2006, p. 197). However, it is recommended to practice "tolerant, cosmopolitan dealing with plurality" (Tetens, 1994, p. 28). Weichhart's concept of the "complementarity idealist" can build on a complementary understanding of plurality, as proposed by the philosopher Gottfried Gabriel. Gabriel models a pluralism of "not relativistic, but complementary design" (Gabriel, 2005, p. 326; see also Schildknecht et al., 2008), i.e. as "perspectives" that "can exist side by side without excluding each other [...] as complementary attitudes" (Gabriel, 1993, p. 190).

This advocacy of a plurality of paradigms had to be based on a "radical reading" (Weichhart, 2018 [published in 2020]) of Kuhn's considerations. Kuhn distinguished 'mature' from 'immature' sciences: In the context of normal scientific development, so-called 'mature' sciences such as physics, chemistry, or biology are characterized by the fact that they are 'monoparadigmatic', whereas 'immature' sciences are 'polyparadigmatic'. In his reconstruction of scientific change, Kuhn had predominantly dealt with paradigms of the natural sciences, to which his diagnosis actually applies in many cases. In the humanities, social sciences, or cultural studies, however, things are different; here several paradigms can indeed coexist. This is correspondingly true for disciplines that are at the boundary and intersection of natural sciences and humanities/ social/cultural sciences. The geographer Gerhard Hard (2003), following Toulmin (1978) therefore distinguished between 'hard' and 'diffuse' disciplines and classified geography as "diffuse," that is, as "constitutionally pre- or polyparadigmatic" (Hard, 2003, p. 179) discipline, insofar as it is precisely defined by *multiple* paradigms. As a further example, sociology can be mentioned, which, for instance, was described by Pierre Bourdieu (1992) and by Kneer and Schroer (2009, pp. 7–18) have determined as a 'multiparadigmatic' discipline. Against the idea that 'mature' sciences are monoparadigmatic in composition, therefore, a 'post-Kuhnian stocktaking' of the structure of contemporary science in geography (cf. Schurz & Weingartner, 1998; Weichhart, 2006) and philosophy (e.g., Lakatos., 1974a; Stegmüller, 1985) resistance arose. The normal case of scientific development (at present, at least), on the other hand, seems to be a situation in which a paradigm or style of thought is "not singular, but plural" (Kneer & Schroer, 2009, p. 7). Thus Peter Weichhart (2006) for human geography twelve, for physical geography five 'individual paradigms' reconstructed and distinguished from each other.

It is also criticized that the sequence of normal scientific development, which is *reconstructed* and *described* by Kuhn from a historical perspective of science, if it is interpreted *normatively* in a fallacious way (see Sect. 2.4.4), "how sciences should be" (Kornmesser & Schurz, 2014, p. 11), cannot do justice to the current factual development of science and its productivity. The appropriation of normative consequences from descriptions of a change within the sciences has also been criticized by Peter Weingart (1999, p. 48) with reference to Kuhn's thesis of paradigm change using the example of 'new' science production in 'Mode 2' (see Sect. 5.4.2):

"It is alarming that 'Mode 2', just like [...] Thomas Kuhn's 'paradigm shift', is interpreted not as an empirical description of a change, but as a normative pattern for its design." ◀

4.3   Theory Dynamics and 'Sophisticated Falsificationism': Imre Lakatos

Popper had already shifted the focus away from the question of the connection between observations and theory and towards the comparison of theories with each other, Kuhn and Popper had reconstructed scientific progress as the progressive development of theories. As with Popper and Kuhn, the Hungarian philosopher Imre Lakatos (1922–1974) also focuses on theories, namely their structures. Against Popper, Lakatos emphasizes the tenacity of theories against falsifications by falsifying observational data, since the theory-ladenness of observation (cf. Textbox 16) and the heuristic, research-enabling and guiding function of theories leads to the fact that familiar and so far successful theories are actually rarely abandoned in practice. Against this background, Lakatos focuses on so-called "research programs": According to his "methodology", these are

"the greatest scientific achievements [...], which can be evaluated on the basis of progressive and degenerative problem shifts; and scientific revolutions consist in the fact that one research program supersedes another (in the course of progress). This methodology provides a new rational reconstruction of science." (Lakatos, 1974b, pp. 279–280)

The term 'research program' highlights, on the one hand, the *pragmatic*, in particular the *heuristic* function of theories, and on the other hand, paradigm shifts can be *rational* reconstructed on the basis of the criteria 'progressive and degenerative'. A research program is 'progressive' if it can lead to the discovery of new phenomena and the number of possible and confirmed predictions increases; 'degenerative' if this does not happen permanently, that is, neither new phenomena are discovered nor the number of confirmed predictions increases. 'Progress' in this sense means "the replacement of a degenerate research program by a progressive research program, the latter being an improvement over the former in that it has proven to be more effective in predicting new phenomena" (Chalmers, 2006 [1996], p. 113).

Lakatos differentiates between 'negative' and 'positive heuristics'. In the sense of a 'negative heuristic', a self-defined 'hard core' of fundamental general hypotheses is held, which first "form the basis from which the program must be developed" (Chalmers, 2006 [1996], p. 108) and second "cannot be shaken by counterexamples" (Poser, 2009, p. 158). This transfers Kuhn's basic idea that a paradigm is unquestionably valid and research-guiding on the concept of research programs. The 'hard core' is surrounded by a 'belt' of 'auxiliary' or 'additional hypotheses' from the scientists, which is supposed to preserve this 'core' from falsification. If observation data contradict a theory, the contradiction is located in the belt and the auxiliary hypotheses are either modified or supplemented by further auxiliary hypotheses. This 'immunizes' the 'hard core' against falsifications—

against Popper. The auxiliary hypotheses are falsified by the Popperian method of 'trial and error'. Otherwise, it would be possible that the logic of falsification by 'modus tollens' (see Sect. 3.6) falsifies the conjunction of the hard core and its belt *as a whole*.

In contrast to the 'negative heuristic', which indicates what is to be omitted (namely the 'falsification of the hard core'), a research programme in the sense of a '*positive* heuristic' contains a

> "Series of suggestions or hints on how to change and develop the 'falsifiable versions' of the research programme and how to modify and refine the 'falsifiable' protective belt." (Lakatos, 1974a, p. 131)

So these are defence strategies of the 'hard core' and the protective belt. In the course of 'progressive problem shifting', the further development of a research programme can be seen as a sequence of theories that are continuously modified. Even if this (against Kuhn) provides a rational criterion for scientific progress, no criterion can be given for whether a (degenerative or progressive) research programme should be abandoned or continued. Because a research programme can both come to new insights after a degenerative phase and, conversely, a research programme can stand still permanently after a progressive problem shift. Ultimately, a practical decision by the community of scholars is required, which cannot be derived or formalised theoretically (cf. Poser, 2009).

As with Kuhn's paradigm shift, a 'rationality gap' also arises here, this time between the different research programmes. Against Kuhn, Lakatos doubts that the standard case of normal science is the dominance of a single, non-competitive paradigm, but on the contrary there are always competing theories, perspectives and research programmes in practice. Since—this is an adaptation of a basic idea by Popper—it is basically not foreseeable in advance which research programme will be the ultimately successful one, he calls for a pluralism of research programmes (cf. Poser, 2009).

4.4   Research Traditions: Larry Laudan

An important extension of the positions of Kuhn and Lakatos is Larry Laudan's approach (1977). Laudan replaces the concept of the research program with that of the *research tradition*. His thesis against Lakatos and Kuhn is that even the "hard core" of research programs or the basic assumptions of a paradigm are by no means immune to change, but can be modified in the course of research work in the face of unavoidable resistance (e.g. conceptual or empirical). A "research tradition" is—like any other tradition—on the one hand relatively stable and resistant to change, or modification; on the other hand, under pressure of adaptation, corresponding adaptations and modifications to changed conditions can occur within traditions. This fundamental possibility of modification affects *all* basic assumptions of the research tradition, that is—in the terminology of Lakatos—both the "hard core" and the "safety belt".

This basic and comprehensive modification option has the consequence that Laudan not only takes into account the discrepancy between theory and observation, but also discrepancies at the level of *basic concepts*. For example, the basic conceptual disputes in geography can be mentioned, whether 'space', 'landscape' or other concepts are research-oriented, without empirical evidence for these concepts or terms. Popper rated a contradiction of theory and observation as a falsification, Kuhn (positively) as a puzzle to be solved or (negatively) as an insurmountable anomaly, Lakatos as a progressive or degenerative problem shift. Laudan, on the other hand, evaluates these empirical problems as an opportunity to change and further develop the theory tradition at the empirical level. As a criterion for such further development, he conceives of scientific progress as the "problem-solving efficiency" (Poser, 2009, p. 167) of a research tradition. This results from the *interplay* of an explanation of the theoretical-conceptual basis as a "progress in conceptual clarity" *and* a "progress in empirical problem-solving ability", which also depends on what can be accepted as a solution (cf. Poser, 2009).

4.5   Methodological Pluralism: Paul Feyerabend

That and why science is not always rational in methodological and research-logical terms, but certainly contains 'irrational' elements, has already been shown by the comments on Popper, Kuhn, Lakatos and Laudan. In Sect. 1.2 it also became clear why science often comes with a quasi-religious optimism about science or even a 'belief in science' (Jaspers, 1975; Tetens, 2013) can go along with it. It is exactly here that the criticism of Paul Feyerabend (1924–1994) starts, when in the "foreword to the German edition" of the first edition of his classic "Against Method" (1976) he criticizes that science is the new religion of our time, that everything and everyone is oriented according to its results and that, like a religion, it has become unquestionable and therefore unassailable (Feyerabend, 1976, pp. 11–27). As an antidote to the quasi-religious rule of science and its "methodical compulsion", which is based on theories, research logics, methodological rules and methods that are petrified in research traditions and apparently unassailable, he propagates a "methodological anarchism". Its maxim is the famous "Anything goes"—in the first edition, however, there is also "(Do what you want)" in brackets directly behind it (Feyerabend, 1976, pp. 35, 45). With this, Feyerabend does not turn against scientific methodology as such, but against its absolutization in religious manner. He turns against a scientific system that educates and socializes scientists, who often present their results in an authoritarian, intolerant and arrogantly priestly manner and make them immune to criticism, results that, according to all previous experience in the history of science, are uncertain and fallible in their validity as results, and in their genesis are due to partly arbitrary and irrational assumptions. Instead, he pleads for a scientific pluralism and for a "science for free people" (Feyerabend, 1978, p. 349). The slogan "Anything goes" is not intended as a carte blanche for scientific arbitrariness and not as a normative rule, but firstly serves the description of a factual scientific operation (see Sect. 1.2), which he wants to criticize, and secondly the call for creativity and autonomy in the choice of methods. With regard to the first aspect, Feyerabend writes:

> "But the catchphrase 'anything goes' is not a methodological rule that I recommend, but a painful description of the situation of my opponents after comparing their rules with scientific (ethical, political) practice." (Feyerabend, 1978, p. 343)

Those who understand this slogan normatively thus commit the already mentioned fallacy of inferring descriptions from normative consequences (see Sects. 2.4.4 and 3.2.2). As far as the second aspect is concerned, this slogan

> "has a very specific and very concrete meaning: a research direction that contradicts the most fundamental principles of thinking of a certain time and that is therefore irrational can cause a new idea of reason to shine in the researcher and thus appear to be highly reasonable at the end." (Feyerabend, 1978, p. 343)

Feyerabend names a essential reason for his skepticism towards science and his anarchism of methods and draws the corresponding conclusions. The decisive reason for this is that

> "there is not a single rule that has not been violated at any time, no matter how plausible and well-anchored in epistemological terms it may be. It becomes clear that such violations are not accidental [...]. On the contrary, one realizes that they are necessary for progress." (Feyerabend, 1983, p. 21)

As examples of consciously intended or unconsciously achieved and necessary for the progress of science rule violations, Feyerabend lists among others the "invention of the theory of atoms in antiquity, the Copernican Revolution, the rise of modern atomic theory" (Feyerabend, 1983, p. 21). From this finding, he derives "anti-rules", "contradict the known rules of scientific procedure" (Feyerabend, 1983, p. 33). These are anti-inductivist and thus anti-positivist 'anti-rules', which first require "developing hypotheses that contradict recognized and well-confirmed *theories*" and second "well-confirmed *facts*" (Feyerabend, 1983, p. 33). Both with theories and facts, the 'theory-ladenness of observation' applies (see Textbox 16). Theories can only be refuted by hypotheses and data that do not stem from the theory-ladenness of one's own theory, but from another theory; therefore, a 'pluralistic methodology' and the comparison of a plurality of 'incompatible' and possibly 'incommensurable' theory alternatives are required. Facts are theory-laden anyway and "there is not a single interesting theory that agrees with all known facts in its field" (Feyerabend, 1983, p. 35).

Ultimately, Feyerabend's position is a plea against method dogmatism and for the free choice of method (s). There is not the one right method or the one and only right method canon in the individual sciences. Researchers are also called upon not only to apply tried and tested methods in the course of normal scientific routine, but also to break such routines and the rules underlying them. Fleck (see

Sect. 5.2) already addressed the 'training of perception' as a necessary prerequisite for scientific observation; Feyerabend generalizes the thought of the 'training', when he states that scientists often behave "like a well-trained pet", which "obeys its master" (Feyerabend, 1983, p. 24). 'Dressed' scientists will hardly be able to break rules creatively and gain new scientific perspectives.

Feyerabend's considerations on the connection between creativity and rule-breaking are in line with current research on creativity. For creativity as "the art of the new" (Abel, 2005) fundamentally resists any application of rules. Creativity is obviously not recursively rule-based—all attempts to bring it to rules, calculations or algorithms have so far failed because it is "neither analyzable nor individualizable by calculation" (Abel, 2005, p. 9). '*Strong* creativity' therefore means "the transformation, the breaking, the replacement of old by new principles, regularities and laws", '*weak* creativity' means "the combinatorial re-arrangement of already existing elements" (Abel, 2007, p. 15). Both forms of creativity are significant, but can only develop if the freedom conditions required for this are actually given in the partly "hostile to human" (Feyerabend, 1983, p. 17) scientific enterprise, including its training and employment conditions.

Textbox 21: Conclusion to: The Contextualization of Science I: Time
This chapter showed that and why Popper's approach can be considered the culmination of the philosophy of science up to those approaches which are characterized by the inclusion of the historical and social contexts of the emergence and validity of scientific knowledge and the functioning of the scientific enterprise. First, positions were presented in this chapter which were developed from the history of science and therefore take the factor 'time' into account. The positions presented share, despite their differences, the attempt to better understand the factual development of sciences and theories in their rationally difficult to grasp dynamics and to avoid or overcome the problems associated with Popper's falsificationism. A difficult question in this context is and was on which basic assumptions sciences cannot do without, whether there is such a thing as a 'hard core' of theories, paradigms or sciences which is to be preserved if possible, or whether this 'hard core' is also subject to development dynamics. The inescapable fact of the diversity of differentiated sciences as well as the diversity of paradigms, theories and methods within a science has furthermore led to the question of practicable and appropriate forms of dealing with these pluralisms. With this question, the question of what role the social context of scientific activity plays came into focus. This question leads to the next chapter, in which the social embeddedness of knowledge and sciences is examined, that is, which relationships can be observed between knowledge and social power relations or between scientific communities and knowledge generation or between (political) world views or ideational systems and the scientific enterprise.

Further Reading

Andersson (1988): Contains a thorough and well-founded discussion of Kuhn, Lakatos and Feyerabend and their criticism of Popper's Critical Rationalism.

Chalmers (2006 [1996]): A standard work of university teaching on the philosophy of science and its key positions and issues.

Poser (2012): A very well-founded, comprehensive and readable introduction to the philosophy of science, which also takes into account recent developments, such as evolutionary approaches.

Original Literature

Chalmers (1999): Again Chalmers, this time with a discussion of the limits of science as well as a criticism of Popper, but also a criticism of the negative consequences of Feyerabend's approach.

Feyerabend (1976): Another still readable classic against the methodical compulsion of repressive structures in the scientific community.

Kuhn (1976): A still readable classic on the theory of paradigms.

Popper (1959): Central work of the philosophy of science, perhaps the central work. And still—relatively—easy to read.

References

Abel, G. (2005). Die Kunst des Neuen. Kreativität als Problem der Philosophie. In G. Abel (Hrsg.), *Kreativität. XX. Deutscher Kongress für Philosophie, 26.–30. September 2005 in Berlin* (Sektionsbeiträge, Bd. 2, S. 1–21). Universitätsverlag der TU Berlin.

Abel, G. (2007). Kreativität – Worin besteht sie und was macht sie so wertvoll? In G. Abel, J. Conzett, U. P. Jauch & P. M. Da Rocha (Hrsg.), *Grenzüberschreitungen im Entwurf* (Architekturvorträge der ETH Zürich, Bd. 5, S. 10–43). gta.

Andersson, G. B. J. (1988). *Kritik und Wissenschaftsgeschichte. Kuhns, Lakatos' und Feyerabends Kritik des kritischen Rationalismus* (Die Einheit der Gesellschaftswissenschaften, Bd. 54). Mohr.

Bauberger, S. (2016). *Wissenschaftstheorie. Eine Einführung.* Kohlhammer.

Bourdieu, P. (1992). *Homo academicus* (Suhrkamp-Taschenbuch Wissenschaft, Bd. 1002). Suhrkamp (französische Originalausgabe 1984).

Carrier, M. (2017). *Wissenschaftstheorie zur Einführung* (Zur Einführung, 4., überarb. Aufl., Bd. 353). Junius.

Chalmers, A. F. (1999). *Grenzen der Wissenschaft.* Springer.

Chalmers, A. F. (2006 [1996]). *Wege der Wissenschaft. Einführung in die Wissenschaftstheorie.* Springer.

Dahrendorf, R. (1980). *Die neue Freiheit. Überleben und Gerechtigkeit in einer veränderten Welt.* Suhrkamp.

Dahrendorf, R. (2008). *Versuchungen der Unfreiheit. Die Intellektuellen in Zeiten der Prüfung* (Beck'sche Reihe, Bd. 1875). Beck.

Düwell, M. (2001). Angewandte Ethik – Skizze eines wissenschaftlichen Profils. In A. Holderegger & J.-P. Wills (Hrsg.), *Interdisziplinäre Ethik. Grundlagen, Methoden, Bereiche. Festgabe für Dietmar Mieth zum sechzigsten Geburtstag* (Studien zur theologischen Ethik, Bd. 89, S. 165–184). Universitätsverlag.

Eisel, U. (1997). Unbestimmte Stimmungen und bestimmte Unstimmigkeiten. Über die guten Gründe der deutschen Landschaftsarchitektur für die Abwendung von der Wissenschaft und die schlechten Gründe für ihre intellektuelle Abstinenz – mit Folgerungen für die Ausbildung in diesem Fach. In S. Bernhard & P. Sattler (Hrsg.), *Vor der Tür. Aktuelle Landschaftsarchitekru aus Berlin* (S. 17–33). Callwey. http://www.ueisel.de/fileadmin/dokumente/eisel/Unbestimmte_Stimmungen/Eisel_Unbestimmte_Stimmungen_fertig.pdf. Zugegriffen: 12. Jan. 2019.

Feyerabend, P. (1976). *Wider den Methodenzwang. Skizze einer anarchistischen Erkenntnistheorie.* Suhrkamp.

Feyerabend, P. (1983). *Wider den Methodenzwang. Against method* (Suhrkamp-Taschenbuch Wissenschaft, Bd. 597). Suhrkamp.

Feyerabend, P. K. (Hrsg.). (1978). *Ausgewählte Schriften* (Bd. 1). Vieweg.

Feyerabend, P. K. (1981). Der Pluralismus als ein methodologisches Prinzip. In P. K. Feyerabend (Hrsg.), *Probleme des Empirismus. Schriften zur Theorie der Erklärung, der Quantentheorie und der Wissenschaftsgeschichte. Ausgewählte Schriften II* (Wissenschaftstheorie Wissenschaft und Philosophie, Bd. 2, S. 7–14). Vieweg+Teubner.

Franck, G. (2007). *Ökonomie der Aufmerksamkeit. Ein Entwurf.* dtv.

Gabriel, G. (1993). *Grundprobleme der Erkenntnistheorie. Von Descartes zu Wittgenstein* (3., durchgesehene. Aufl.). Schöningh.

Gabriel, G. (2005). Orientierung – Unterscheidung – Vergegenwärtigung. Zur Unverzichtbarkeit nicht propositionaler Erenntnis für die Philosophie. In G. Wolters & M. Carrier (Hrsg.), *Homo Sapiens und Homo Faber. Epistemische und technische Rationalität in Antike und Gegenwart. Festschrift für Jürgen Mittelstraß* (S. 323–334). de Gruyter.

Gehlen, A. (2016). *Moral und Hypermoral. Eine pluralistische Ethik.* Klostermann.

Hard, G. (2003). Studium in einer diffusen Disziplin. In G. Hard (Hrsg.), *Dimensionen geographischen Denkens. Aufsätze zur Theorie der Geographie* (Osnabrücker Studien zur Geographie, Bd. 23, S. 173–230). V & R Unipress.

Heidegger, M. (1963). *Holzwege* (4. Aufl.). Klostermann.

Honneth, A. (1992). *Kampf um Anerkennung. Zur moralischen Grammatik sozialer Konflikte.* Suhrkamp.

James, W. (Hrsg.). (2005 [1909]). *Das pluralistische Universum. Vorlesungen über die gegenwärtige Lage der Philosophie.* Wissenschaftliche Buchgesellschaft.

Jaspers, K. (1975). *Was ist Philosophie? Ein Lesebuch.* Piper.

Kneer, G., & Schroer, M. (Hrsg.). (2009). *Handbuch Soziologische Theorien.* VS Verlag.

Kornmesser, S., & Büttemeyer, W. (2020). *Wissenschaftstheorie. Eine Einführung.* Metzler.

Kornmesser, S., & Schurz, G. (2014). Die multiparadigmatische Struktur der Wissenschaften: Einleitung und Übersicht. In S. Kornmesser & G. Schurz (Hrsg.), *Die multiparadigmatische Struktur der Wissenschaften* (S. 11–46). Springer VS.

Kuhn, T. S. (1976). *Die Struktur wissenschaftlicher Revolutionen* (Suhrkamp-Taschenbuch Wissenschaft, Bd. 25, zweite revidierte und um das Postskriptum von 1969 ergänzte Aufl.). Suhrkamp.

Kühne, O. (2008). *Distinktion – Macht – Landschaft. Zur sozialen Definition von Landschaft.* VS Verlag.

Kühne, O. (2017). *Zur Aktualität von Ralf Dahrendorf. Einführung in sein Werk* (Aktuelle und klassische Sozial- und Kulturwissenschaftlerinnen). Springer VS.

Kühne, O. (2018). Die Landschaften 1, 2 und 3 und ihr Wandel. Perspektiven für die Landschaftsforschung in der Geographie – 50 Jahre nach Kiel. *Berichte. Geographie und Landeskunde, 92*(3–4), 217–231.

Kühne, O., Weber, F., & Jenal, C. (2018). *Neue Landschaftsgeographie. Ein Überblick (Essentials).* Springer VS.

Lakatos, I. (1974a). Falsifikation und die Methodologie wissenschaftlicher Forschungsprogramme. In I. Lakatos & A. Musgrave (Hrsg.), *Kritik und Erkenntnisfortschritt* (Abhandlungen des Internationalen Kolloquiums über die Philosophie der Wissenschaft, Bd. 4, S. 89–189). Vieweg.

Lakatos, I. (1974b). Die Geschichte der Wissenschaften und ihre rationalen Rekonstruktionen. In I. Lakatos & A. Musgrave (Hrsg.), *Kritik und Erkenntnisfortschritt* (Abhandlungen des Internationalen Kolloquiums über die Philosophie der Wissenschaft, Bd. 4, S. 271–312). Vieweg.

Laudan, L. (1977). *Progress and its problems. Towards a theory of scientific growth.* University of California Press.

Popper, K. R. (1959). *The logic of scientific discovery.* Harper & Row.

Popper, K. R. (1984). *Auf der Suche nach einer besseren Welt. Vorträge und Aufsätze aus dreißig Jahren.* Piper.

Popper, K. R. (1997). *Lesebuch. Ausgewählte Texte zur Erkenntnistheorie, Philosophie der Naturwissenschaften, Metaphysik, Sozialphilosophie* (2. Aufl.). UTB GmbH.

Poser, H. (2009). *Wissenschaftstheorie. Eine philosophische Einführung.* Reclam.

Poser, H. (2012). *Wissenschaftstheorie. Eine philosophische Einführung* (2., überarb. u. erw. Aufl.). Philipp Reclam jun.

Schildknecht, C., Teichert, D., & van Zantwijk, T. (Hrsg.). (2008). *Genese und Geltung. Für Gottfried Gabriel.* Mentis.

Schurz, G., & Weingartner, P. (Hrsg.). (1998). *Koexistenz rivalisierender Paradigmen. Eine postkuhnsche Bestandsaufnahme zur Struktur gegenwärtiger Wissenschaft.* Westdeutscher.

Stegmüller, W. (1985). *Probleme und Resultate der Wissenschaftstheorie und analytischen Philosophie* (Theorie und Erfahrung, Bd. 2). Springer (Zweiter Halbband: Theorienstrukturen und Theoriendynamik).

Ströker, E. (1998). Wissenschaftstheorie. In A. Pieper (Hrsg.), *Philosophische Disziplinen. Ein Handbuch* (Reclam-Bibliothek, Bd. 1643, S. 437–456). Reclam.

Tetens, H. (1994). *Geist, Gehirn, Maschine. Philosophische Versuche über ihren Zusammenhang* (Universal-Bibliothek, Bd. 8999). Reclam.

Tetens, H. (1999). Wissenschaft. In H. J. Sandkühler (Hrsg.), *Enzyklopädie Philosophie* (S. 1763–1773). Meiner.

Tetens, H. (2013). *Wissenschaftstheorie. Eine Einführung.* Beck.

Toulmin, S. E. (1978). *Menschliches Erkennen I: Kritik der kollektiven Vernunft.* Suhrkamp.

Vilsmaier, U., & Lang, D. J. (2014). Transdisziplinäre Forschung. In H. Heinrichs & G. Michelsen (Hrsg.), *Nachhaltigkeitswissenschaften* (S. 87–113). Springer Spektrum.

Weichhart, P. (2006). Humangeographische Forschungsansätze. In W. Sitte & H. Wohlschlägl (Hrsg.), *Beiträge zur Didaktik des „Geographie und Wirtschaftskunde"-Unterrichts* (Materialien zur Didaktik der Geographie und Wirtschaftskunde, 4., unveränd. Aufl., Bd. 16, S. 182–198). Institut für Geographie und Regionalforschung.

Weichhart, P. (2018 [2020 erschienen]). Die Landschaft der Landschaften. *Berichte. Geographie und Landeskunde, 92*(3–4), 203–216.

Weingart, P. (1999). Neue Formen der Wissensproduktion: Fakt, Fiktion und Mode. *TATuP-Zeitschrift für Technikfolgenabschätzung in Theorie und Praxis, 8*(3–4), 48–57.

Welsch, W. (1996). *Vernunft. Die zeitgenössische Vernunftkritik und das Konzept der transversalen Vernunft.* Suhrkamp.

The Contextualization of Science II: Social Affinity

5

Science is also carried out in social contexts—as already hinted at in the previous chapter. Whether universities, research institutes, research departments in companies or even in the study rooms of private scholars: the production, distribution and reception of knowledge is embedded in social contexts. In this chapter we present essential relationships between knowledge and science in their social context and then conclude with the question of which world-view positions lie at the bottom of scientific approaches to the world (can), but first we will again lay down relevant terms for the following chapter in Textbox 22.

Textbox 22: Essential terms for Chap. 5

Dichotomy and polarity: Dichotomy describes two mutually exclusive and incompatible, strictly separated, but interrelated statements (linguistically: either … or …). Polarity, on the other hand, describes two statements between which there are different intensities of transition (linguistically: both … and …).

Globalization: Increasing global interweaving of economy, politics, society and culture, but also ecology (e.g. loss of biodiversity or global climate change). Often, 'globalization' focuses on economic processes, such as the increase in world trade, foreign direct investment, the importance of transnational corporations, the internationalization of financial markets, etc. (Beck, 2007; Rademacher et al., 1999).

Institution: In the social sciences, an 'institution' is understood to be a complex of regulations that guide the behaviour of people and thus reduce arbitrariness and behavioural insecurity on the one hand, but also restrict freedom of action on the other (e.g. marriage, the education system, etiquette; Häußling, 2018).

© The Author(s), under exclusive license to Springer Fachmedien Wiesbaden
GmbH, part of Springer Nature 2022
O. Kühne and K. Berr, *Science, Space, Society*,
https://doi.org/10.1007/978-3-658-39140-9_5

Class society: With the abolition of privileges of the estates (estate society) and the introduction of legal equality, the class society emerged. Karl Marx is considered the founder of modern class theory. For him, the ownership of the means of production is the central difference between the class of capitalists and the workers (who are forced to sell their labor power because they do not own the means of production; Dahrendorf, 1957).

Power and domination: In his classical definition, Max Weber (1976 [1922], p. 28) describes power as "any chance to impose one's own will on another person in a social relationship, regardless of what this chance is based on." Domination, on the other hand, is—according to Weber (1976 [1922], p. 541)—a "special case of power": In contrast to power, he understands domination as a chance for a certain command to find obedience. In comparison to the more general power, domination is therefore more specific and can be better regulated (Dahrendorf, 1979).

Milieu society: In the social sciences, 'milieu' describes a subgroup of society that shares similar principles of life, values and mentalities. The society shaped by milieus is no longer stratigraphic (vertical, that is, divided into 'upper' and 'lower' classes, like the class and stratified society), but also 'horizontal', on a 'vertical' layer. The current societies in North America and Central and Western Europe can be subsumed under this term (Hradil, 2018).

Social polarization: This is spoken of when the differences in society increase. Mostly, this is aimed at the economic differences, but polarization can also be found in relation to education, social networks, political participation, etc. Polarization also finds its expression spatially, for example in stigmatized quarters (Kraemer & Bittlingmayer, 2001; Kronauer & Siebel, 2013).

Layered or stratified society ('Schichtgesellschaft'): Social 'layer' describes a subset of society in which members are in a similar social situation. While class society is characterized by the (especially economic) division, layered society is characterized by gradual transitions (Hradil, 1995). In Germany, this expression is very popular to describe the social conditions in the 1950s to 1980s.

5.1 The Social Embeddedness of Knowledge: Sociology of Knowledge

The creation and distribution of knowledge is neither social nor regional, the discussion of "educationally disadvantaged groups" or "knowledge regions" illustrates this. Knowledge can be vital for one person, completely worthless for another. Where I find water in desert regions is usually of no practical use to most residents of Central Europe, but essential for the residents of the corresponding desert region. In certain milieus, certain knowledge can bring prestige, other knowledge can be ridiculed. A detailed knowledge of the writings of Immanuel Kant, their interpretations and further developments increases social prestige in an education-friendly milieu, while knowledge of all match results including goal sequences and line-ups of MSV Duisburg since the existence of professional football is likely to have the opposite effect. In a fan club of MSV Duisburg (a traditional soccer club from the Ruhr area, which also played internationally, but today belongs to the third Bundesliga), the distribution of social recognition is likely to be reversed. The sociological sub-discipline that deals with the reciprocal relationship between "knowledge" and "social" is the sociology of knowledge (Schützeichel, 2012). The central position of the sociology of knowledge is "the departure from the traditional model of the autonomous, individual knowledge actor or a Cartesian knowledge subject" (Schützeichel, 2012, p. 17). Questions about the genesis of knowledge in societies are given great topicality by developments such as the disintegration of generally accepted interpretation and evaluation patterns, the emergence of new areas of knowledge, the expansion of the central importance of knowledge beyond Western society, the cultural differentiation of societies, the rapid obsolescence of knowledge, the formation and dissemination of "conspiracy myths", "the general attribution of knowledge and convictions to social interests, referred to as 'ideologization'" (Schützeichel, 2012, p. 17; in relation to Karl Mannheim).

The 'classical sociology of knowledge' was developed in Germany in the first three decades of the 20th century, although the founder of scientific sociology, Auguste Comte (1798–1857), had highlighted the importance of knowledge for social processes and Karl Marx had also pointed to the class dependence of knowledge, whereby the recognized knowledge (of the bourgeoisie) became ideology through the exploitation of the proletariat (Knoblauch, 2006). The term 'sociology of knowledge' probably goes back to Max Scheler (1924, 1926), although Wilhelm Jerusalem (1925 [1909]) had already dealt extensively with the 'sociology of cognition' and Karl Mannheim (1922, 1931) had formulated a differentiated sociology of knowledge research program. The representatives of the 'classical sociology of knowledge' dealt with the question of how knowledge (and its distribution) and social structures are related to each other. For Karl Mannheim's interpretation of the sociology of knowledge, the focus was on ideologies, that is, the dependence of the world view on social structures and the attempt to legitimize one's own point of view. According to him, knowledge is 'intertwined', not only in terms of social classes, but also other groups develop their own stocks of knowledge and styles of thinking, from which a particular world view results (this can be seen, for example, in the sciences, see Chap. 4). With the modernization of society, in this case the differentiation of society, the number of perspectives increases accordingly. Mannheim (1931) formulated it as the task of the sociology of knowledge, first, to view social world views from a distance (e.g. what world views are there? For example, the conservative, liberal and socialist; see Sect. 5.4), on the one hand, to understand the social binding of knowledge (for example: who has why which knowledge?) and, on the other hand, to reveal the particular validity of social knowledge (e.g.: for whom which knowledge why is binding? More detailed: Knoblauch, 2006; Maasen, 2015; Schützeichel, 2012).

A milestone in social research far beyond the sociology of knowledge is the book "The Social Construction of Reality. A Treatise in the Sociology of Knowledge" (Berger & Luckmann, 1966). The authors formulated—based on the phenomenological sociology of Alfred Schütz (1960 [1932]; Schütz, 1971a [1962])—not only the basics of a scientific constructivism, but shifted the focus of sociological considerations of knowledge away from ideologies and scientific knowledge, towards the formation of everyday knowledge (more detailed Knoblauch, 2006). With the sociological investigation of the origin of knowledge in the social world ('knowledge of first order'), a knowledge of 'second order' arises, that is "constructions of those constructions which are formed in the social field by the actors whose behavior the scientist observes and tries to explain in accordance with the procedural rules of his science" (Schütz, 1971a [1962], p. 7). Social constructivism pursues the research program, which—also empirically—"is concerned with the question of which reality interpretations gain social binding" (Kneer, 2009, p. 5). The process of 'construction' does not designate "an intentional action, but a culturally mediated subconscious process" (Kloock & Spahr, 2007 [1986], p. 56). Abstractions in the form of prior knowledge about the world flow into every perception (Schütz, 1971b), with the consequence that "there is nowhere such a thing as pure and simple facts" (Schütz, 1971a [1962], p. 5) is. Perception is not only a result of "a very complicated interpretation process, in which present perceptions are related to previous perceptions" (Schütz, 1971a [1962], pp. 123–124), but also in a conceptual framework used by humans and in categories that already exist in our culture (Burr, 2005). The knowledge with which we face the world (including ourselves) is thus socially defined and transmitted to the individual in the process of socialization. The typical communication between people takes place by means of the use of language. Typifications are not "self-contained isolated interpretation schemes, but rather interconnected and graded" (Schütz & Luckmann, 2003 [1975], p. 125), which means that they are changeable. This also makes the sociality of knowledge clear: Only in a coordinated interplay with other people does knowledge form for actors, based on the mediation by these other people and based on the social institutions emerging between people, which are then subjected to an 'objectivation' (Knoblauch, 2013). It thus becomes a convention that is accepted as objectively given and not questioned: "The world is accepted as self-evident and 'real' at least as long as it is not questioned, as long as it is not problematized" (Werlen, 2000, p. 39). The knowledge about the world (and oneself) is divided into everyday knowledge (a social common sense of what is really and how to judge it) and special knowledge stocks. In the context of social differentiation and the increase in knowledge, these special knowledge stocks not only increase quantitatively, but also mean that society's dependence on certain experts increases. The 'knowledge society' has arisen (Weingart, 2001). Before we turn to the sociology of science (Sect. 5.3) closely linked to it, we will deal with two authors who critically deal with the social consequences of the increase in the importance of expertise: Michel Foucault and Pierre Bourdieu.

Foucault (1983 [1976], p. 114) states an insoluble connection between knowledge and power, for him power is "the will to knowledge". The symbiosis of power and knowledge is characterized on the one hand by the fact that no power relations can exist without knowledge relations, on the other hand no knowledge relations without power relations. A central element of this connection is the transformation of power into the will of the dominated itself (Foucault, 2019 [French original 1975]). Here he uses the metaphor of the panopticon prison (a form of prison in which the guards have a central position from which they can potentially see into all cells and all inmates). According to Foucault, the panoptic gaze leads to the constitution of humans as subjects (Hillebrandt, 2000, p. 120): "It is only through the subjectivation of the individual through its own self that the individual is able to recognize what the disciplinary gaze captures when it is directed at the individual." The individual thus learns to submit to the will of power voluntarily, "so that it forces the prisoner to approach an ideal, a behavioral norm, a model of obedience" (Butler, 2001, p. 82). This makes the prisoner the "principle of his own submission" (Foucault, 1977, p. 260). In the course of social differentiation and the specialization of knowledge, power is increasingly less tied to a person (such as an absolutist ruler), to a point (the absolutist capital) or to certain groups of people (politicians, scientists, managers, etc.), but becomes an omnipresent web in which (with differentiations) everyone exercises power over everyone, while Foucault (Foucault, 1983 [1976], p. 113) describes power as "the variety of power relations that populate and organize an area". Power becomes a game, "which in incessant fighting and disputes transforms, strengthens, inverts these power relations; the supports that these power relations find in each other by connecting to systems—or the shifts and contradictions that isolate them from each other" (Foucault, 1983 [1976], p. 113). Since these power relations are always in symbiosis with knowledge relations, one cannot be understood without the other. The essential expression of these power-knowledge symbioses is discourse. In this, firstly, it is regulated what does not belong to discourse, for example, what may not be talked about, secondly, the internal control, by defining rules of who is allowed to communicate in what way (for example, as a commentator or author), and, thirdly, the restriction of the subjects involved in the discourse, by, for example, requiring certain qualifications as "entrance tickets" for participation in the discourse (Foucault, 1996; Foucault, 2007 [orig. French 1971]; we will deal with the topic of discourse in more detail later: Sect. 6.3; an introduction to Foucault's work can be found in: Kammler, 2014; Kleiner, 2001; Sarasin, 2016).

For Pierre Bourdieu, the question is at the center of how society reproduces itself. For him, habitus is the central concept of the production and reproduction of "cultural capital" (knowledge, educational titles, cultural objects; see Textbox 25). For Bourdieu (1996), habitus is the social that has been incorporated into the body, so habitus is "a system of boundaries" (Bourdieu, 1982, p. 33), because "who knows the habitus of a person, he feels or knows intuitively what behavior of this person is denied" (Bourdieu, 1982, p. 33). The inscription of the social into the subject leads to the fact that "the social actors are spontaneously ready to do what society demands from them" (Wayand, 1998, p. 226), a motive that we already know from Foucault, but which is further differentiated here on the stratigraphy of society. In particular, the habitus of the lower classes is for Bourdieu an element of the endurance of the social power relations (Bourdieu, 2001). The willingness of lower classes to follow submission is—as Bourdieu (2001)—laid out by the educational system. The school "imparts to the children of the dominated classes the respect for the ruling culture, without providing them access to it" (Fuchs-Heinritz & König, 2005, p. 42), so in the field of education, the culture of lower classes is devalued, because as soon as the children of lower classes "offer their language there, they get bad grades; they lack the proper pronunciation, the proper syntax, etc." (Bourdieu, 1982, p. 49). The social mechanism of securing domination through the distribution of cultural capital is effective, not least because it— as Bourdieu (2005 [1983b]) states—is unrecognized and underestimated: "The dominated even find taste in his negative state. So poverty becomes a self-chosen lifestyle. Coercion or oppression is experienced as freedom" (Han, 2005, p. 56). So in disadvantaged families, the belief prevails that "talent and diligence are the only decisive factors for school success" (Bourdieu, 2005 [1983b], p. 16), even if their own children were driven out of higher education institutions. This belief is an example of what Bourdieu (2000) calls 'Doxa', that is, a belief that does not know itself as such. Accepted interpretations are perpetuated, (stereo) typed and moralized. Doxa regulates as a 'class-typical collective unconscious' the behavior in an indirect way, by people orienting themselves to the self-evidently valid social order, which is only questioned by crisis experiences, but then new Doxa are generated by explicit discussion and associated legitimation by arguments (Maasen, 2015), a topic that can be continued almost seamlessly in the following sociological considerations of science.

5.2   On the Genesis of Scientific Facts: Ludwik Fleck

The physician and microbiologist Ludwik Fleck (1896–1961) already drew attention in the 1930s to the fact that scientific knowledge is bound to social contexts on the basis of medical research (1980 [1935], 1983, 2011). Accordingly, he can be considered a pioneer of modern sociology of science (Neun, 2017). However, his groundbreaking studies were only re-evaluated, appreciated and taken into account after the Second World War as part of the aforementioned development of a "path into pragmatism". They can be read as very early, specifically sociological studies of science, which is why they are discussed in this chapter.

According to Fleck, medical "scientific facts" and findings do not result from unmediated observation, but from trained active and selective seeing within the framework of group-typical learned ways of seeing and thinking (Fleck, 1983, pp. 46–58). Fleck is thus opposed to a "veni-vidi-vici-epistemology" (Fleck, 1980 [1935], p. 114), which generates knowledge "like a conqueror of the type of Julius Caesar": "One wants to know something, one makes the observation or the experiment—and one already knows it" (Fleck, 1980 [1935], p. 111). But whoever stands in front of the microscope for the first time and wants to "see" or "recognize" something, "looks for similarities with the known, thus overlooking the new, incomparable, specific. He also has to learn to see" (Fleck, 1983, p. 47). It is only to *learn* what can be seen in the microscope, but then a learned 'figure' stands out clearly from other perceptual figures. This "immediate Gestalt perception" (Fleck,

1980 [1935], p. 121) is the result of discipline-specific habituation, conditioning and disciplining procedures within the framework of a "kind of collective experience" (Fleck, 1980 [1935], p. 57). The *"developed immediate Gestaltsehen* "thus solves the still" *unclear initial look"* (Fleck, 1980 [1935], p. 121). Scientists are thus introduced as part of professional training within a "thinking collective" in a profession and science-typical "thinking style". Fleck describes a "thinking collective" as a

> *"community of people who are in exchange of ideas or in mental interaction, so we have in him the bearer of historical development of a field of thought, a certain body of knowledge and cultural level, thus a special thinking style".* (Fleck, 1980 [1935], pp. 54–55)

A "thinking style" is by no means always rational, but partly based on irrational assumptions, in particular "moods" with which the "readiness for selective perception and for corresponding directed action" is given:

> "We can therefore *thinking style as a directed perception, with corresponding mental and physical processing of the perceived, define"*. (Fleck, 1980 [1935], p. 130)

Thinking styles are not only selective, they also exert pressure and coercion on scientists, because "he determines 'what cannot be thought otherwise'" (Fleck, 1980 [1935], p. 130). Such a thinking style primarily determines the thinking, acting and research of the scientists and only makes possible a discipline-specific aspectual and directed perception of certain observational data, which present themselves as "shapes" and "facts" in the course of such "perceptual training". Only "tradition, education and habituation [call] *a readiness for style-appropriate, i.e. directed and limited perception and action* forth" (Fleck, 1980 [1935], p. 111). Only then can the necessary personal "experience" of a scientist, which "brings an irrational, logically unlegitimatable element into knowledge" (Fleck, 1980 [1935], p. 125), be curbed and disciplined, even if this irrationality cannot be completely overcome.

In addition, Fleck rejects the idea of a linear or even teleological progress of science, theories and concepts always also depend on "earlier thinking styles" (Fleck, 1980 [1935], p. 131) and they are subject to a developmental history, insofar as thinking styles, concepts and theories have to "fit" into their historical environment (English *to fit*: "fitness" = adaptability, suitability) (Fleck, 1980 [1935], pp. 31-35). As in Darwin's theory of evolution, this process is indeed anchored in developmental history, but by no means predetermined: The course (and progress) of the sciences remains open in terms of direction and content.

5.3   Sociology of Science

The "knowledge society" already mentioned in Sect. 5.1 has an orientation towards scientific knowledge, accordingly science has a great importance in the generation and legitimation of knowledge, it also provides access to professional careers via qualification, also beyond the sciences. Finally, as Bourdieu (1992, p. 71) points out, the sciences have a "specific power" in the "struggle of ideas [...], that is, as an idea that is true and acknowledged," which gives "those who have scientific knowledge—about the social world—or seem to have it, the monopoly on the legitimate standpoint" (Bourdieu, 1992, p. 71), which in turn often results in a "self-fulfilling prophecy" (Bourdieu, 1992, p. 71). However, this position (increasingly) is weakened by "alternative" world views, such as conspiracy theories, but also by the orientation of science towards (economic) exploitation interests and efficiency criteria (cf. Kaiser & Maasen, 2010; Knoblauch, 2006; Münch, 2011; Weingart, 2010; see Sect. 1.2 and 5.4.2 in more detail).

In view of these increasingly complex relationships between science and (the rest of) society, the sociology of science has developed in recent decades, which, like the theory of science and the history of science, is also a meta-science. So it is a "(sub-)discipline whose object is science itself" (Weingart, 2015, p. 11).

In addition to the already mentioned in Sect. 5.2 work of Fleck's formative work in the sociological view of sciences is Robert K. Merton's (1973) "The sociology of science". He understood "science as a social subsystem with the function of providing evidence-based knowledge generated by scientific methods" (Weingart, 2015, p. 87; Hasse, 2012; see also Hofmann & Hirschauer, 2012; Weingart, 2015). In doing so, he focused—in the logic of systems theory (see also Sect. 6.3 and 6.4)—on the question of which specific rules a democratic science functions, in view of the scientific corruptibility in the Nazi regime in Germany. For him, four norms of scientific behavior were characteristic, as they were developed since the 17th century (more detailed in: Weingart, 2001, 2015; see also Sect. 3.1.2):

1. Universalism, as the independent assessment of truth claims from the social position and personal characteristics of the carriers. For such an assessment, for example, religious affiliation, ethnic origin or position in the scientific system are decisive.
2. Communism, as the recognition that scientific knowledge is the result of social cooperation with other scientists, individual recognition is measured by what has been contributed individually to the advancement of knowledge.
3. Disinterestedness, as a special moral integrity that is reflected in the rejection of using unauthorized means for one's own benefit.
4. Organized skepticism, as an instruction to make a judgment only when the available facts have been checked on the basis of logical and empirical criteria and one's own attitude has been reflected.

These norms continue to shape the sciences today and were intensified in some cases as early as the 1970s, for example in the form of the (sometimes quite problematic) *peer review* (see Textbox 23) or state evaluation as well as the sanctioning of data manipulation. The principles mentioned are also "the basis for the functional differentiation of modern science and the development of a largely self-contained communication system with its own specialized languages" (Weingart, 2010, p. 144), which self-organize and withdraw from considerations of social utility. With Niklas Luhmann (2002 [1990]; for a more detailed discussion of his theory of society, see: Sect. 6.3) this can be interpreted as an expression of the manifestation of a social functional system that, like other functional systems (such as politics, the economy, the judiciary, etc.), also fulfills a specific task for society as a whole, namely the acquisition of "unfamiliar, surprising knowledge" (Luhmann, 2002 [1990], p. 216). The new knowledge is generated along the binary code of 'true'/'false' (the binary code of the economy is 'have'/'not have'; see Luhmann, 1984, 1986) and manifests itself in its own specialist communications, whose addressees are the experts, not the public (Luhmann, 2002 [1990]). The autonomy of science can ultimately lead to a loss of trust in the self-regulation of science and attempts at political control of science (for this relationship, see Sect. 5.4.2).

Textbox 23: Peer Review Procedures

The peer review system is designed to "protect the communication community of scholars from nonsensical, false and fraudulent contributions" (Weingart, 2015, p. 33); see also Weingart (2001). In the scientific system, it is used in three contexts: first, to check the quality of scientific work before it is published (e.g. in journals); second, to assess grant proposals in order to use research funds efficiently; third, to assess research results with regard to their implications, which form the basis for political decision-making (Weingart, 2015). In this way, the system takes on a double function: within the sciences, it should create "confidence in the reliability and reciprocity of scientific communication" (Weingart, 2015, p. 25), while towards the public it should create "confidence in the reliability of the knowledge produced" (Weingart, 2015, p. 25), ultimately securing (financial) resources. However, a key problem with the system is that it is assumed that every peer is on the same level of (sub-) discipline and that there is a consensus in the assessment of work, which is however hardly possible even in closely related fields and (especially in the social and humanities) there are often fundamental differences in the assessment of work, as a result of theoretical or methodological preferences (Neidhardt, 2010; Reinhart, 2012; Weingart, 2005). The great social importance of peer review is thus made clear by the fact that it has a door-opening function to three (scarce) media of generating recognition: first, publications, only those who (in renowned journals, book series, etc.) are published have the chance to be noticed and cited (i.e. to gain attention within the profession); second, third-party funding (grant funding beyond the budget of one's own research organisation); third, public attention and influence on political processes (which in turn affects points 1 and 2 (Kühne, 2008; Weingart and Winterhager, 1984), these relationships are discussed in more detail in Sect. 5.4.2). ◀

An alternative to this institutionalist sociological perspective was developed in the 1970s (see: Hofmann and Hirschauer, 2012): David Bloor (1991 [1976]) and Barry Barnes (2013 [1974]) at the University of Edinburgh developed the *Strong*

Programme of the *Sociology of Scientific Knowledge* (SSK). In contrast to their predecessors in the sociology of knowledge and science, they saw not only erroneous, but also true knowledge as socially determined (Maasen, 2015). They understood "scientific rationality itself and its methods as 'social institutions' and thus as social constructions" (Hofmann & Hirschauer, 2012, p. 89). Bloor (1991 [1976]) formulates the program of this variety of science research in four points:

1. It is causally oriented, in that it deals with those conditions which lead to knowledge, not limited to social causes.
2. It is oriented impartially, in terms of truth and falsity, rationality and irrationality, success or failure. Both sides of these dichotomies need explanation.
3. It is oriented symmetrically, in terms of its style of explanation. The same kinds of causes, for example, would explain both true and false beliefs.
4. They are reflexively oriented, in principle their patterns of explanation must also be applicable to sociology itself.

"With this perspective, science becomes the subject of sociological observation like any other social context, it loses its privileged position. Here the difference to Merton's institutionalist perspective becomes clear as well: While the latter tries to explain under which conditions science (preferably) arrives at true knowledge, Bloor wants to investigate under which conditions knowledge is established as scientific knowledge and then as 'true' or 'false'" (Hofmann & Hirschauer, 2012, p. 90).

The social constructivist sociology of science, also known as 'laboratory constructivism', goes one step further, because according to this approach even the objects of scientific knowledge are created by the practices of scientists (Maasen, 2015). The approach of Bruno Latour, Steve Woolgar and Karin Knorr-Cetina was based on the observation of the work of natural scientists in laboratories, who do not simply find 'nature' (Knorr, 1980; Knorr-Cetina, 2002a, b; Latour & Woolgar, 2013 [1979]; Fig. 5.1), but produce it under 'laboratory conditions': "This aims at a concept of knowledge that not only sees natural scientific results as historically and socially embedded, but also as concretely constructed in the laboratory" (Knorr-Cetina, 2002b, p. 22). The creation of such scientifically produced knowledge is based on a chain of decisions that have an impact on the results, what is measured with which methods and apparatus, how many data are generated and how is the methodology evaluated? Not only the internal laboratory decisions are contextualized (for example in relation to the resources available for the research), but also externally in relation to the publishability of results (after all, the manuscripts have to go through a *peer review*), which means a transformation from innovative to acceptable formats (Knorr-Cetina, 2002a). What is considered acceptable is in turn controlled by the specific field: the differentiation of the sciences has shaped knowledge cultures as "practices, mechanisms and principles that, bound together by kinship, necessity and historical coincidence, determine in a field of knowledge how we know what we know" (Knorr-Cetina, 2002a, p. 11).

Fig. 5.1 The laboratory (here: the laboratory for soil science and geoecology in the department of geosciences at the Eberhard Karls University of Tübingen) is the center of making 'nature' available in scientific research, here comparable data are determined by means of standardized methods, which are then subjected to subsequent evaluation and interpretation and evaluation. All these steps are socially contextualized, standards for determining the data (here starting from the sampling), the methods of their evaluation up to the professional access of interpretation and evaluation, the procedure is subject to social (in this case: scientific) conventions. (Photo: Timo Sedelmeier)

The sedimentation, summarization and formation of typical knowledge, which is rarely questioned with regard to its constitution and construction, which extends over generations of scientists, is illustrated by Bruno Latour with the example of a group of researchers bent over maps in the transition area between rainforest and savannah in Brazil: "Take away the maps, mix up the cartographic conventions, delete the thousands of hours invested in the Atlas of Radambrasil, disturb the radar of the airplanes, and our four researchers would be lost in the landscape. They would be forced to start all the exploration, location, triangulation and surveying work of their hundreds of predecessors from scratch" (Latour, 2002 [1999], p. 41).

Constructivist approaches are often accused of wanting to relativize (scientific) facts. However, "the sociology of science in general and laboratory studies in particular have 'shown that the hardness and immovability of facts are not to be thought of without their social construction'" (Hofmann & Hirschauer, 2012,

p. 98). Such an understanding of the material dimension, the individual and social construction of the world opens up possibilities of dealing with the world, beyond representational epistemologies, as they can be found, for example, in actor-network theory, assemblage theory or neo-pragmatism (see Sects. 6.10 and 6.11).

5.4   Science and Politics: From World Views and the Mutual Dependence of Science and Politics Today

The relationship between political world and scientific world has been mentioned several times in the previous sections, but not elaborated in more detail. We will make up for this in the following. First, the question will be addressed how different ideational systems (or world views) are expressed and differ from each other, connected with different interpretations and evaluations of spaces. Subsequently, the changing relationship between society, in particular politics, and science in the transition from Mode 1 to Mode 2 science will be addressed.

5.4.1   Political Ideational Systems and Their Implications on the Evaluation of Spaces: Socialism, Liberalism and Conservatism

The proximity of knowledge(s) and political ideation systems has been reflected in the history of the sociological reflection of knowledge(s) by different sides, for example by Karl Mannheim or David Bloor (see also Hofmann & Hirschauer, 2012). Political positions also find their way into the space sciences, not only in normative statements about current and future developments, but also in the choice of research questions, in terminology and scientific-theoretical anchoring, whereby the "philosophical background" is often not explicitly named and can only be inferred from the context. In this respect, we will now outline the essential features of socialism, liberalism and conservatism, as well as their space-scientific implications, following Karl Mannheim (1931, 1952). Liberals, conservatives and socialists not only differ "in their interpretations of historical change" (Berlin, 1995 [1969], p. 80), but also with regard to the question of "which are the elementary needs, interests and ideals of people and who represents these ideals to the greatest extent and over what period of time" (Berlin, 1995 [1969], p. 80; see also: Kühne et al., 2021).

The central concept of liberalism (hence also the name) is liberty. Ralf Dahrendorf (2007b, p. 26) initially understands liberty as the "absence of coercion", more concretely, people are "as free as they can make their own decisions. In the state of freedom we find conditions that reduce coercion to a minimum. The goal of liberalism or the politics of freedom is that there is a maximum of liberty under given restrictions" (Dahrendorf, 2007b, p. 26). The axioms of a freely born, equal in rights to others, naturally good and rational human being as an individual form the basis of liberalism (Bauer & Wall-Strasser, 2016; Leonhard, 2001; Pennington, 2002; Schaal & Heidenreich, 2006). Liberalism is a world view in which the individual plays a central role: It should be able to develop itself to the greatest possible extent without social constraints and in its own responsibility (which also means the duty to earn its own livelihood), whereby it should be able to choose an alternative freely according to its own convictions on the basis of a maximum number of alternatives (political, economic, cultural, etc.). Liberalism has an optimistic attitude towards the future, i.e. the future will be better through progress (Leonhard, 2001), the present, though not perfect, is better than the past (Popper, 2019 [1987]), accordingly Liberalism is "necessarily a philosophy of change" (Dahrendorf, 1979, p. 61). Society is—according to the liberal view—not subject to any higher (e.g. divine) order, it also does not develop teleologically towards a goal (Popper, 2011 [1947], 2019 [1987]), as is the case, for example, with a Marxist view. The task of society is to provide the individual with security in his pursuit of happiness. The classical tasks of a liberal state are defined briefly by Mises (1927, p. 33): "Protection of property, liberty and peace". In the political context, liberalism is associated with the "defence of certain individual rights and freedoms such as freedom of expression, freedom from discrimination on the grounds of race, gender or nationality, procedural rights (e.g. the right to defence) and political rights to democratic participation and participation in elections" (Rivera López, 1995, p. 17).

Freedom has—according to Isaiah Berlin (1995 [1969])—two dimensions, a negative and a positive one. The "negative freedom" outlines the absence of coercion, threat and violence, also called "freedom from". The "positive freedom" ("freedom to"), refers to the ability to act autonomously, "the 'positive' meaning of 'freedom' is derived from the individual's desire to be his own master. I want my life and my decisions to depend on me and not on any external powers" (Berlin, 1995 [1969], p. 211). This distinction facilitates the understanding of different currents of liberalism. The "classical liberalism" follows rather the negative understanding of freedom, by wanting to limit the state to the tasks of security and preferring an economic system that is as little restricted by the state as possible, in order to increase individual freedoms. The liberalism that relies more on the positive concept of freedom assumes that freedom does not arise formally through the guarantee of basic rights, but substantively through "to be fulfilled civic rights [...] social rights, such as the right to protection against unforeseeable hardship or the right to adequate old-age provision or the right to education" (Dahrendorf, 1983, p. 104), which in turn extends the tasks of the state.

Just as liberalism developed, early conservatism also developed in critical engagement with the French Revolution and its ideas, albeit rejectingly. The reference idea of conservatism (derived from the Latin *'conservare'*, German 'preserve', 'preserve') is tradition, it forms the central orientation, order and reference framework. Thus, conservatism points to the ideas of rationality and enlightenment, but not affirmatively as liberalism, but rejectingly (cf. Greiffenhagen, 1971; Lenk, 1989; Ottmann, 2008; Schoeps, 1981). This position outlines the first dilemma of conservatism: It is constitutively bound to that which it rejects (Greiffenhagen, 1971; Trepl, 2012). The second dilemma of conservatism is that it is forced to follow the logic of the Enlightenment in order to defend what is actually self-evident to it, institutions such as religion, family and nation, require justification by recourse to reflective reason (1989; Schoeps, 1981). The normative understanding of human coexistence fundamentally differentiates conservatism from liberalism, not the 'society of independent individuals', but a community in the sense of an organic connection of people is the guiding idea here (1971; Lenk, 1989; Trepl, 2012; Voigt, 2009b). The individuals should "serve the whole like organs in the organism, and that too at the place assigned to them respectively" (2012, p. 141). Accordingly, the concept of freedom of conservatism also differs from that of liberalism: "Freedom therefore means in this conception the adaptation to the higher order of the whole" (1999, p. 23), not the maximization of individual life chances. Instead of the liberal right to freedom of expression in order to determine the most suitable arguments, the conservative "does not argue with every person about what is right [...], in order to convince him" (2012, p. 145). He rather takes a "paternalistic attitude" (2012, p. 145) towards "those who would not understand it anyway". This paternalistic position can be understood as an expression of the conservative understanding of authority, secured by tradition, it is not—as in liberalism—to be regulated and limited, but rather seen as a guarantee for the preservation of the community (2004; Lenk, 1989; Voigt, 2009b).

Even conservative ideas can be differentiated. Eppler (1975) distinguishes between structural conservatism and value conservatism. Structural conservatism advocates for the preservation (of traditional) privileges and power relations, while value conservatism is directed at preserving a livable environment, a solidary society, and the dignity of the individual. The position of value conservatism makes it receptive to the environmental movement, which has been gaining strength since the 1970s, but also to social democratic and trade union positions, as well as to liberals (Euchner et al., 2015). Another distinction can be made in relation to the continental European and Anglo-American lines of conservatism with regard to the importance of the state. While the former affirms the (even authoritarian) state as a guarantee for the "natural" order, the latter evaluates the state as a symbol of anonymous rule and guardianship, beyond the securing of internal and external security (here there are clear parallels to classical liberalism, while both differ significantly in their evaluation of religion, which classical liberalism understands as a private matter; see Doeker, 1973; Greiffenhagen, 1971; Shell, 1986).

Socialism (from Latin '*socialis*') picks up the ideas of the French Revolution, with rejection of the liberal and conservative world view. It can be represented as a general term for different theories and ideas, which have the "priority of the 'society' resp. the 'social'" (Bärsch, 1981, p. 170) in common, in contrast to the individual. From the prioritization of the 'social community' follows the rejection of privilege for individuals, which arise from tradition or the pursuit of economic gain (Bärsch, 1981). While classical socialism aims for a collectivization of production means, communism strives for a transfer of the "means of production as well as [the] consumer goods into the public property (distribution of goods)" (Bärsch, 1981, p. 172; cf. also Fainstein, 2010). Socialism has an orientation towards the future, which is to be achieved in revolutionary socialism through the formation of a 'revolutionary elite': "What guided them [the 'revolutionary elite'] was the conviction that it was their task to free the exploited and oppressed" (Becker, 2013, o. S.). Also here a paternalistic attitude becomes clear, although the goal of socialism is to replace the 'external leadership' by 'revolutionary elites' in egalitarian socialist societies through the 'internal leadership' of the individual in the sense of socialism. This is the central task of education, here the 'false' (self-interested) should be replaced by a 'correct' (solidary) consciousness (Bärsch, 1981), which makes a central difference to the understanding of education in liberalism: "The liberals want to stylize the worker into a citizen, integrate him into the linguistic, political and intellectual traditions of the bourgeoisie, while the socialists try to use education to create class solidarity and class consciousness, also politically" (Knoll, 1981, p. 92).

For a socialist understanding of social development, the formulation of a goal of this development is essential (teleology). This teleology is clearly visible in Marxism, which formulates a sequence of stages of social development. Accordingly, social development is driven by economic contradictions: the contradiction between production relations and productive forces comes to a head in a revolutionary transition to the next stage (Fig. 5.2), a development lawfulness which is referred to as 'historical materialism'. This is also how the 'self-destruction' of capitalism is understood, because "the market economy already contains the causes of its own downfall" (Herzog, 2013, p. 109), because the principle of private ownership leads to the pauperization of the working masses (Bärsch, 1981), in its own country right up to the global scale. From this perspective, reformist social efforts are also rejected which, on the way to socialism, understand reforms within the market economy (e.g. more participation in the workplace for workers,

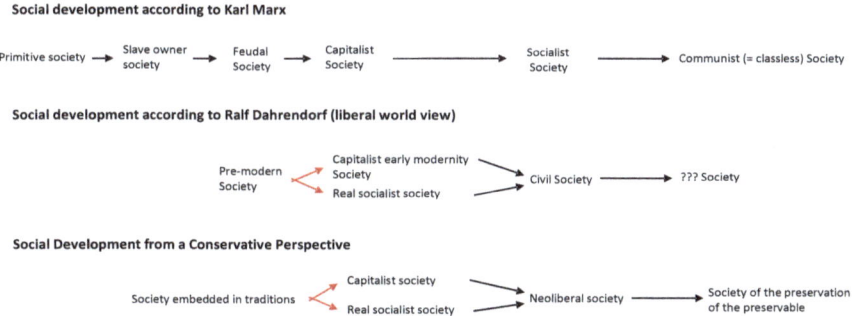

Fig. 5.2 The different stages of social development according to socialist (here Karl Marx), liberal (here Ralf Dahrendorf) and conservative understanding. For Marx, history runs teleologically towards the communist society, whereas for Dahrendorf social future is open. According to the conservative understanding, a society is aimed at after the current 'neoliberal society' which 'preserves what is worth preserving' (own representation according to: Kühne, 2017)

higher wages, etc.) as 'management and manipulation of the capitalist crisis' (see, for example, Harvey, 2005, 2008). The acquisition of 'bourgeois freedoms' (voting rights, freedom of speech, freedom of coalition, etc.), which liberalism advocates, would thus remain incomplete until people were no longer economically obliged to sell their labour power (Harvey, 2009; Misik, 2012). So Marxist liberals see liberal democracy as a 'hostage of capitalist interests', because state institutions are corrupted by economic elites (e.g. Agnoli, 1968). The reduction of social development, but also the assumption of the law of nature-like predictability of social development, has in turn led to the accusation that Marxism is essentially essentialist in its arguments (e.g. Dahrendorf, 1971; Kühne, 2017; Laclau & Mouffe, 1985). The efforts to translate the idea of socialism into state action gave rise to state or real socialism, in which a pronounced system of existence security (pronounced health care, housing, food) was opposed to a pronounced economic efficiency (low productivity and high environmental pollution) and a system of political repression. Thus, for Ralf Dahrendorf, real socialism is, next to the capitalist early modern society, an alternative way to the liberal civil society. In comparison to socialist teleology, he sees the difference in terms of social development in that the social future is open and not predetermined (see Fig. 5.2; Dahrendorf, 1972, 2007a; Popper, 1996, 2003 [1945], 2018 [1984]).

In recent decades, approaches have been developed to which the prefix 'Neo' is prefixed, but there is by no means general consensus about their use: the development of Neo-Marxism (since the Second World War) rejects the developments of Real Socialism in Eastern Central and Eastern Europe on the one hand (as 'Western Marxism'), on the other hand it no longer follows the teleological schema of Marxism. Neo-conservatism has been developed in the United States since the end of the 1960s. The values of 'classical conservatism', such as family, homeland, tradition, etc., are supplemented by the ideas of economic liberalism (see also the section on Anglo-American conservatism above). He sees the connection of political democracy and market economy as the highest level of social development. The word 'neoliberalism', on the other hand, is associated with two completely different concepts. On the one hand, 'neoliberalism' is understood to mean a liberal current that arose in the 1930s (the so-called 'Freiburg School'). Its core concerns are the modernization of classical liberalism, in which the state sets a clear competition regime to protect people from the self-destructive effects of the market (so-called 'ordoliberalism') and which was reflected in the 'social market economy' in Western Germany after the Second World War. On the other hand, the word 'neoliberalism' describes a (particularly socialist) diagnosis of present-day society. This is characterized by deregulation of state tasks, economization and globalization of all areas of life, performance pressure and increasing polarization of society.

In recent decades, the scientific study of world view preferences in relation to space has been intensified (for example, Eisel, 1982, 2009; Kirchhoff & Trepl, 2009a; Kühne, 2015; Vicenzotti, 2006, 2011; Kirchhoff, 2019b; Kühne, 2011; Voigt, 2009a, b). These are to be compared along the space types 'wilderness', 'rural space', 'suburbium' and city:

- Wilderness is contradictorily connoted from a liberal perspective: on the one hand it is regarded as a symbol for a dangerous, pre-societal state of nature of struggle, on the other hand it is also understood as a place of freedom in which the individual develops his abilities in cultivation (here the US-American founding myth becomes clear; see also Pregill & Volkman, 1999), in order to make an unproductive space a space of economic productivity. Internal as well as external wilderness is "regarded by conservatism as a sphere of instinctual drives. It is the temptation to resist, to tame and to leave behind" (Vicenzotti, 2011, p. 140; see also Kötzle, 1999). Wilderness can, however, also be understood as a symbol of a paradisiacal origin or as a symbol of 'innocent youth'. From a socialist perspective, 'wilderness' is connoted with the 'primeval state' of society. This state has been overcome.
- Rural areas, often interpreted as 'cultural landscape', are from a conservative perspective "an expression, ideal and symbol of successful cultural development" (Vicenzotti, 2011, p. 147), here is the successful synthesis of 'land and people' in a 'superorganism' (Eisel, 1982, 2004; Rodewald, 2001; Vicenzotti, 2011). They are an expression "of a perfection that corresponds to both the nature of the community (national character) and the living space" (Trepl, 2012, p. 156). Both liberalism and socialism see these traditional rural areas as "a more advanced stage in comparison to wilderness", but "still below the level of development of the city" (Vicenzotti, 2011, p. 116). Both world views associate rural life with political, social and technical backwardness: Traditional rural communities are associated with social control and a lack of education, irrational agricultural land use. Karl Marx spoke of the 'idiocy of rural life' (Ipsen, 1992). Marxism saw the working class as the carrier of socialism, so the real socialism aimed, on the one hand, to resettle rural population in the cities, on the other hand to urbanize rural areas and to industrialize rural land use (Esser, 1998; cf. Domański, 1997; Fierla, 1999; Jaehne, 1972).
- Not only socialism, but also liberalism regards the city as the symbolic place of the preferred state of society, "the place where the state of nature is overcome and entry into bourgeois society is carried out" (Vicenzotti, 2011, p. 121), it is the "place of productive channeling of passions" (Vicenzotti, 2011, p. 122; see also Eisel, 1982). Here, competition is not fought by violence, but in the economic channel. From a conservative perspective, the (large) city is described as the place of seduction by purchasable lust, as the place of moral depravity, unnaturalness and artificiality, the city itself is "imagined as a self-giving, self-opening, devouring female figure" (Löw, 2008, p. 198). In the metropolis, the barbarism of the proletariat meets the overcivilization of the bourgeoisie (Vicenzotti, 2011; for the Anglo-Saxon world: Muir, 1998).
- Suburban spaces are regarded from both a conservative and a socialist perspective as an expression of an excessive striving for ownership, a socially normed apparent voluntary submission of the human being to the rules of the financial market in capitalist societies, as well as the retreat into the private sphere of the single-family home instead of a communal exchange (e.g. Belina, 2009; Bourdieu, 1998; Soja, 2000). From a conservative perspective, the loss of 'iden-

tity' of suburban settlements is also lamented here, their 'uniformity' as well as the expansion of urban lifestyles into formerly rural areas (cf. Vicenzotti, 2012; Hunt, 2016). From a liberal point of view, suburban areas are regarded as an expression of the striving for ownership, independence from communal constraints and privacy, which also allows for individual spatial design.

Not only the interpretation and evaluation of social developments, but also of spaces is—as shown—therefore strongly dependent on which world view is taken. These basic views are—also in scientific texts—often not explicitly named. The (implicit) dependence on world-view interpretations and evaluations is not limited to cultural and social sciences, but also to natural sciences, for example many ecological theories, such as the ecosystem approach, are related to the approach of ecosystem services, with the further connotation of a 'neoliberalization' in the sense of an economization of nature (among many: Kirchhoff, 2019a; Kühne & Duttmann, 2019; Leibenath, 2017; Voigt, 2009a).

5.4.2   The Transition from Mode 1 to Mode 2 Science

In recent decades (with the beginnings going back to the 19th century), the relationship between science and politics/society has changed fundamentally (Bender, 2004; Gibbons et al., 1994; Nowotny et al., 2001; Nowotny, 2005): The mode of scientific production in the form of a strict separation of basic and applied research, the clear separation of science and society (Mode 1), has been replaced by a form of scientific production which is referred to as Mode 2. However, critics point out that the allegedly new connection between scientific knowledge and social benefit stands in a historical context in which the "Mode 2" does not appear as a new phenomenon at all. Science has always operated "under the double obligation of knowledge and benefit" (Carrier et al., 2007, p. 24), it has always also been "practice-stabilizing knowledge" (Mittelstraß, 2004, p. 259) for society understood and handled. For example, modern natural science was "at all times full of tension on [these] two goals" (Carrier et al., 2007, p. 24), which, however, were not, as initially hoped, to be achieved. In 'Mode 2', mixed forms of application-oriented basic research are developed, (Elias et al., 2014), private research funding strengthens the application-oriented, to the detriment of the scientific treatment of basic questions, in the race for external research funding (third-party funding) submit themselves to the rules of economic logic, by trying to improve their position in rankings, in order to be more attractive for third-party funders and renowned researchers, also the competition private research institutes against universities increases, which in turn are spread with university-educated staff. The result is an increasing contextualization of science/research in non-scientific networks (in detail at: Weingart, 2001). Latour (2002 [1999], p. 31) sees here the transition from science to research: "Science had certainty, coolness, reserve, objectivity, distance and necessity, research on the other hand seems to carry all the opposite characteristics: It is uncertain, open-ended, involved in the lower problems of money, instruments and know-how and can not so easily distinguish between hot and cold, subjective and objective, human and non-human." Gibbons et al. (1994) and Nowotny (2005) state that the transition from Mode 1 to Mode 2 is associated with a fundamental epistemological change: Science is no longer concerned with the exploration of natural laws, but rather research in interdisciplinary application contexts produces 'social robust knowledge' (cf. also Viehöver, 2005), less 'truth' becomes the criterion for the evaluation of scientific results, but 'suitability'.

The production of knowledge in Mode 2 is—as mentioned—significantly context-dependent in four dimensions (Fig. 5.3; Latour, 2002 [1999]):

- Mobilization of the world. This refers to means that require the involvement of non-human beings in the discourse (see also Sect. 5.3 and 6.10). Examples include technical instruments, data collection, surveys or expeditions, with which information about the world outside of science is gathered. This loop also "affects the sites where all the mobilized objects of the world [in expeditions, surveys, etc.; note O.K.] are gathered and stored together" (Latour, 2002 [1999], p. 122).
- Autonomy. This refers to the work of a discipline, profession or clique to become independent of alternative evaluation criteria. This creates its own evaluation and reference system.
- Alliances. This refers to the need for people outside of science to be interested in specific scientific topics and disciplines (Latour, 2002 [1999], p. 125): "You

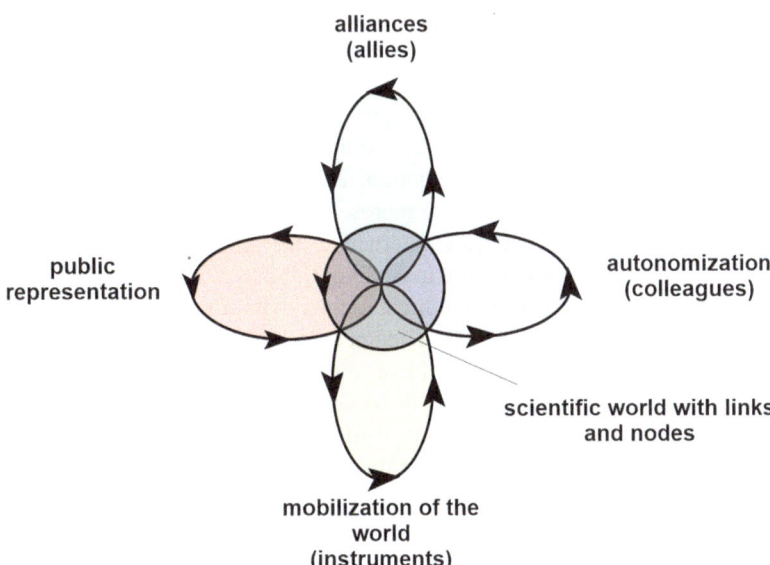

Fig. 5.3 Model of the integration of science and research into the social context (according to: Latour, 2002)

have to interest military men in physics, industrialists in chemistry, kings in cartography, teachers in pedagogy and deputies in political science. Without such an effort to interest, the other loops will be as futile as traveling with your finger on the map."

- Public representation. This refers to the regulation of the relationship between sciences and journalists and "the woman and the man on the street" (Latour, 2002 [1999], p. 127). Research must be presented publicly in order to be noticed (and to secure resources).

At the center of these loops, science/research occupies the position of a node (which results from the focus on science, if such a mapping were developed with the focus on 'politics', this would be the node between (other) loops). Success (of a (sub-) discipline, a scientist, a university, a research area, etc.) depends on the fact that a sufficient connection of these loops succeeds in mobilizing material and immaterial resources. In this respect, the function of scientists is increasingly changing: The production of innovative knowledge is no longer a guarantee of success, financial resources must be mobilized for the production of knowledge, it must be accepted in one's own community (i.e. not too innovative), it must be compatible with neighboring disciplines and publicly communicated in order to in turn mobilize new resources.

The transition from Mode 1 to Mode 2 means on the one hand an extension of the accountability basis of science, namely ultimately towards society, no longer only towards one's own colleagues (Nowotny, 2005), on the other hand this 'demystification' also means an increase of science on social, in particular political decisions, "and even if only in the sense that it gives them the additional legitimacy of rationality and 'objectivity'" (Weingart, 2003, p. 92). Scientific knowledge is thus less established, tested, distributed and protected from the logic of the 'science system' (Luhmann, 2002 [1990]), but its development serves more the usefulness for justification, legitimation and social conviction activity (Bloor, 1982). Especially with the successful mobilization of the public, there is the danger of the "colonization" (Weingart, 2003, p. 98) of politics by individual experts or a small group of these, who represent a single perspective on the subject of communication and do not allow alternative interpretations and evaluations to arise (discursive closure; see section in detail 6.3). Such a form of 'colonization' in turn represents "a constant potential threat to the other, primary basis of legitimacy [...], namely the will of the people documented by elections" (Weingart, 2003, p. 92). On the other hand, the control of knowledge production by politics and economy through the granting of third-party funds can also be regarded as 'colonization' (Giroux, 2015; Moore et al., 2011; Morrissey, 2015), because "knowledge production is oriented towards the market, politics, media. In the sense of this mutuality of expectations, *scientization* of society always also means *economization, politicization* and *mediatization* of science" (Weingart, 2001, p. 124).

Before such (certainly mutual) 'colonizations' can take place, a translation between different logics is necessary: From the perspective of actor-network theory (see detailed Sect. 6.10), living and non-living objects interact with each other in networks. These networks arise through 'translations' of initially uncon-nected elements and resources such as their contextual relationships (Weingart, 2015; Callon, 1984), for example, ecological or political problems are translated into scientific ones, scientific questions have to be translated into political or eco-nomic ones: Due to communication in different codes (true/false in science; hav-ing/not having power in politics), it is necessary to translate political problems into scientific research programs, the scientific answers in turn have to be translated into the language of politics (cf. Luhmann, 1997, 2002 [1990]; Weingart, 2001, 2003; Weingart et al., 2008). Here (in spite of all 'colonization') there is the prob-lem that expertise is transgressive, that means "that all experts have to exceed their scientific competence because they are asked questions that are not their own" (Nowotny, 2005, p. 37; cf. also Holzinger, 2004; Levidow, 2005). In other words: Science (as an observation instance of the world) exceeds the (professionally) safe territory as soon as normative questions are asked. But to which it is forced in Mode 2—and (sometimes) understands how to use it for its own purposes: by skillfully using the public representation, science (or its more prominent represent-atives) is able to bring issues onto the political agenda, only to be commissioned by politics again to formulate proposals for solving problems that were formulated themselves: "The success of scientists in producing knowledge and thus confer-ring social validity on it cannot be explained by the truth of this knowledge and its persuasiveness. Rather, it requires the skillful manipulation of the relevant network of heterogeneous units, technical artifacts and natural objects in order to secure their 'support' for one's own goals" (Weingart, 2015, p. 72).

In addition, there are further differentiations: For example, the scientific system is characterized by the transformation of the dichotomy of experts and laypeople into a field of different polarities, in which, within science, the scientist, who is among the most respected experts in his subdiscipline, is at best an interested layman in a neighboring subdiscipline. On the other hand, with the expansion of education since the 1960s, the abilities of laypeople to acquire expert knowledge have increased, sometimes in relation to the collection and evaluation of scientific data (*citizen sci-ence*; see Bonney et al., 2009; Finke, 2014; Silvertown, 2009), but with the aim of strengthening their own individual or collective ideological interests vis-à-vis conflicting interests (a current example: the physical manifestations of the energy transition, which is being fought over with bitterness between proponents and opponents—using scientifically generated knowledge –; among many: Jenal, 2018; Kühne et al., 2019; Leibenath & Otto, 2012; Weber et al., 2017; Weber, 2018).

These processes of the delinking of science and society, more critically: the mutual 'colonization' take place—as already mentioned several times—against the background of a society that is becoming more complex, with more complex feedbacks to non-human living and non-living objects and their constellations. The increase in complexity leads to an increase in the value of uncertain knowledge, of uncertainty, non-knowledge and "hypothetical risks" (Fischer, 2005, p. 111)—in spite of an increase in research funding. Hypothetical risks can be—based on the classical risk formula (risk as the product of damage and probability of occurrence), with Fischer (2005, p. 111) as those risks which "both the amount of damage and the probability of occurrence are unknown". In addition, there is currently a higher potential for the population as a whole to be affected (cf. Beck, 1986).

In addition, science's handling of non-knowledge is very specific: Science usually constructs non-knowledge as a precursor to knowledge as specified non-knowledge, by limiting contingency, by weighing risks, or as not-yet-knowledge, that is, non-knowledge that can be assumed to be accessible by science (Luhmann, 2002 [1990]; Merton, 1987; see for the distinction between scientific knowledge already Sect. 3.1.2). If non-knowledge is neither specific nor soon convertible into knowledge, it escapes observation by science. In turn, such types of scientific non-knowledge can be "thematized by media, social movements or political institutions" (Böschen, 2005, p. 247). An investigation of this non-knowledge is carried out—following the Mode 2-logic—by stimulating the 'science' functional system through the awarding of subsidies. The handling of still uncertain knowledge is particularly controversial within the (sub-)disciplines, even within a scientific-theoretical understanding, with the interplay of expertise and counter-expertise (Nennen & Garbe, 1996) not being uniformly understood as an expression of an attempt to secure knowledge or as a sign of increased non-knowledge (after all, new knowledge always produces new questions of (non-)knowledge; in the sense of Nowotny, 2005), but often as non-intended self-questioning (in the sense of Beck, 1986). That the influence of society by science is not risk-free is shown by Funtowicz and Ravetz (1990) using the example of the early communication of experts about anthropogenic climate change: The mechanisms of stabilization of the temperature of the Earth's atmosphere are complex and multi-feedback, weather and weather conditions are correspondingly volatile and the standard measurement periods for calculating the climate are correspondingly long (climate data is obtained in 30-year observation intervals), so that in the 1980s there were still no secure findings on the development of air temperature, the climate models were still relatively crude, so that there was a high probability that the forecasts of the experts would not materialize. This in turn would have led to a loss of public legitimacy of science, a renunciation of positioning in turn would have been associated with the accusation that it would not meet its public obligation to name risks (Weingart, 2001). If the relevant non-knowledge is converted into knowledge, the translation of scientific knowledge into political action is not without risk. Often, the further pursuit of a political logic is understood as a failure (whether due to inability or unwillingness) of decision-makers to "implement scientific truth" correctly. Here, a return to Mode 1 communication is often observed, "still very much shaped by the Enlightenment pattern of natural scientists, for example according to the motto: 'We know where it's going, and first of all you have to come to a minimum level of scientific knowledge, then we can talk further'" (Nowotny, 2005, p. 40).

Textbox 24: Interim Conclusion on: The Contextualization of Science II: Social Embeddedness

This chapter has shown that knowledge does not arise solely in social contexts, but is also selected and imparted. This applies not only to everyday knowledge, but also to scientific knowledge. The "constructivist turn" in the sociology of science could lead to the conclusion that scientific knowledge is levelled with non-scientific knowledge (as Karin Knorr-Cetina has done), but it can also be concluded that the justification of the decisions leading to the formulation of scientific results is required. This justification also has another context: With the increasing mediatization, the availability of information increases, methods of dealing with this information gain in importance, as does the conscious handling of the context-dependence of (not only) presented information. The clear presentation and justification of decisions in research and even in the process of acquiring knowledge reduces the risk of an unreflective ideologization (in the sense of Karl Mannheim). In addition, the fact that the sciences have their own system of reputation generation, the securing of their own standards and, last but not least, that other parts of society have expectations of the sciences with regard to the generation of usable knowledge speaks in favour of an independent position of the sciences in society.

A overview of the genealogy of scientific concepts and thinkers from antiquity to the present, as they are presented in this book (Chaps. 3, 4 and 5), can be found in Fig. 5.4.

Further Reading

Kirchhoff and Trepl (2009b): A very instructive introduction to the cultural and social determinants of essential spatial categories.

Maasen (2015): An instructive introduction to the sociology of knowledge with some remarks on the sociology of science.

Weingart (2015): A now classic introduction to the sociology of science.

Original Literature

Berger and Luckmann (1966): The classic work of the sociology of knowledge, with which the constructivist paradigm was shaped in the cultural and social sciences.

Luhmann (2002 [1990]): Due to Luhmann's own terminology, not an easy book to read, but it provides informative insights into the function of science in society.

Knorr-Cetina (2002a): A grippingly written book that social-scientifically represents the process of the emergence of scientific knowledge.

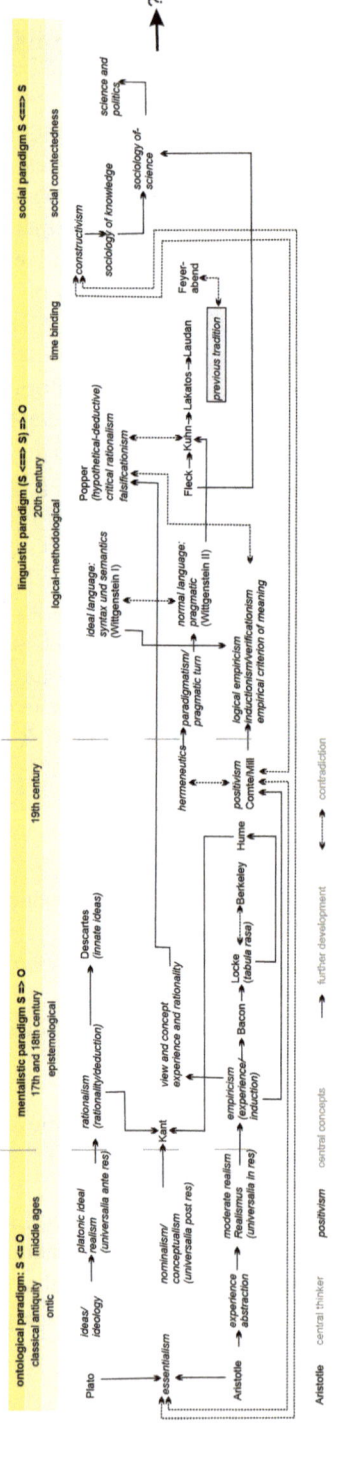

Fig. 5.4 A genealogy of scientific concepts and thinkers (own representation)

References

Agnoli, J. (1968). *Die Transformation der Demokratie*. Europäische Verlaganstalt.

Barnes, B. (2013 [1974]). *Scientific knowledge and sociological theory*. Routledge.

Bärsch, C.-E. (1981). Sozialismus. In J. H. Schoeps, J. H. Knoll & C.-E. Bärsch (Hrsg.), *Konservativismus, Liberalismus, Sozialismus. Einführung, Texte, Bibliographien* (Uni-Taschenbücher Politologie, Neuere Geschichte, Soziologie, Bd. 1032, S. 140–249). Fink.

Bauer, L., & Wall-Strasser, S. (2016). Liberalismus/Neoliberalismus. *Politik und Zeitgeschehen, 4*, 15–37.

Beck, U. (1986). *Risikogesellschaft. Auf dem Weg in eine andere Moderne* (Edition Suhrkamp, Bd. 1365). Suhrkamp.

Beck, U. (2007). *Weltrisikogesellschaft. Auf der Suche nach der verlorenen Sicherheit* (Edition Zweite Moderne). Suhrkamp.

Becker, W. (2013). *Macht ohne Maß und kein Ende? Katholizismus, Kapitalismus (Imperialismus) und Kommunismus*. Engelsdorfer.

Belina, B. (2009). Kriminalitätskartierung – Produkt und Mittel neoliberalen Regierens, oder: Wenn falsche Abstraktionen durch die Macht der Karte praktisch wahr gemacht werden. *Geographische Zeitschrift, 97*(4), 192–212.

Bender, G. (2004). Modus 2 – Wissenserzeugung in globalen Netzwerken? In U. Matthiesen (Hrsg.), *Stadtregion und Wissen. Analysen und Plädoyers für eine wissensbasierte Stadtpolitik* (S. 87–96). VS Verlag.

Berger, P. L., & Luckmann, T. (1966). *The social construction of reality. A treatise in the sociology of knowledge*. Anchor Books.

Berlin, I. (1995 [1969]). *Freiheit. Vier Versuche*. Fischer.

Bloor, D. (1982). Durkheim and Mauss revisited: Classification and the sociology of knowledge. *Studies in History and Philosophy of Science London, 13*(4), 267–297.

Bloor, D. (1991 [1976]). *Knowledge and social imagery*. University of Chicago Press.

Bonney, R., Cooper, C. B., Dickinson, J., Kelling, S., Phillips, T., Rosenberg, K. V., & Shirk, J. (2009). Citizen science: A developing tool for expanding science knowledge and scientific literacy. *BioScience, 59*(11), 977–984. https://doi.org/10.1525/bio.2009.59.11.9.

Böschen, S. (2005). Reflexive Wissenspolitik. Formierung und Strukturierung von Gestaltungsöffentlichkeit. In A. Bogner & H. Torgersen (Hrsg.), *Wozu Experten? Ambivalenzen der Beziehung von Wissenschaft und Politik* (S. 241–265). VS Verlag.

Bourdieu, P. (1982). *Leçon sur la leçon*. Les Éditions de Minuit.

Bourdieu, P. (1992). *Homo academicus* (Suhrkamp-Taschenbuch Wissenschaft, Bd. 1002). Suhrkamp (französische Originalausgabe 1984).

Bourdieu, P. (1996). Die Praxis der reflexiven Anthropologie. Einleitung zum Seminar an der École des hates études en sciences sociales. Paris, Oktober 1987. In P. Bourdieu & L. Wacquant (Hrsg.), *Reflexive Anthropologie* (S. 251–294). Suhrkamp.

Bourdieu, P. (1998). *Der Einzige und sein Eigenheim*. VSA.

Bourdieu, P. (2000). *Les structures sociales de l'économie*. Seuil.

Bourdieu, P. (2001). *Meditationen: zur Kritik der scholastischen Vernunft*. Suhrkamp.

Bourdieu, P. (2005 [1983b]). Politik, Bildung und Sprache [1997]. In P. Bourdieu (Hrsg.), *Die verborgenen Mechanismen der Macht. Schriften zu Politik und Kultur 1* (S. 13–30). VSA.

Burr, V. (2005). *Social constructivism*. Routledge.

Butler, J. (2001). *Psyche der Macht. Das Subjekt der Unterwerfung*. Suhrkamp.

Callon, M. (1984). Some elements of a sociology of translation: Domestication of the scallops and the fishermen of St Brieuc Bay. *The Sociological Review, 32*(1), 196–233.

Carrier, M., Krohn, W., & Weingart, P. (Hrsg.). (2007). *Nachrichten aus der Wissensgesellschaft. Analysen zur Veränderung der Wissenschaft*. Velbrück-Wissenschaft.

Dahrendorf, R. (1957). *Soziale Klassen und Klassenkonflikt in der industriellen Gesellschaft*. Enke.

Dahrendorf, R. (1971). *Die Idee des Gerechten im Denken von Karl Marx*. Verlag für Literatur und Zeitgeschehen.

Dahrendorf, R. (1972). *Konflikt und Freiheit. Auf dem Weg zur Dienstklassengesellschaft*. Piper.

Dahrendorf, R. (1979). *Lebenschancen. Anläufe zur sozialen und politischen Theorie* (Suhrkamp-Taschenbuch, Bd. 559). Suhrkamp.

Dahrendorf, R. (1983). *Die Chancen der Krise. Über die Zukunft des Liberalismus*. Deutsche Verlags-Anstalt.

Dahrendorf, R. (2007a). *Auf der Suche nach einer neuen Ordnung. Vorlesungen zur Politik der Freiheit im 21. Jahrhundert* (Krupp-Vorlesungen zu Politik und Geschichte am Kulturwissenschaftlichen Institut im Wissenschaftszentrum Nordrhein-Westfalen, 4. Aufl., Bd. 3). Beck.

Dahrendorf, R. (2007b). Freiheit – eine Definition. In U. Ackermann (Hrsg.), *Welche Freiheit. Plädoyers für eine offene Gesellschaft* (S. 26–39). Matthes & Seitz.

Doeker, G. (1973). Konservatismus in den Vereinigten Staaten von Amerika. *Der Staat, 12*, 369.

Domański, B. (1997). *Industrial control over the socialist town: Benevolence or exploitation?* Praeger Publishers.

Eisel, U. (1982). Die schöne Landschaft als kritische Utopie oder als konservatives Relikt. Über die Kristallisation gegnerischer politischer Philosophien im Symbol „Landschaft". *Soziale Welt, 33*(2), 157–168.

Eisel, U. (2004). Politische Schubladen als theoretische Heuristik. Methodische Aspekte politischer Bedeutungsverschiebungen in Naturbildern. In L. Fischer (Hrsg.), *Projektionsfläche Natur. Zum Zusammenhang von Naturbildern und gesellschaftlichen Verhältnissen* (S. 29–44). Hamburg University Press.

Eisel, U. (2009). *Landschaft und Gesellschaft. Räumliches Denken im Visier* (Raumproduktionen: Theorie und gesellschaftliche Praxis, Bd. 5). Westfälisches Dampfboot.

Elias, F., Franz, A., Murmann, H., & Weiser, U. W. (Hrsg.). (2014). *Praxeologie. Beiträge zur interdisziplinären Reichweite praxistheoretischer Ansätze in den Geistes- und Sozialwissenschaften* (Materiale Textkulturen, Bd. 3). de Gruyter.

Eppler, E. (1975). *Ende oder Wende. Von der Machbarkeit des Notwendigen*. Kohlhammer.

Esser, B. (1998). Das Selbstverständnis einer Nation. Von der ersten Teilung bis zum Ende des Sozialismus. *Geographische Rundschau, 50*(1), 12–16.

Euchner, W., Stegmann, F. J., Langhorst, T., Jähnichen, T., & Friedrich, N. (2015). *Geschichte der sozialen Ideen in Deutschland: Sozialismus – Katholische Soziallehre – Protestantische Sozialethik. Ein Handbuch*. Springer.

Fainstein, S. S. (2010). *The just city*. Cornell University Press.

Fierla, I. (1999). *Repetytorium z geografii gospodarczej*. Polskie Wydawnictwo Ekonomiczne.

Finke, P. (2014). *Citizen Science. Das unterschätzte Wissen der Laien*. Oekom.

Fischer, R. (2005). Regulierter Rinderwahnsinn. Die Reform der wissenschaftlichen Politikberatung innerhalb der Europäischen Union. In A. Bogner & H. Torgersen (Hrsg.), *Wozu Experten? Ambivalenzen der Beziehung von Wissenschaft und Politik* (S. 109–130). VS Verlag.

Fleck, L. (1980 [1935]). *Entstehung und Entwicklung einer wissenschaftlichen Tatsache. Einführung in die Lehre vom Denkstil und Denkkollektiv* (Wissenschaftsforschung). Suhrkamp (Mit einer Einleitung herausgegebn von Lothar Schäfer und Thomas Schnelle).

Fleck, L. (1983). *Erfahrung und Tatsache. Gesammelte Aufsätze* (Suhrkamp-Taschenbuch Wissenschaft, Bd. 404). Suhrkamp (Mit einer Einleitung herausgegeben von Lothar Schäfer und Thomas Schnelle).

Flick, U. (2011). *Triangulation*. Springer Fachmedien.

Foucault, M. (1977). *Überwachen und Strafen. Die Geburt des Gefängnisses* (Suhrkamp-Taschenbuch Wissenschaft, Bd. 184). Suhrkamp.

Foucault, M. (1983 [1976]). *Der Wille zum Wissen. Sexualität und Wahrheit* (Suhrkamp-Taschenbuch Wissenschaft). Suhrkamp taschenbuch wissenschaft.

Foucault, M. (1996). *Diskurs und Wahrheit. Berkeley-Vorlesungen 1983*. Merve.

Foucault, M. (2007 [frz. Original 1971]). *Die Ordnung des Diskurses. Mit einem Essay von Ralf Konersmann*. Fischer Taschenbuch.

Foucault, M. (2019 [frz. Original 1975]). *Überwachen und Strafen. Die Geburt des Gefängnisses*. Suhrkamp.

Fuchs-Heinritz, W., & König, A. (2005). *Pierre Bourdieu. Eine Einführung*. UVK-Verlagsgesellschaft.

Funtowicz, S. O., & Ravetz, J. R. (1990). *Uncertainty and quality in science for policy*. Kluwer Academic Publishers.

Gibbons, M., Limoges, C., Nowotny, H., Schwartzmann, S., Scott, P., & Trow, M. (1994). *The new production of knowledge. The dynamics of science and research in contemporary societies*. Sage Publications.

Giroux, H. A. (2015). Public intellectuals against the Neoliberal University. In N. K. Denzin & M. D. Giardina (Hrsg.), *Qualitative inquiry-past, present, and future. A critical reader* (S. 194–221). Left Coast Press.

Greiffenhagen, M. (1971). *Das Dilemma des Konservatismus in Deutschland*. Piper.

Han, B.-C. (2005). *Was ist Macht?* Reclam.

Harvey, D. (2005). *A brief history of neoliberalism*. Oxford University Press.

Harvey, D. (2008). The right to the city. *New Left Review, 27*(53), 23–40.

Harvey, D. (2009). *Cosmopolitanism and the geographies of freedom* (The Wellek library lectures). Columbia University Press.

Hasse, J. (2012). *Atmosphären der Stadt. Aufgespürte Räume*. Jovis.

Häußling, R. (2018). Institution. In J. Kopp & A. Steinbach (Hrsg.), *Grundbegriffe der Soziologie* (12. Aufl., S. 191–193). Springer Fachmedien.

Herzog, L. (2013). *Freiheit gehört nicht nur den Reichen. Plädoyer für einen zeitgemäßen Liberalismus* (C. H. Beck Paperback, Bd. 6127). Beck.

Hillebrandt, F. (2000). Disziplinargesellschaft. In G. Kneer, A. Nassehi & M. Schroer (Hrsg.), *Soziologische Gesellschaftsbegriffe. Konzepte moderner Zeitdiagnosen* (UTB für Wissenschaft Uni-Taschenbücher Soziologie, 2. Aufl., Bd. 1961, S. 101–126). Fink.

Hofmann, P., & Hirschauer, S. (2012). Die konstruktivistische Wende. In S. Maasen, M. Kaiser, M. Reinhart & B. Sutter (Hrsg.), *Handbuch Wissenschaftssoziologie* (S. 85–99). Springer.

Holzinger, M. (2004). *Natur als sozialer Akteur. Realismus und Konstruktivismus in der Wissenschafts- und Gesellschaftstheorie* (Forschung Soziologie, Bd. 197). VS Verlag.

Hradil, S. (1995). Schicht, Schichtung und Mobilität. In H. Korte & B. Schäfers (Hrsg.), *Einführung in Hauptbegriffe der Soziologie* (6. Aufl., S. 145–164). VS Verlag.

Hradil, S. (2018). Milieu, soziales. In J. Kopp & A. Steinbach (Hrsg.), *Grundbegriffe der Soziologie* (12. Aufl., S. 319–322). Springer Fachmedien.

Hunt, R. (2016). *Huts, bothies and buildings out-of-doors: An exploration of the practice, heritage and culture of ,out-dwellings'in rural Scotland*. PhD thesis, University of Glasgow.

Ipsen, D. (1992). Stadt und Land – Metamorphosen einer Beziehung. In H. Häußermann, D. Ipsen, R. Krämer-Badoni, D. Läpple, M. Rodenstein & W. Siebel (Hrsg.), *Stadt und Raum. Soziologische Analysen* (2. Aufl., S. 117–156). Centaurus.

Jaehne, G. (1972). *Landwirtschaft und Landwirtschaftliche Zusammenarbeit im Rat für gegenseitige Wirtschaftshilfe Comecon*. Duncker & Humblot.

Jenal, C. (2018). Ikonologie des Protests – Der Stromnetzausbau im Darstellungsmodus seiner Kritiker(innen). In O. Kühne & F. Weber (Hrsg.), *Bausteine der Energiewende* (S. 469–487). Springer VS.

Jerusalem, W. (1925 [1909]). Soziologie des Erkennens. In W. Jerusalem (Hrsg.), *Gedanken und Denker. Gesammelte Aufsätze* (o.S.). Wien: Wilhelm Braumüller.

Kaiser, M., & Maasen, S. (2010). Wissenschaftssoziologie. In G. Kneer & M. Schroer (Hrsg.), *Handbuch Spezielle Soziologien* (S. 685–705). VS Springer.

Kammler, C. (2014). Einführung: Konzeptualisierungen der Werke Foucaults. In C. Kammler, R. Parr, U. J. Schneider & E. Reinhardt-Becker (Hrsg.), *Foucault-Handbuch. Leben – Werk – Wirkung* (S. 9–11). Metzler.

Kirchhoff, T. (2019a). Ökosystemdienstleistungen. In O. Kühne, F. Weber, K. Berr & C. Jenal (Hrsg.), *Handbuch Landschaft* (S. 807–822). Springer VS.

Kirchhoff, T. (2019b). Politische Weltanschauungen und Landschaft. In O. Kühne, F. Weber, K. Berr & C. Jenal (Hrsg.), *Handbuch Landschaft* (S. 383–396). Springer VS.

Kirchhoff, T., & Trepl, L. (2009a). Landschaft, Wildnis, Ökosystem: zur kulturbedingten Vieldeutigkeit ästhetischer, moralischer und theoretischer Naturauffassungen. Einleitender Überblick. In T. Kirchhoff & L. Trepl (Hrsg.), *Vieldeutige Natur. Landschaft, Wildnis und Ökosystem als kulturgeschichtliche Phänomene* (Sozialtheorie, S. 13–68). transcript.

Kirchhoff, T., & Trepl, L. (Hrsg.). (2009b). *Vieldeutige Natur. Landschaft, Wildnis und Ökosystem als kulturgeschichtliche Phänomene* (Sozialtheorie). transcript.

Kleiner, M. S. (Hrsg.). (2001). *Michel Foucault. Eine Einführung in sein Denken.* Campus.

Kloock, D., & Spahr, A. (2007 [1986]). *Medientheorien. Eine Einführung* (UTB). Fink.

Kneer, G. (2009). Jenseits von Realismus und Antirealismus. Eine Verteidigung des Sozialkonstruktivismus gegenüber seinen postkonstruktivistischen Kritikern. *Zeitschrift für Soziologie, 38*(1), 5–25. https://doi.org/10.1515/zfsoz-2009-0101.

Knoblauch, H. (2006). *Wissenssoziologie.* UVK-Verlagsgesellschaft/UTB.

Knoblauch, H. (2013). Wissenssoziologie, Wissensgesellschaft und die Transformation der Wissenskommunikation. *Aus Politik und Zeitgeschichte,63*(18–20), 9–16.

Knoll, J. H. (1981). Liberalismus. In J. H. Schoeps, J. H. Knoll & C.-E. Bärsch (Hrsg.), *Konservativismus, Liberalismus, Sozialismus. Einführung, Texte, Bibliographien* (Uni-Taschenbücher Politologie, Neuere Geschichte, Soziologie, Bd. 1032, S. 87–139). Fink.

Knorr, K. (1980). Die Fabrikation von Wissen. Versuch zu einem gesellschaftlich relativierten Wissensbegriff. In N. Stehr & V. Meja (Hrsg.), *Wissenssoziologie. Kölner Zeitschrift für Soziologie und Sozialpsychologie* (Bd. 22, S. 226–245). Westdeutscher. [Themenheft].

Knorr-Cetina, K. (2002a). *Die Fabrikation von Erkenntnis. Zur Anthropologie von Wissenschaft.* Suhrkamp.

Knorr-Cetina, K. (2002b). *Wissenskulturen. Ein Vergleich naturwissenschaftlicher Wissensformen* (Suhrkamp-Taschenbuch Wissenschaft, Bd. 1594, Dt. Erstausg, 1. Aufl.). Suhrkamp.

Kötzle, M. (1999). Eigenart und Eigentum. Zur Genese und Struktur konservativer und liberaler Weltbilder. In S. Körner, T. Heger, A. Nagel & U. Eisel (Hrsg.), *Naturbilder in Naturschutz und Ökologie* (Landschaftsentwicklung und Umweltforschung, Bd. 111, S. 19–36). TU Berlin.

Kraemer, K., & Bittlingmayer, U. H. (2001). Soziale Polarisierung durch Wissen. In P. A. Berger & D. Konietzka (Hrsg.), *Die Erwerbsgesellschaft. Neue Ungleichheiten und Unsicherheiten* (S. 313–329). VS Verlag.

Kronauer, M., & Siebel, W. (Hrsg.). (2013). *Polarisierte Städte. Soziale Ungleichheit als Herausforderung für die Stadtpolitik.* Campus.

Kühne, O. (2008). *Distinktion – Macht – Landschaft. Zur sozialen Definition von Landschaft.* VS Verlag.

Kühne, O. (2011). Die Konstruktion von Landschaft aus Perspektive des politischen Liberalismus. Zusammenhänge zwischen politischen Theorien und Umgang mit Landschaft. *Naturschutz und Landschaftsplanung, 43*(6), 171–176.

Kühne, O. (2015). Weltanschauungen in regionalentwickelndem Handeln – die Beispiele liberaler und konservativer Ideensysteme. In O. Kühne & F. Weber (Hrsg.), *Bausteine der Regionalentwicklung* (S. 55–69). Springer VS.

Kühne, O. (2017). *Zur Aktualität von Ralf Dahrendorf. Einführung in sein Werk* (Aktuelle und klassische Sozial- und Kulturwissenschaftlerlinnen). Springer VS.

Kühne, O., & Duttmann, R. (2019). Recent challenges of the ecosystems services approach from an interdisciplinary point of view. *Raumforschung und Raumordnung Spatial Research and Planning.* https://doi.org/10.2478/rara-2019-0055.

Kühne, O., Weber, F., & Berr, K. (2019). The productive potential and limits of landscape conflicts in light of Ralf Dahrendorf's conflict theory. *Società Mutamento Politica, 10* (19), 77–90. https://oajournals.fupress.net/index.php/smp/article/view/10597. Zugegriffen: 22. Juni 2020.

Kühne, O., Berr, K., Schuster, K., & Jenal, C. (2021). *Freiheit und Landschaft. Auf der Suche nach Lebenschancen mit Ralf Dahrendorf.* Wiesbaden, in Vorbereitung: Springer.

Laclau, E., & Mouffe, C. (1985). *Hegemony and socialist strategy. Towards a radical democratic politics.* Verso.

Latour, B. (2002 [1999]). *Die Hoffnung der Pandora. Untersuchungen zur Wirklichkeit der Wissenschaft.* Suhrkamp.

Latour, B., & Woolgar, S. (2013 [1979]). *Laboratory Life: The construction of scientific facts.* Princeton University Press.

Leibenath, M. (2017). Ecosystem services and neoliberal governmentality – German style. *Land Use Policy, 64,* 307–316.

Leibenath, M., & Otto, A. (2012). Diskursive Konstituierung von Kulturlandschaft am Beispiel politischer Windenergiediskurse in Deutschland. *Raumforschung und Raumordnung, 70*(2), 119–131. https://doi.org/10.1007/s13147-012-0148-0.

Lenk, K. (1989). *Deutscher Konservatismus.* Campus.

Leonhard, J. (2001). *Liberalismus. Zur historischen Semantik eines europäischen Deutungsmusters* (Veröffentlichungen des Deutschen Historischen Instituts London/ Publications of the German Historical Institute London). R. Oldenbourg.

Levidow, L. (2005). Expert-based policy or policy-based expertise? Regulating GM crops in Europe. In A. Bogner & H. Torgersen (Hrsg.), *Wozu Experten? Ambivalenzen der Beziehung von Wissenschaft und Politik* (S. 86–108). VS Verlag.

Löw, M. (2008). Wenn Sex zum Image wird. Über die Leistungsfähigkeit vergeschlechtlichter Großstadtbilder. In D. Schott & M. Toyka-Seid (Hrsg.), *Die europäische Stadt und ihre Umwelt* (S. 193–206). Wissenschaftliche Buchgesellschaft.

Luhmann, N. (1984). *Soziale Systeme. Grundriß einer allgemeinen Theorie.* Suhrkamp.

Luhmann, N. (1986). *Ökologische Kommunikation. Kann die moderne Gesellschaft sich auf ökologische Gefährdungen einstellen?* Westdeutscher.

Luhmann, N. (1997). *Gesellschaft der Gesellschaft.* Suhrkamp.

Luhmann, N. (2002 [1990]). *Die Wissenschaft der Gesellschaft* (4. Aufl.). Suhrkamp.

Maasen, S. (2015). *Wissenssoziologie* (2., komplett überarb. Aufl.). transcript.

Mannheim, K. (1922). *Die Strukturanalyse der Erkenntnistheorie* (Kant-Studien Ergänzungshefte, Bd. 57). Reuter & Reichard.

Mannheim, K. (1931). Wissenssoziologie. In A. Vierkandt (Hrsg.), *Handwörterbuch der Soziologie* (S. 659–680). Enke.

Mannheim, K. (1952). *Ideologie und Utopie.* Schulte-Bulmke.

Merton, R. K. (1973). *The sociology of science. Theoretical and empirical investigations.* Chicago: University of Chicago Press.

Merton, R. K. (1987). The fragments from a sociologist's notebooks: Establishing the phenomenon, specified ignorance, and strategic research materials. *Annual Review of Sociology, 13,* 1–29.

von Mises, L. (1927). *Liberalismus.* Verlag von Gustav Fischer.

Misik, R. (2012). *Halbe Freiheit. Warum Freiheit und Gleichheit zusammengehören* (edition suhrkamp digital). Suhrkamp.

Mittelstraß, J. (2004). Theoria. In J. Mittelstraß (Hrsg.), *Enzyklopädie Philosophie und Wissenschaftstheorie* (Bd. 4, S. 259–260). Metzler.

Moore, K., Kleinman, D. L., Hess, D., & Frickel, S. (2011). Science and neoliberal globalization: A political sociological approach. *Theory and Society, 40*(5), 505–532.

Morrissey, J. (2015). Regimes of performance: Practices of the normalised self in the neoliberal university. *British Journal of Sociology of Education, 36*(4), 614–634.

Muir, R. (1998). Reading the landscape, rejecting the present. *Landscape Research, 23*(1), 71–82. https://doi.org/10.1080/01426399808706526.

Münch, R. (2011). *Akademischer Kapitalismus. Zur politischen Ökonomie der Hochschulreform.* Suhrkamp.

Neidhardt, F. (2010). Selbststeuerung der Wissenschaft. In D. Simon, A. Knie, S. Hornbostel & K. Zimmermann (Hrsg.), *Handbuch Wissenschaftspolitik* (S. 280–292). VS Springer.

Nennen, H.-U., & Garbe, D. (Hrsg.). (1996). *Das Expertendilemma. Zur Rolle wissenschaftlicher Gutachter in der öffentlichen Meinungsbildung.* Springer.

Neun, O. (2017). Zum Verhältnis von Ludwik Flecks und Karl Mannheims Wissenssoziologie. In M. Endreß, K. Lichtblau & S. Moebius (Hrsg.), *Zyklos 3. Jahrbuch für Theorie und Geschichte der Soziologie* (S. 71–89). Springer Fachmedien.

Nowotny, H. (2005). Experten, Expertisen und imaginierte Laien. In A. Bogner & H. Torgersen (Hrsg.), *Wozu Experten? Ambivalenzen der Beziehung von Wissenschaft und Politik* (S. 33–44). VS Verlag.

Nowotny, H., Scott, P., & Gibbons, M. (2001). *Re-thinking science. Knowledge and the public in an age of uncertainty.* Polity.

Ottmann, H. (2008). *Geschichte des politischen Denkens* (Neuzeit, Bd. 3). Metzler (Teilband 3: Die politischen Strömungen im 19. Jahrhundert).

Pennington, M. (2002). A Hayekian liberal critique of collaborative planning. In M. Tewdwr-Jones & P. Allmendinger (Hrsg.), *Planning futures. New directions for planning theory* (S. 187–205). Routledge.

Popper, K. R. (1996). *Alles Leben ist Problemlösen. Über Erkenntnis, Geschichte und Politik.* Piper.

Popper, K. R. (2003 [1945]). *Die offene Gesellschaft und ihre Feinde* (Der Zauber Platons, Bd. 1, 8. Aufl.). Mohr Siebeck.

Popper, K. R. (2011 [1947]). *The open society and its enemies.* Routledge.

Popper, K. R. (2018 [1984]). *Alle Menschen sind Philosophen.* Piper (Herausgegeben von Heidi Bohnet und Klaus Stadler).

Popper, K. R. (2019 [1987]). *Auf der Suche nach einer besseren Welt. Vorträge und Aufsätze aus dreißig Jahren.* Piper.

Pregill, P., & Volkman, N. (1999). *Landscapes in history. Design and planning in the eastern and western traditions.* Wiley.

Rademacher, C., Schroer, M., & Wiechens, P. (Hrsg.). (1999). *Spiel ohne Grenzen? Ambivalenzen der Globalisierung.* Westdeutscher.

Reinhart, M. (2012). *Soziologie und Epistemologie des Peer Review* (Schriftenreihe Wissenschafts- und Technikforschung, Bd. 10). Nomos.

Rivera López, E. (1995). *Die moralischen Voraussetzungen des Liberalismus.* Alber.

Rodewald, R. (2001). *Sehnsucht Landschaft. Landschaftsgestaltung unter ästhetischem Gesichtspunkt* (2. Aufl.). Chronos.

Sarasin, P. (2016). *Michel Foucault zur Einführung* (6., erg. Aufl.). Junius.

Schaal, G. S., & Heidenreich, F. (2006). *Einführung in die Politischen Theorien der Moderne* (UTB Politikwissenschaft, Bd. 2791). Budrich.

Scheler, M. (Hrsg.). (1924). *Versuche zu einer Soziologie des Wissens.* Dunker & Humbolt.

Scheler, M. (1926). *Die Wissensformen und die Gesellschaft.* Der Neue-Geist-Verlag.

Schoeps, J. H. (1981). Konservativismus. In J. H. Schoeps, J. H. Knoll & C.-E. Bärsch (Hrsg.), *Konservativismus, Liberalismus, Sozialismus. Einführung, Texte, Bibliographien* (Uni-Taschenbücher Politologie, Neuere Geschichte, Soziologie, Bd. 1032, S. 11–86). Fink.

Schütz, A. (1960 [1932]). *Der sinnhafte Aufbau der sozialen Welt. Eine Einleitung in die Verstehende Soziologie* (2. Aufl). Julius Springer. (Originalarbeit erschienen 1932).

Schütz, A. (1971a [1962]). *Gesammelte Aufsätze 1. Das Problem der Wirklichkeit.* Martinus Nijhoff.

Schütz, A. (1971b). *Gesammelte Aufsätze 3. Studien zur phänomenologischen Philosophie.* Martinus Nijhoff.

Schütz, A., & Luckmann, T. (2003 [1975]). *Strukturen der Lebenswelt.* UTB.

Schützeichel, R. (2012). Wissenssoziologie. In S. Maasen, M. Kaiser, M. Reinhart & B. Sutter (Hrsg.), *Handbuch Wissenschaftssoziologie* (S. 17–26). Springer.

Shell, K. L. (1986). *Der amerikanische Konservatismus.* Kohlhammer.

Silvertown, J. (2009). A new dawn for citizen science. *Trends in Ecology & Evolution, 24*(9), 467–471.

Soja, E. W. (2000). *Postmetropolis. Critical studies of cities and regions*. Blackwell.

Trepl, L. (2012). *Die Idee der Landschaft. Eine Kulturgeschichte von der Aufklärung bis zur Ökologiebewegung*. transcript.

Vicenzotti, V. (2006). Kulturlandschaft und Stadt-Wildnis. In I. Kazal, A. Voigt, A. Weil, & A. Zutz (Hrsg.), *Kulturen der Landschaft. Ideen von Kulturlandschaft zwischen Tradition und Modernisierung* (Landschaftsentwicklung und Umweltforschung, Bd. 127, S. 221–236). Technische Universität Berlin.

Vicenzotti, V. (2011). *Der ‚Zwischenstadt'-Diskurs. Eine Analyse zwischen Wildnis, Kulturlandschaft und Stadt*. transcript.

Vicenzotti, V. (2012). Gestalterische Zugänge zum suburbanen Raum – Eine Typisierung. In W. Schenk, M. Kühn, M. Leibenath & S. Tzschaschel (Hrsg.), *Suburbane Räume als Kulturlandschaften* (Forschungs- und Sitzungsberichte, Bd. 236, S. 252–275). Selbstverlag.

Viehöver, W. (2005). Der Experte als Platzhalter und Interpret moderner Mythen. Das Beispiel der Stammzellendebatte. In A. Bogner & H. Torgersen (Hrsg.), *Wozu Experten? Ambivalenzen der Beziehung von Wissenschaft und Politik* (S. 149–171). VS Verlag.

Voigt, A. (2009a). ‚Wie sie ein Ganzes bilden'– analoge Deutungsmuster in ökologischen Theorien und politischen Philosophien der Vergesellschaftung. In T. Kirchhoff & L. Trepl (Hrsg.), *Vieldeutige Natur. Landschaft, Wildnis und Ökosystem als kulturgeschichtliche Phänomene* (Sozialtheorie, S. 331–348). transcript.

Voigt, A. (2009b). *Die Konstruktion der Natur. Ökologische Theorien und politische Philosophien der Vergesellschaftung* (Sozialgeographische Bibliothek, Bd. 12). Steiner.

Wayand, G. (1998). Pierre Bourdieu: Das Schweigen der Doxa aufbrechen. In P. Imbusch (Hrsg.), *Macht und Herrschaft. Sozialwissenschaftliche Konzeptionen und Theorien* (S. 221–237). VS Verlag.

Weber, F. (2018). *Konflikte um die Energiewende. Vom Diskurs zur Praxis*. Springer VS.

Weber, F., Jenal, C., Roßmeier, A., & Kühne, O. (2017). Conflicts around Germany's *Energiewende*: Discourse patterns of citizens' initiatives. *Quaestiones Geographicae, 36*(4), 117–130. https://doi.org/10.1515/quageo-2017-0040.

Weber, M. (1976 [1922]). *Wirtschaft und Gesellschaft. Grundriß der verstehenden Soziologie*. Mohr Siebeck.

Weingart, P. (2001). *Die Stunde der Wahrheit? Zum Verhältnis der Wissenschaft zu Politik, Wirtschaft und Medien in der Wissensgesellschaft*. Velbrück Wissenschaft.

Weingart, P. (2003). *Wissenschaftssoziologie* (Einsichten). transcript.

Weingart, P. (2005). *Die Wissenschaft der Öffentlichkeit. Essays zum Verhältnis von Wissenschaft, Medien und Öffentlichkeit*. Velbrück Wissenschaft.

Weingart, P. (2010). Wissenschaftssoziologie. In D. Simon, A. Knie, S. Hornbostel & K. Zimmermann (Hrsg.), *Handbuch Wissenschaftspolitik* (S. 118–129). VS Springer.

Weingart, P. (2015). *Wissenschaftssoziologie*. transcript.

Weingart, P., & Winterhager, M. (1984). *Die Vermessung der Forschung. Theorie und Praxis der Wissenschaftsindikatoren*. Campus.

Weingart, P., Engels, A., & Pansegrau, P. (2008). *Von der Hypothese zur Katastrophe. Der anthropogene Klimawandel im Diskurs zwischen Wissenschaft, Politik und Massenmedien* (2., leicht veränd. Aufl.). Budrich.

Werlen, B. (2000). *Sozialgeographie. Eine Einführung*. Haupt.

Conceptions of Space and Theories of Spaces

6

When we talk about "space" or "landscape", we often mean certain object constellations, but "space" (or "landscape") are also complex concepts, we have learned what we can describe as "space" (or "landscape") and what not. Before we deal with different concepts of space, however, we want to introduce an analytical categorization with which we can explain the different theories—and not least compare them. This categorization is based on the considerations of a thinker whose ideas on the theory of science we have presented in detail (see in particular Sects. 3.6 and 4.1): Karl Popper. Here we now use his three-world theory (Textbox 25).

Textbox 25: Karl Popper's Three-World Theory as a Reference Framework for Space Research

To illustrate the different dimensions of 'space', we use Karl Popper's Three Worlds Theory (Niemann, 2019; Popper, 1973, 1979, 2019 [1987]) (see also: Weichhart, 1993, 1999; Kühne, 2020). This approach provides access to spatial concepts and ideas without the need for them to be already known, as is the case with other space concepts (such as Weichhart's, 1999, 2018a, b or Hard's, 2003) (for example, Weichhart's space$_3$ as a construction of immaterial relations and relationships). The access to space used here is therefore an analytical tool for understanding space concepts, not a categorization of space concepts (as is the case with Weichhart or Hard and Bartels).

Karl Popper refers to 'World 1' as living and non-living bodies, to 'World 2' as conscious content, that is, individual thoughts and feelings, like "maybe also the subconscious experiences" (Popper, 2018 [1984], p. 82). He refers to 'World 3' as "all planned or desired *products* of human mental activity" (Popper, 2019 [1987], p. 17; emphasis in original), such as mathematical statements, but also largely shared understandings of space, landscape, theory, and of

© The Author(s), under exclusive license to Springer Fachmedien Wiesbaden GmbH, part of Springer Nature 2022
O. Kühne and K. Berr, *Science, Space, Society*,
https://doi.org/10.1007/978-3-658-39140-9_6

course the Three Worlds Theory itself. For Popper, 'reality' therefore consists of three worlds that are interrelated. However, certain 'things' can be assigned to several worlds, for example a building is both part of 'World 1' (as a material object) and 'World 3' (as a carrier of cultural and social meanings). The Three Worlds Theory can be seen as Popper's attempt to develop an alternative to three widely held understandings of the world (Popper, 2019 [1987]):

1. To the materialistic world view, which only recognized the 'world 1' as 'real',
2. to the immaterialistic world view, which explained the 'world 2' alone as 'real',
3. to the dualistic world view, which held the 'worlds 1 and 2' for 'real'.

He supplemented the two worlds 1 and 2 with world 3, "that is, the world that anthropologists call 'culture'" (Popper, 2019 [1987], p. 18).

Poppers 3-world theory was criticized in particular with regard to the thesis of the independence or autonomy of "world 3" and the bridging function of "world 2" between worlds 1 and 3. The argument against the independence of "world 3" is: "Nothing is more dependent and vulnerable, less autonomous than the 'world' of culture" (Bunge, 1984, p. 215). The contents of "world 3"—so the objection—must be either materialized ("world 1") or thought ("world 2"), otherwise they would not exist. That "world 2" is the indispensable bridge between worlds 1 and 3 leads to the so-called "body-soul problem" or "brain-mind problem", the central question of which is how mental ("soul", "res cogitans", "world 2") can influence physical ("body", "res extensa", "world 1") and how both worlds can interact. Popper doubts that this question will ever be clarified and that the ideal of a complete understanding may have to be abandoned (Popper & Eccles, 1977, Chap. P2, Sect. 10). This question thus leads to an aporia.

The approach does not mean—as shown by the example of the building—a strict separation of the three worlds. Rather, it offers an analytical framework for dealing with hybridities and interactions, making it suitable for the research of spaces in their relations to society, individuals and materialities (Kühne, 2018b; Weichhart, 1999). World 2 is connected to world 3 in that the individual is introduced into the social stock of knowledge, interpretation and evaluation patterns (socialization), it also has the opportunity to anchor new knowledge, interpretation and evaluation patterns in society (innovation; Fig. 6.1). With world 1, the world is connected by its observation, but it also receives a structuring framework for its activities (after all, consciousness cannot detach itself from its corporeality; in more detail: Kühne, 2020 and below in space).

In terms of the consideration of space, living and non-living bodies—in spatial arrangement—form the physical substrate of "Space 1" as part of "World 1". "Space 2" accordingly comprises the individual concepts, sensations, emotions, norm concepts, experiences, etc. of and relating to space (as part of "World 2"). "Space 3" refers to the social constructions of space and the social conventions relating to patterns of interpretation and evaluation of space (Fig. 6.2).

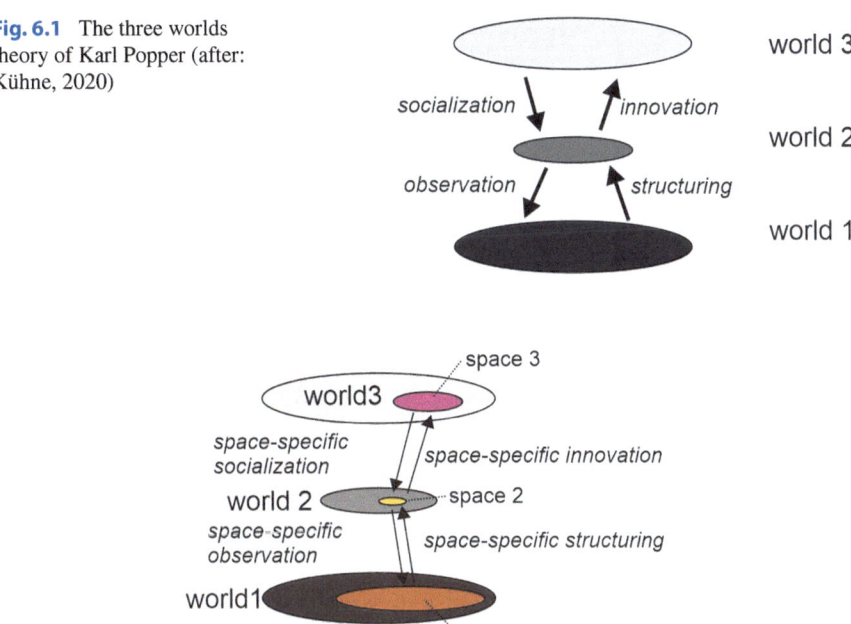

Fig. 6.1 The three worlds theory of Karl Popper (after: Kühne, 2020)

Fig. 6.2 The derivation of the three spaces from Popper's three worlds. Spaces each represented a subset of the respective world on the different levels (It could be argued that matter is always spatially arranged, inasmuch as World 1 and Space 1 are identical. However, this approach is strongly dependent on the space understanding represented here (container space understanding; Sect. 6.1). Other approaches (in particular constructivist; see Sect. 6.5) assume that spaces arise through the observation of objects, this view is visualized here.) (own representation)

The influence of the three spaces always takes place via 'space 2'. For example, from 'space 3' to 'space 2', because our individual ideas and feelings of and about 'space' are only partially based on our own immediate experience of 'space 1'. Rather, we learn which social ideas, interpretations and evaluations of space, i.e. of 'space 3', exist and how we use them (see, inter alia, Greider & Garkovich, 1994; Kühne, 2019c; Nissen, 1998). 'Space 3' acts on 'space 1' via the mediation of 'space 2': people manifest their (mostly socially mediated) ideas materially, for example by building houses, as 'one' (space 3) builds houses. The mediation of 'space 2' in the relationship between 'space 1' and 'space 3' is necessary because only the individual can fall back on a body (which in turn is also part of the 'world 1'). In the simplest case, 'space 2' acts almost as a transmission disk of the social ideas, interpretations and evaluations of 'space 3', on 'space 1'. But the reflective human being is able to formulate alternative interpretations, evaluations and ideas and possibly also to anchor them in 'space 3', i.e. to change them (see, in general: Dahrendorf, 1979, 2007).

Fig. 6.3 Basis for the visualizations of the different theories carried out in the following (own representation)

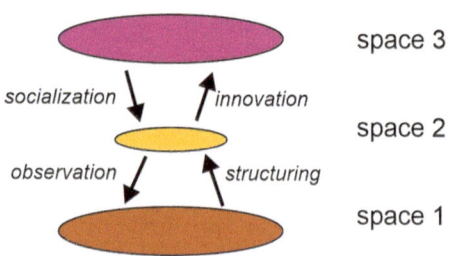

To illustrate which of the three spatial levels and which relations the respective theories focus on, a graphical representation is made on the basis of Fig. 6.3. For the sake of greater clarity, we have omitted to show that the respective spatial levels are parts of the respective worlds, as well as omitting to mention that the relations are related to spaces (see Fig. 6.2). In the following, the levels and relations that are not in the focus of the respective theory are shown to be strongly transparent, while those that have a certain meaning are shown to be weakly transparent. ◀

In everyday language, 'space' is understood as something self-evident and object-given, it is quasi "an attribute of material nature" (Läpple, 1991, p. 36), which surrounds us, in which we move, which we can also change. 'Space' is therefore clearly assigned to 'World 1' in Popper's sense (see Textbox 21). But 'space' is ultimately the result of an individual (World 2) and social (World 3) abstraction. If humans develop their ideas of space in relation to themselves, so-called 'egocentric spaces' (Vorwerg, 2013; see also Smith, 1984; Fig. 6.4), this is expressed in old units of measure such as ell, foot, acre and daywork. The etymological origin of the word 'space' also makes this connection clear: 'space' originally meant clearing, which is cut into the wilderness for the purpose of making it arable (Bollnow, 1963; Läpple, 1991). The resulting idea of objects to each other is more strongly influenced by places instead of spaces (see Textbox 26). In the following, different abstract ideas of space and spaces will now be discussed, as they are discussed today in geography, philosophy, cultural studies and sociology, but also partly (without discussion) form the basis of space science work. The abstract dealing with spaces has intensified in the past three decades, so the understanding

Fig. 6.4 Visualization of the space focusing 'egocentric spaces', spaces 1 and 3 form only a frame for the individual construction of spaces (own representation)

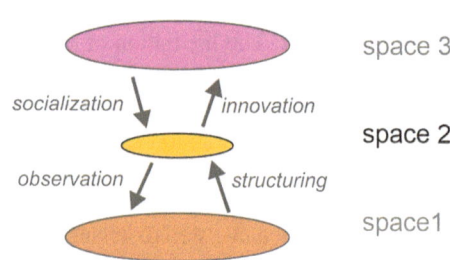

of spaces and their genesis from different scientific perspectives has been illuminated, for example from humanities (Dünne & Günzel, 2006; Rau, 2017), from geographical (Egner, 2010; Escher & Petermann, 2016; Fliedner, 2015; Weichhart, 2018a, b) or sociological (Dangschat, 2014; Läpple, 1992; Löw, 2001; Löw et al., 2008) perspective.

Textbox 26: Place and Space in Medieval Europe

In medieval Europe, the idea of a space independent of human consciousness did not dominate yet. Rather, the idea of a hierarchical and "perfect and strictly structured" (Bojadžiev, 2003, p. 74) ensemble of different places (Foucault, 1991, p. 66) prevailed: "sacred places and profane places; protected places and open, defenseless places; urban and rural places: for the life of humans". The places on earth were opposed to the heavenly and super-heavenly places. In contrast to the idea of a 'space', the reference to places therefore dominated in the Middle Ages, as is also shown by the low importance of territorial government organization in the Middle Ages, for example, places of medieval rule were palaces, cities and castles (cf. Schneider, 2004). Foucault (1991) therefore calls the medieval space a "space of localization". ◄

The following space-theoretical concepts differ in their focus on the different spaces (1, 2 and 3) or interpret the relations between them differently. We begin with understandings that primarily focus on one of the three spaces, then move on to approaches that focus on multiple spaces, then those that focus on the relationships between spaces, and finally those that advocate (partial) redefinition of the spaces. Essential terms for this chapter are introduced in Textbox 27.

Textbox 27: Essential Terms for Chap. 6

Exploratory research: This is designed to research objects that are not yet well known. So it's primarily about generating hypotheses, not testing them. In particular, qualitative methods of empirical social research are used (Flick, 2007; Lamnek, 2010).

Inclusive and exclusive thinking: While exclusive thinking is based on producing unambiguities, inclusive thinking understands ambiguities as a resource. In spatial planning, exclusive thinking is manifested, for example, in the strict spatial separation of functions, such as living and working, while inclusive planning is concerned with integrating different functions (Kühne, 2006; Sloterdijk, 1987).

Corporatism: In the original sense, corporatism (Latin: *corporatio*, corporation) means the connection of people of the same profession or status (such as guilds). In political science, 'corporatism' refers to the incorporation of organized interests into politics as well as "their participation in the formulation and execution of political decisions" (Voelzkow, 2000, p. 185).

Reification (or reification): 'Reification' means 'objectification'. In philosophy, this means that an abstraction is treated as if it were a concrete object. In geography (and social sciences), it is understood that social phenomena are materially inscribed, for example in physical space (Bourdieu, 1991; Löw, 2001; Meyer & Miggelbrink, 2018).

6.1 'Space 1' Understandings: Container or Container Space, the Relational Space and the Space-Time Continuum

With the understanding of the container or container space, an abstraction was made with respect to ego-centric space understandings. Space is understood as a three-dimensional limited area, "a kind of container, into which one can put something and [which] is equipped with objects" (Egner, 2010, p. 98) and in which science investigates structures and processes (Herrmann, 2010). The idea of the container space is based on the concepts of classical physics, which goes back to the Greek antiquity in its basic principles, but was finally conceived in Europe between the 13th and 17th centuries. It represents a simplification of Newton's space understanding, because Newton had conceived the absolute space as an infinite space, not as a closed container (e.g. Sturm, 2000; Schroer, 2006). Space is described as a reality independent of consciousness, the understanding of living in a container space is conveyed in socialization and educational processes and "is undoubtedly a culturally necessary achievement for the constitution of many spaces in order to be able to classify objects, oneself or other people in a grid. This ordering activity is based and strengthens the idea of 'living in space'" (Löw, 2001, p. 63). Space becomes a clearly definable "and more determinable section of the visible, material earth surface" (Egner, 2010, p. 98). Even today, the container space concept based on 'Space 1' is dominant in many space sciences, for example in natural space research, when, for example, geomorphological processes are investigated 'in space', but also in classical settlement and (cultural) landscape research, when it is assumed that certain settlement forms 'spread in space' or that 'cultural landscapes', understood in a positivist sense, could be delimited as 'real units' in that certain objects with distinguishable features would occur in a certain spatial extent, such as house, settlement or field forms. The concept of container space or container space has a special meaning in spatial planning, here 'a space' (often defined by administrative boundaries) is planned by determining which objects are allowed to be placed here in which arrangement and which are not.

The idea of the relational order space can be seen as the reverse of the logic of the container space understanding: Objects are not in a space, they rather form spaces through their arrangement. Space is thus not an object, but a property. The consequence of this: "Without things there is no space" (Weichhart, 2008, p. 79; similarly also: Läpple, 1992; Egner, 2010; Ipsen, 2002). The focus of this

Fig. 6.5 Visualization of the space focusing in the container space concept and in the relational space, spaces 2 and 3 have no explicit meaning (own representation)

understanding is—as with the container space understanding—on the dimension of 'World 1', since here too the world of material objects is focused. A relational understanding of space is then used in the space sciences when relationships between objects come into focus, for example in the planning assignment of centrality or in the relational economic geography when activities and relationships (for example in the form of networks or the integration into company alliances) come into focus (for example at: Bathelt & Glückler, 2003; Boggs & Rantisi, 2003).

The ideas of the container space as well as of the relational order space emphasize the static structure of 'Space 1' (Fig. 6.5), the idea of the space-time continuum means a dynamization, by attributing processes a constitutive meaning in the creation of spaces. The basis for this idea of space is Albert Einstein's theory of relativity, which replaces the ideas of absolute space on the one hand and absolute time by the idea of space and time as constitutively interrelated magnitudes. Space is understood as a relational structure between constantly moving bodies (cf. Löw, 2001): "Every change in 'space' is a change in 'time', every change in 'time' is a change in 'space'" (Elias & Schröter, 1994, p. 75). As a constantly revisible arrangement of objects, space is *"dependent on the observer's reference system"* (Löw, 2001, p. 34; Fig. 6.6). This makes it clear: The concept of the space-time continuum expands the focus from 'space 1' to 'space 2'. In the field of space research, the relationality of space and time is focused, for example, where 'the human being' sets himself or herself in his or her physicality in relation to arrangements of other material objects—using time, for example in 'Time Geography' (Hägerstrand, 1970), which examines the relationship between space (overcoming) and time (use) (see also: Kramer, 2012). Evolutionary economic geography

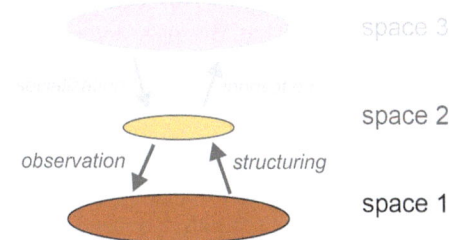

Fig. 6.6 Visualization of the space focusing in the time-space continuum, in which space 2 has a certain meaning. Space 3 has no explicit meaning (own representation)

(Boschma & Frenken, 2006) also focuses on the dynamics of the economy with its spatial feedback loops, "which implies an emphasis on the contingency of spatial development paths" (Schamp, 2012, p. 121).

6.2   The Focus 'Space 2': Space as a Priori in Kant's Understanding of Space

Immanuel Kant takes a different path of understanding space and time (Kant, 1959 [1781]). He assumes that space and time are a priori concepts, which lie at the empirical concepts or precede them. This means that they are sensual conditions of the possibility of experience and knowledge, namely as a priori (experience-preceding) sensual forms of intuition in the subject of knowledge (Kant, 1959 [1781], B 33–73). Kant justified this thesis with two arguments each. The first argument pair is directed against empiricism and its denial of the a priori nature of space and time. The second pair is directed against rationalism and its rejection of the intuitive character of space and time. Central aspects of Kant's understanding of space are (see also: Koriako, 2005):

1. Space (Kant, 1959 [1781], B 38) and time (Kant, 1959 [1781], B 46) are not *empirical* concepts: They cannot originate from experience, since they already underlie every intuition. They are always already assumed when an adjacent, behind or next to each other or a before or after of objects is perceived.
2. Space (Kant, 1959 [1781], B 38–39) and time (Kant, 1959 [1781], B 46), however, are *necessary* concepts that underlie all intuitions *a priori*. This thesis is proven by the fact that space and time can be represented without any object or any appearance, but not that there is no space and no time. So space and time are necessary a priori concepts.
3. Space (Kant, 1959 [1781], B 39) and time (Kant, 1959 [1781], B 47) are also not aprioric *concepts*, because concepts are general ideas that refer to independent examples, but space and time are not general, but something individual. There is only *one* space with its dependent subspaces and only *one* time with its dependent subtimes.
4. Space (Kant, 1959 [1781], B 40) and time (Kant, 1959 [1781], B 47) are rather *intuitions* a priori, because they are presented as infinite given magnitudes. They contain infinitely many ideas *in* themselves. A concept, on the other hand, can only contain *certain* ideas (characteristics) *in* itself and perhaps infinitely many *under* itself. So space and time cannot be concepts, but they are *intuitions a priori*.

As ways of intuition, space and time are thus not contingent properties attached to things, but the fundamental, necessary and inalienable sensual conditions of the possibility to view things in this way and not otherwise. In this understanding, space and time are therefore dependent on the subject of cognition ('space 2') and

Fig. 6.7 Visualisation of Kant's focus on space, in which spaces 1 and 3 play a more supporting role in the emergence of space 2 (own representation)

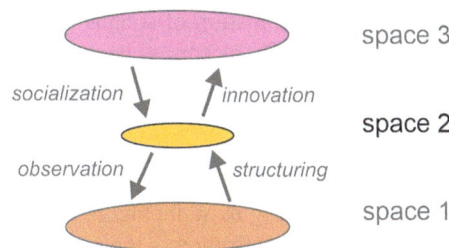

thus, as ways of cognition that precede experience, not an objective reality external to consciousness ('space 1'; Fig. 6.7).

6.3 'Space 3': The Social Production of Space—Spaces in Radical Constructivism and Discourse Theory

Similar to the focus on 'space 2' in Kant, in the social construction of spaces, 'space 1' is not understood as the constitutive level of space. Unlike Kant, in this understanding, it is not individual consciousness, but social ideas of space, that is, 'space 3' that is the constitutive dimension of space. According to this view, spaces are spatialized social structures (Paasi, 1998, 1999), they are results of constitution processes taking place in society, processes that are coupled to communication, particularly language (Jørgensen & Phillips, 2002). The construction processes, as well as their (temporary results) are contingent (Wardenga, 2002), which has the following consequences: First, they are not constantly reproduced in social practice, they lose their existence. Second, because of their contingency, they are always constructible in different ways.

From a radical constructivist perspective, following Niklas Luhmann (see Helmut Klueter, 1986), space is "initially understood as a medium of perception, and increasingly as a medium of communication" (Egner, 2010, p. 99). Space is thus a medium "of the measurement and calculation of objects" (Luhmann, 1995, p. 179; see also Egner, 2008). Space is thereby generated, "that locations can be identified independently of the objects they occupy in each case" (Luhmann, 1995, p. 180; see also Hard, 2002; Lippuner, 2008; Pott, 2007). Luhmann (1995) understands locations as the medium of space, objects as the forms of space. This is how location differences mark the medium, object differences the forms of the medium (Luhmann, 1995). Accordingly, the elements of the medium 'space' are arranged, "which is assumed as a kind of neurological basic structure in perception and communication, but itself only perceptible by means of its forms, that is, by means of object differences" (Lippuner, 2011, p. 323). Space arises through differences (e.g. through distinguishable objects) and through communication. Language is the central medium for the coupling of psychic and social systems (here the approach is extended to "World 2"). Language can be understood "in the sense of a loose coupling of elements as a set of sounds or words […]. Only through the combination of sounds into words and sentences does the medium of language (based on its forms) become recognizable as such and meaningful communication possible" (Lippuner, 2008). "Objects" are—if one follows a Luhmannian concept of space as a medium of communication—understood as semantic units of communication, i.e. they are by no means things given in the external world independent of communication processes (Luhmann, 2001 [1997]; Pott, 2007; Redepenning & Wilhelm, 2014). Places, in turn, can be understood as specific forms that are formed in the medium of space. These are positions that can be distinguished from other positions (Pott, 2007; Fig. 6.8). Since the different social subsystems (politics, economy, science, etc.) are subject to different logics, they also construct space in very different ways (Redepenning & Wilhelm, 2014; see in particular Sect. 6.4).

A somewhat different perspective on World 3 is provided by discourse theory, which focuses less on how space is created in communication processes, but rather on the fact that space construction is not uncontentious, space constructions compete with each other. Discourses as more or less closed systems of interpretation and evaluation compete for hegemony over alternative discourses (among many: Glasze & Mattissek, 2009; Langer & Wrana, 2013; as the basis for this approach: Laclau & Mouffe, 1985). To illustrate this with an example: In the past two decades, the debate about the spatial consequences of the energy transition has increased in importance, with two competing discourses emerging in terms of the landscape consequences, 'energy transition destroys beautiful and familiar landscape' vs. 'energy transition produces landscapes of sustainable development' (among many: Kühne & Weber, 2015; Leibenath & Otto, 2012; Weber, 2017). Another field of research of a discourse-theoretical approach can be found in settlement research, for example the discourses around the 'neoliberal city' (Mattissek, 2008, 2010) or the hybridization of city and country (Weber, 2018). It becomes clear that also from this perspective elements from 'space 1' and 'space 2' are dealt with, but materiality is not understood as an independent reality, but only actualized when it is the subject of communication (Glasze, 2013; Weber, 2019b), individuals are not understood as persons or actors, but rather as 'subject positions' in discourses (Weber, 2019a; Fig. 6.9)

Fig. 6.8 Visualization of the space focus of radical constructivism. Spaces 1 and 2 are merely products of communication (own representation)

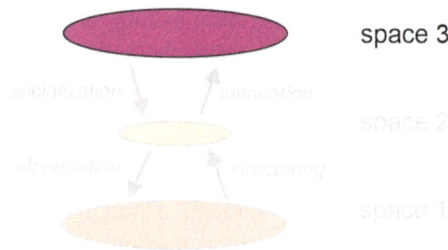

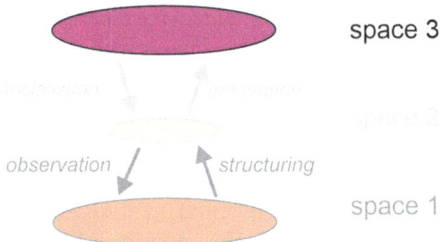

space 3

observation   structuring

space 1

Fig. 6.9 Visualization of the space focusing of discourse theory. Constitutive is—as with radical constructivism space 3, but space 1 has a certain importance as an object of discourse, while space 2 is hardly given any importance (own representation)

6.4   The Dominance of 'Space 3' Over 'Space 1': The Inscription of System Logics and the Domestication of Spaces

A step towards the domestication of space is the subjugation of 'space 1' to a logical structure or order structure. 'Space 1' is laid over as a logical structure, which "an act of ordering" (Egner, 2010, p. 98) called, for example by grids, topographical and thematic maps. The form of 'space 3' stands for "immaterial relations, something thought" (Egner, 2010, p. 98). These types of order of 'space 1' in turn form a essential basis for the intervention of 'space 3' in 'space 1' (by means of 'space 2', which is largely ignored in the accesses to space presented here). For example, the drainage of swamps, the straightening of rivers, the construction of railway lines requires an exact survey of 'space 1' (see Blackbourn, 2007 for details).

In his "Essay on Space", sociologist Dieter Läpple (1991) deals with the specific logics of the enrollments of "Space 3" in "Space 1" based on political scientist Elmar Altvater (1987). He uses a systems theory perspective for this. The starting point is the diagnosis that with the modernization of society, its differentiation into subsystems went hand in hand, such as the subsystems of the economy, society, politics and culture (Parsons, 1991 [1951]). These social subsystems are concerned with the solution of specific social problems. The economic system, for example, regulates scarcity, and the political system the general orientation of society (cf., for example, Parsons, 1951; Luhmann, 1984). Läpple (1991, p. 198) states that these social subsystems "also develop *a specific spatial manifestation each*". Following Altvater (1987), Läpple (1991) refers to these manifestations as "functional spaces", since specific functions are carried out here for society (such as the production of food, the staging of cultural events, the representation of political power). These "functional spaces" have "*different spatial extensions or areas of influence*, and they develop, according to their respective function specialization, also *different space-forming or space-structuring tendencies*" (Läpple, 1991, p. 199). The individual social subsystems usually do not manage to organize sections of "Space 1" (here the container space perspective greets), exclusively according to their own systemic logic (e.g. the economic one), so that "these functional spaces overlap accordingly according to their respective manifestations" (Läpple, 1991, p. 199). For example, the enforcement of economic calculation in agriculture is controlled by agricultural law, in construction by building law. The consequence: In "Space 1" an "*complex and contradictory configuration* of economic, social, cultural and political functional spaces, which, although they each have their specific development dynamics, are nevertheless in a mutual relationship and tension" (Läpple, 1991, p. 199; Fig. 6.10).

The sociologist Rudolf Stichweh (1998, 2003) also deals with the relationship between 'space 3' and 'space 1' from a systems theory perspective. While Läpple focuses more on the basic forms of the inscription of social sub-systems in 'space 1', Stichweh (2003) deals with the strategies of social control of 'space 1' in the course of socio-cultural evolution. His concern is to "understand the structural effects of the operational enactments of a system" (Stichweh, 2003), that is, inscriptions of social sub-systems in space 1, whereby he identifies five strategies (Stichweh, 2003):

1. The substitution of natural by artificial circumstances. Physical structures created by humans, take the place of natural given circumstances, "which do not seem to be influenced by communication and social action" (Stichweh, 2003). In other words: roads replace the individual search for a way through forests, the canalization of rivers makes them (as is usually desired) navigable all year round, residential buildings replace caves etc.
2. The superimposition of physical spaces by social spaces (in the terminology of our volume: superimposition of 'space 1' by 'space 3'). The restrictions associated with the physical presence of humans in 'space 1' are minimized, so mountains in the information age are a much smaller obstacle to the dispatch of messages than at the time of the stagecoach. Physical-spatial copresence is no longer a condition for social proximity (for example in video conferences or in long-distance relationships), which means "that our neighbor can be com-

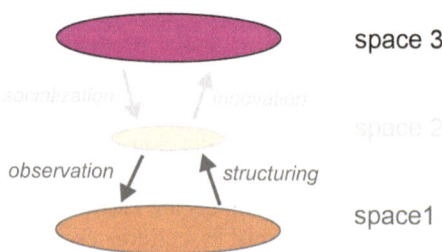

Fig. 6.10 Visualization of the space focus of the theory of the inscription of system logics and the domestication of spaces. Constitutive is the level of space 3, space 1 has a rather passive function, which is adjusted according to social logics. Space 2 remains without meaning (own representation)

pletely foreign to us, while our most intimate partner lives at the other end of the world" (Rosa, 2013).
3. The invisibilisation of a spatially ordered system that actually exists and cannot be structurally overridden. This refers to strategies and techniques for using physical space ('1') that make the structuring effect of this usage invisible" (Stichweh, 2003). For example, mobile phone usage "requires a precise decomposition of the earth into cells and segments and neighbourhood relations between these cells and segments" (Stichweh, 2003). This can be seen as an extension of the logical structure or order structure development mentioned at the beginning of this section, as we know it from cartography.
4. The substitution of spatial orders with functional orders. In this case, functions that were previously located in space exactly (e.g. markets, retail stores) are carried out via alternative routes (e.g. e-commerce). From this, Stichweh (2003, p. 99) concludes in system theoretical terminology: "The functional reinterpretation of spatial distinctions takes place in ever new forms". This also means that specific locations can lose their previous function, making them functionless on the one hand, but also filled with functions that were previously located elsewhere or not located at all, for example in the re-use of old industrial objects, such as cultural institutions, housing, software forge, etc.
5. The strategy of domestication of space ('1'). This strategy does not aim at the substitution of 'space 1', but its control. Precise spatial orders are expressions of a network of localization like predictable changes of location. The peculiarity of their own name is taken away from buildings, localization is done by unambiguous and also locatable without local knowledge addresses (name, street, house number, country code, zip code, settlement name).

In his consideration of spaces, Rudolf Stichweh draws an emancipation of 'space 3' from 'space 1' in the context of social modernization. 'Space 1' may still exist, but it becomes less and less relevant for social functions.

6.5   The Interaction of 'Space 2' and 'Space 3': The Social Constructivist Understanding of Space or Landscape

Within the space-related sciences, the social constructivist perspective has been able to establish itself in the German-speaking world as well as beyond, especially in landscape research. This may be due to the fact that, with regard to the dependence of vision on 'landscape', which always includes a space 3 perspective (landscape as an aesthetic construct) in addition to a space 1 perspective, the objective component is more pronounced in the German-speaking world than the aesthetic one (Berr & Kühne, 2020; Drexler, 2013; Müller, 1977; Olwig, 2002; Schenk, 2017). Even though predecessors of social constructivist landscape theory can be traced back to the middle of the 20th century in English landscape research (2005 [1955]), on which Denis Cosgrove (1984, 1993) built, today a theoretical framework based on the phenomenological sociology of Alfred Schütz (1960

[1932], 1971a [1962], 1971b) and its further development by Peter Berger and Thomas Luckmann to social constructivism (Berger & Luckmann, 1966) dominates (among many: Aschenbrand, 2017; Bruns & Kühne, 2015; Fontaine, 2017; Greider & Garkovich, 1994; Kühne, 2018c; Stemmer, 2016).

If it is therefore to be assumed that "the social reality is socially constructed, in which acting subjects and social structures constitute each other" (Risse-Kappen, 1995, p. 174), then an essential aspect of social constructionist landscape or general space research is already outlined: the questions of the relationships between landscape 3 and 2 (space 3 and 2), because, like all other sign systems, social interpretation and attribution patterns of landscape have to be learned by the individual that is socializing: "There is no naive relationship to the landscape before all society. The naive cannot see the landscape, because he has not learned its language" (Burckhardt, 2006, p. 20). On the other hand, social constructionist landscape research also asks how individual ideas can change the social construction of landscape. Landscapes are—in this view—individually projected into space 1 (landscape 2) on the basis of socially shared interpretation and evaluation patterns. Constitutive for landscape (or general: space) is therefore the level landscape 3/ space 3.

Now landscapes 1, 2 and 3 are subject to a 'triple landscape change' (Kühne, 2018a [2020 published], 2020). In particular, rapidly occurring changes in the physical space are individually subjected to interpretations and evaluations that are based on culturally, socially and spatially differentiated patterns of interpretation and evaluation (Bruns, 2016; Bruns & Münderlein, 2019), i.e. landscape 3 is not uniform, but contains different patterns of interpretation and evaluation. Three modes of social construction of landscape can be traced: the 'native normal landscape', the 'common-sense understanding' and the 'expert special knowledge' (Kühne, 2018a [2020 published]). While the normative understanding of the mode of the native normal landscape is that of the stability of physical structures, the common-sense understanding is strongly influenced by the spatial implementation of largely socially shared ideas of a beautiful, typical and 'good' landscape. In contrast, the 'expert special knowledge' is so strongly influenced by professional specifics that the evaluations of the change of physical landscape foundations are very divergent (what appears to the architectural critic as a 'milestone of design practice', the urban sociologist may consider as a 'manifestation of gentrification'; see also Sect. 5.3). Changes in the physical foundations are therefore subject to a very differentiated interpretation and evaluation—depending on which landscape mode they are considered and to what extent their own (also economic) interests are affected. With landscapes 2 and 3 also subject to change (in detail: Kühne, 2018a).

Landscape 1 is only updated in the feedback relationships between Landscape 2 and 3 if it is symbolically occupied or its foundations (i.e. the material foundations of the symbolic occupation) are subject to a change that lies below the threshold of perception. Changes to and within Landscape 1 then acquire particular relevance if they contradict individual or social landscape norms (e.g. the construction of

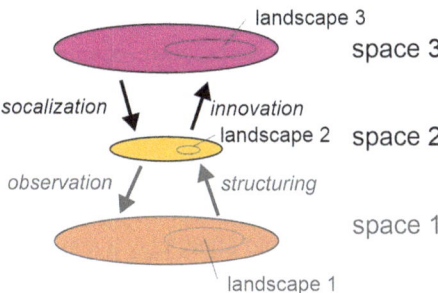

landscape 3
space 3

socalization ↓   ↑innovation
                  landscape 2   space 2

observation ↓   ↑ structuring

space 1

landscape 1

Fig. 6.11 Visualisation of the spatial focus of the social constructivist approach to spaces—and as a subset of spaces on different levels, including landscapes. This approach focuses on the interactions between space 2 and 3 (or landscapes 2 and 3). Space 1 (landscape 1) only has meaning if certain objects are symbolically occupied (own representation)

wind turbines in the 'native normal landscape' or a commercial area in stereotypical landscape beauty expectations); social conflicts about landscape are the result, which are often fought with great (verbal) hardness (among many others: Jenal & Berr, 2019; Kühne & Jenal, 2020b; Kühne & Weber, 2018 [online first 2017]; Michler et al., 2019; Wolf, 2020; Fig. 6.11).

6.6   The Individual Experience: 'Space 2' Between 'Space 1' and 'Space 3'—Phenomenology

Phenomenology can be understood as the study and description of phenomena, whereby phenomena are understood to mean all units, things and events that present themselves to the world (Moran, 2000; Tilley, 1997, 2005) and present themselves to the subject: "Phenomenology involves the understanding and descripton of things as they are experienced by a subject" (Tilley, 1997). Phenomenological thinking is strongly shaped by the thinking of Edmund Husserl (1913), Maurice Merleau-Ponty (1945), Martin Heidegger (Heidegger, 2005 [1927]), as well as the Alfred Schütz (1971a, b) already mentioned in connection with social constructivism. The starting point of the phenomenological world view are sensory experiences. Behind the perceptible appearances, an "essence" of the things is sought in essentialist tradition (Sokolowski, 2000). Phenomenology can certainly be seen in the tradition of romantic science (Wylie, 2019), which was concerned with creating a unity of cognitive, moral and intuitive-aesthetic ideas (Eisel, 2009).

Phenomenology follows a third way by neither proceeding empirically-inductively (like positivism) nor theoretically-deductively (like, for example, autopoietic system theory). The starting point of phenomenological engagement are thinking, speaking and acting. Based on a concrete case (independent of whether it is imaginary or 'real'), something essential and fundamental is intuitively concluded. This fundamental principle refers to the experience of the world. The phenomenological space researcher thus becomes a "storyteller" (Tuan, 1989b, p. 240): "His or her description is inexplicably mixed with exegesis and interpretation, because ordinary language not only invites interpretative conjunctions (because, because, because, because, therefore, etc.), but is also very rich in words that point to relationships beyond their literal meaning." Compared to the artistic approach to phenomena, however, phenomenological engagement is based on explicit concepts, the understanding of which it formulates (cf. Sokolowski, 2000; Moran, 2002). The goal of a phenomenological engagement with the world is a subjective gain in knowledge, in which perception and affect play a major role: "A perception is a style of visibility, of being visible, a configuration of light and matter that exceeds the perceptions of a seeing subject, penetrates into them and extends beyond them. An affect is an intensity, perhaps a field of awe, irritation or cheerfulness, that exceeds the feelings and emotions of a subject that sees, penetrates into them and extends beyond them" (Wylie, 2005, p. 236).

The relationships between subject and object are not only understood relationally from a phenomenological perspective (Chemero, 2003; Gibson, 1979), but rather lead to a dissolution of the dichotomy of subject and object (which also makes phenomenology more than representational, Sect. 6.10, can be assigned, but are already being dealt with here, since other theories of space refer to phenomenology, Sects. 6.7 and 6.9). On the one hand, 'subjects' change through their involvement with 'objects' (DeMarrais et al., 2004; Rebay-Salisbury, 2013), on the other hand, knowledge and experience are so intensively written into the human body, they are anchored firmly in the body and can hardly be articulated on a cognitive basis anymore (Sørensen & Rebay-Salisbury, 2012). Berleant (1997, p. 109) summarises: "A landscape, an environment, more so, is embodied experience". The importance of emotional commitment in relation to cognitive is emphasised here: "Emotions are [...] closely linked to material culture, places in the landscape, as well as human actions, practices and rituals" (Rebay-Salisbury, 2013, p. 63). Accordingly, spaces are not primarily a social (space 3) or individual (space 2) construction, but rather these form the starting point for a mental and physical involvement in space 1 (Ingold, 2002; Fig. 6.12). A key research area for phenomenological space research are therefore atmospheres, which develop between subject and object, against the background of social patterns of interpretation and commitment (see in detail: Hasse, 2012; Kazig 2007, 2019; Thibaud, 2003). Here again the affinity to social constructivism becomes apparent (see Textbox 28).

Fig. 6.12 Visualization of space focusing from a phenomenological perspective. In the center are the interactions between space 2 and space 1, the references to space 3 and space 3 itself are less focused (own representation)

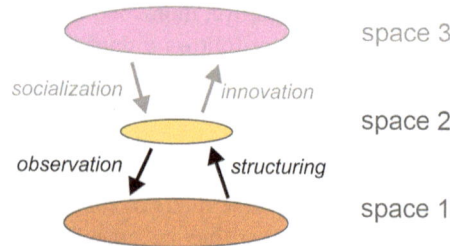

Textbox 28: Social Constructivism and Phenomenology—Similarities and Differences

Current social constructivist and phenomenological research, in particular by Alfred Schütz, show a 'relationship of kinship'. Based on this relationship, it can be seen how the development paths of theoretical approaches to the world can develop differently (Kühne, 2019a): Social constructivism has established itself as a social theory, while phenomenology is primarily a philosophical approach to the world. Accordingly, social processes, in particular the inter-action with the individual, the construction of spatial arrangements, are in the foreground of interest in social constructivist space-related research (i.e. spaces 2 and 3). In contrast, phenomenological space-related research is more focused on the effects and meanings of physical spaces for individual people (relation space 1 and 2, which is influenced by space 3). Even the levels of focus differ: Social constructivist research is based on the social and individual construction processes of space or landscape and thus focuses more on cognitive processes. Phenomenological research refers more to individual practices, appropriations, meanings and emotional attachments, with a strong presence of the 'transient phenomenon of atmosphere' (Kazig, 2007, 2013), located between space 1 and space 2 (more detailed: Kühne, 2019a). ◄

6.7   The Relations of Space 2 to Spaces 1 and 3: Space as a Result of Action—Benno Werlen

Geographer Benno Werlen (1995, 1997, 2000)—based on the structuration the-ory by Anthony Giddens (1984)—advocates a recursive understanding of space between spaces 1, 2 and 3. This structuration theory deals with the mutual influ-ence of individuals and society through action. According to the structuration the-ory, social reality is constituted by competent actors. In their actions, the actors refer to social structures, whereby action is both structured and structuring.

   The action-theoretical social geography—following the structuration theory—puts the focus on human action and deepens the spatial perspective. It is not con-cerned with how humans 'act in space', for it is not relevant to think in terms of space containers, but how spaces are constructed through actions. Compared to the structuration theory, 'theory of everyday regionalization', which Benno Wer-len develops, regionalization is not simply understood as territorialization (in the sense of the creation of delimited territorial units, such as states), but as a practice of binding to and forming the world (i.e. the assignment of meaning, but also the social construction of 'space'). The 'theory of everyday regionalization' focuses on the relationships between space 2 and space 1 as well as between space 2 and space 3, that is, how humans act under the conditions of the material world and those of the socio-cultural contexts. Thus, on the one hand, the acting subjects relate the worlds 1 and 3 to themselves in their everyday world actions, on the

Fig. 6.13 Visualization of the spatial focus from an action-theoretical perspective. Centered on space 2, the interactions with space 1 and 3 in particular are focused on, less so these themselves (own representation)

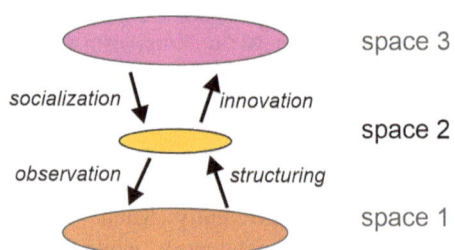

other hand, they shape the world 1 and 3. This also makes it clear that action-theoretical social geography regards space 2 as the central starting point of its considerations and how space 1 and space 3 enable or prevent action (Fig. 6.13); here the proximity to the phenomenological approach becomes apparent (see Sect. 6.6). A outstanding importance is given to the investigation of everyday situations in the action-theoretical social geography. In the "everyday geography-making" the earth's surface is designed in material and symbolic terms. Following a relational concept of space, space 1 is related to social actors (Wardenga, 2002). The space thus created in material and symbolic terms becomes a "factual reality" (Egner, 2010) for the actors, it is accepted unquestioningly as 'normality' and not questioned.

The social framework of the structuring of individual actions sees Werlen (1997) differentiated in everyday regionalizations of the economy, where production takes place and consumption is carried out, the society, in which norms are inscribed in material arrangements and how political control over persons is exercised. In addition, there are the "everyday geographies of information", which are concerned with the generation and distribution of knowledge, while the "everyday geographies of symbolic appropriation" deal with attributions of meaning and emotional occupation of—generated—spaces, for example in the form of "home".

6.8   The Powerful Inscriptions of Society in Space 1: The Space Theory of Pierre Bourdieu

A key starting point for the French sociologist Pierre Bourdieu (1930–2002) is the distinction between social and geometric space (as a container space) by Sorokin (1959). He distinguished between social space in a horizontal (in groups) and vertical differentiation (the position that individual people occupy in groups). However, this space concept leaves out that "there is no social space that does not also constitute itself through material aspects, and vice versa, there is no material space that is not interpreted socially" (Funken & Löw, 2007, p. 86). Pierre Bourdieu (1991) connects this physical and social space with the appropriated physical space, which he also calls reified social space. The physical space, for him a kind of available substrate for social processes, is therefore not at the center of his considerations (Schroer, 2006). 'Social space' is a metaphor for society for Bourdieu, but unlike Sorokin he focuses on the distribution of power and the dynamics in social space through accumulation or loss of symbolic capital (see Textbox 29), for example, the loss of a job usually means not only a loss of economic capital, but also a loss of social networks as social capital. These dynamics are also manifested in the appropriated physical space: This space connects society and physical space: In the appropriated physical space, the relations of the social space become physically manifest (Bourdieu, 1991, p. 29): "The social space realized (or objectified) on a physical level manifests itself as a distribution of different types of goods and services in physical space, as well as physically localized individual actors and groups (in the sense of bodies or corporations bound to a permanent location) with different chances of appropriating these goods and services, as well as physically localized individual actors and groups". The central element of Bourdieu's theory of space is the question of the distribution of social power (in social space), which is inscribed in physical spaces (Rau, 2017). The inscriptions of society in space 1 take place as a result of the different distribution of power of the individual fields (Bourdieu & Wacquant, 1996). Fields are understood as objective, historically grown relations between positions, which in turn are based on certain forms of power or capital. These social fields "form fields of force, but also fields of struggle, on which the preservation or change of the power relations is fought" (Bourdieu, 1985, p. 45). The fields, for example the political, the scientific, the economic, the university field and others have their own functional laws (see also Fuchs-Heinritz & König, 2005; Schroer, 2006), here there are certain similarities of the conception of social fields to the system-theoretical interpretation of the world (Schroer, 2006; see Sects. 6.2 and 6.3), but the social subsystems (especially with Luhmann) are conceived of as exclusivist (the economic logic excludes the political one), Bourdieu conceives his fields as social special areas, which "are structurally superimposed by the order configuration of the social space" (Kneer, 2004, p. 40). Another difference to a radical-constructivist (but discourse-theoretical) understanding of space is the importance of materiality in Bourdieu (Deffner & Haferburg, 2014).

Textbox 29: Symbolic Capital According to Pierre Bourdieu

Symbolic capital appears in the form of economic, social and cultural capital according to Pierre Bourdieu (1979 [original French 1972] and 1987 [1979]). These capitals have scarcity and desirability in common. He understands economic capital to mean material possessions that can be exchanged for money. Social capital "is described as a relational good inherent in social relationships and as a resource of different social structures with different social reach for individuals and corporate actors or communities" (Maischatz, 2010), that is, it arises from reliable social networks. He subdivides cultural capital into three sub-capitals (Bourdieu, 2005 [1983a]): under the objectified form he understands physical manifestations of human activity (books, technical equipment, works of art; thus inscriptions from world 3 into world 1). Under the incorporated form he understands the knowledge and skills inscribed in the subject and thus bound to the physical existence of the acting person (education, cultural skills; thus the inscriptions from world 3 into world 2). Under the institutionalized form he understands society-defined and secured documentation of the availability of incorporated capital (e.g. certificates; thus the documentation of the successful inscriptions from world 3 into world 2). ◀

The intersection of materiality (whether as physical space or as human body) with the concept of social fields has great potential for social science space research (Deffner & Haferburg, 2014; Dirksmeier, 2015; Kühne, 2008): In the appropriated physical space—given the knowledge of the corresponding logics of spatial manifestation, particularly of the different fields—the possibilities of disposal over social, economic and cultural capital can be read. The appropriated physical space also allows conclusions to be drawn about the habitus (to be understood, for example, as the totality of preferences, habits, gestures of a person that shape their taste and at the same time refer to their social background) of the residents (or of those who have manifested themselves in the physical space), because: "It is the habitus that makes the habitat" Bourdieu (1991). The habitus acts as a link between the human being (in his physical corporeality) and the designed place (*place; see* Textbox 30) in the now appropriated physical space (cf. also Deffner & Haferburg, 2014; Dirksmeier, 2007). Here (for example in comparison to action theory, Sect. 6.7) the influence of social space and appropriated physical space on the individual prevails. Thus, it is possible to "read out" "the social from the physical structures of space: Whether it is the Kabyle house, the structure of the school or urban space—always social structures are inscribed in these specific spaces; they tell, so to speak, of the power relations that are expressed through them" (Schroer, 2006, p. 89; Fig. 6.14).

Textbox 30: Place and Space

In particular, in Anglo-Saxon geography, the distinction between *space* and *place* is of great importance. While *space* refers to an abstract geometric space, *place* denotes a holistic and concrete place. Accordingly, *space* is primarily researched and partly modeled using quantitative methods, based on a positivist paradigm. *Place*, on the other hand, is accessed using qualitative methods, in particular under a social constructivist or phenomenological theoretical frame-

Fig. 6.14 Visualization of space focusing from the perspective of Bourdieu's space theory. Space 3 and space 1 take on a central importance, less space 2, this level is strongly determined by the influences from spaces 1 and 3 (own representation)

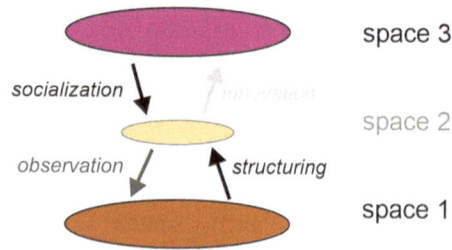

work. The question of *sense of place* and the feeling of belonging to a place (*place attachment*; Hubbard et al., 2005; Tuan, 1989a) is of particular importance here. This concept is characterized by the assignment of the everyday to the local (*place*), a position that Massey (2006) criticizes in that it ignores the relational embeddedness of *places* in supra-local contexts. Against the background of everyday movement in global virtual spaces (such as *social media*), an extension of the concept seems sensible here. ◀

6.9   The Shift in Perspective from the Separate Consideration of Space 1 and Space 3 to an Integrated Thirdspace—Edward Soja

The US-American geographer Edward Soja (1940–2015) dealt with the spatial interweaving of social postmodernization (see Textbox 31). In his analyses, which follow the Marxist French philosopher Henri Lefèbvre (1974), he distinguishes between the three epistemologies of *First-, Second-* and *Thirdspace* (for an introduction to Lefèbrve's understanding of space, see: Ronneberger & Vogelpohl, 2014). He gained his conception of the three epistemologies essentially through his engagement with the agglomeration of Los Angeles, in particular with Orange County, where he diagnosed a dissolution of classical-modern forms of urbanity, which at best only connected to modern experiences of the urban through "merely stage-like facilities" (Krahmer, 2017) and ultimately only simulated urbanity (see in detail: Kühne, 2012, 2015).

Textbox 31: Postmodern

The discussion of postmodernizations came up in the 1970s. Essential aspects of a transition from modern to postmodern thinking are the abandonment of modern dichotomous views (city-country, man-woman, good-bad, beautiful-ugly) and the recognition of hybrids (e.g. settlements that can neither be described as clearly urban or rural, the diverse gender), a gain in importance of the emotional over the cognitive, the upgrading of the aesthetic and the historical, the abandonment of the modernist principle of *form follows function* in favor of a variety of *form follows fiction, form follows fear, form follows finesse* and *form follows finance* (Ellin, 1999). In terms of space, postmodernization is also characterized by a rejection of a comprehensive spatial order in favor of economic and social fragmentation processes (Dear, 2000, 2005; Kubsch, 2007; Kühne, 2006, 2012; Lyotard, 1979; Vester, 1993; see Fig. 6.15 for an overview). ◀

The epistemologies of the *firstspace* encompass the material manifestations of social action (in our chosen terminology: the inscriptions of World/Space 3 into Space 1). The appropriation of the earth, the relationships between nature and

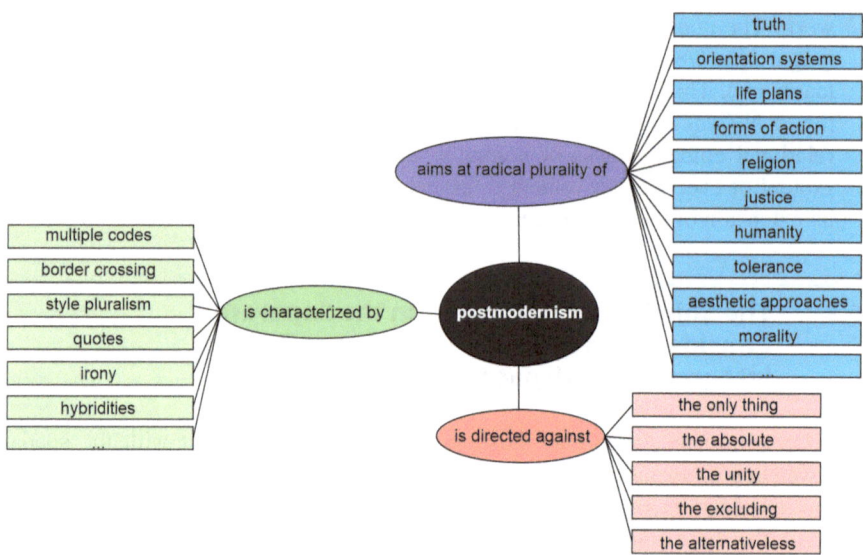

Fig. 6.15 Essential features of postmodernism (expanded according to: Kubsch, 2007)

culture, architecture are here reconstructed as a naive accumulation of spatial elements (Soja, 1996, 2003, 2014). *Firstspace* is—according to Kirchberg (2015)—"the materialized product of planned and unplanned physical and social practice, which we immediately perceive and which guides us through space". The epistemologies of *firstspace* thus privilege "objectivity and materiality and aim for a formal space science" (Soja, 1996). The epistemologies of *secondspace* deal with the spatial concepts that underlie the physical structures (in our chosen terminology: Space 3). These are in turn expressions of existing and historically grown power relations. These are particularly imprinted in normative aesthetic ideas of elites (in particular politicians, investors, architects and planners) (Soja, 1996). *Thirdspace* he conceives, however, as a "lived combination of spatial images and socially produced space" (Fröhlich, 2003), the *thirdspace* "consists of material objects, but all of them are also in the perceptual world of man and socially meaningful" (Hard, 2008). *Thirdspace* represents for Soja (1996) a "cautious deconstruction and heuristic reconstruction of the dualism of the *Firstspace-secondspace* duality" (Soja, 1996). This hybridizes Space 1 and Space 3 (which we will further discuss in Sect. 6.10), while in the aftermath of the 'lived space' by Lefèbre he also refers to a phenomenological tradition (see Sect. 6.6) of space devotion. With the epistemology of the *thirdspace* is a conceptual approach to the recursive connections of physical space, identity construction, the physical manifestations of social power structures, etc. The attempt to shift the focus from the epistemologies of *first*—and *secondspace* to those of *thirdspace* (Soja, 1996, 2003) can be understood as Soja's (2003) attempt to "restore the fundamental triangle of historicity, sociality, and spatiality to a balance." Soja understands his concept of *Thirdspace*

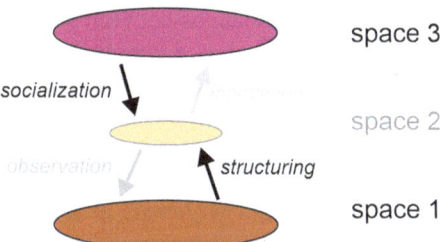

Fig. 6.16 Visualization of space focus from the perspective of Soja's space theory. Space 3 and space 1 play a central role, less space 2. As with Bourdieu, this level is strongly determined by the influences from spaces 1 and 3, but Soja attaches less importance to the observation of space 1 than Bourdieu (here the stronger influence of neo-Marxist thinking on Soja is evident; own representation)

as an extension of the scientific view from the perceived space (*Firstspace*), that is, the "world of direct, immediate spatial experience of empirically measurable and cartographically accessible phenomena" (space 1), but also the mental space (*Secondspace*), which focuses on the "cognitive, constructed and symbolic 'worlds'" (Soja, 2003; space 3) to the entanglements of spaces 1, 2 and 3 (Fig. 6.16).

6.10   More-than-Representational Approaches: Assemblage Theory and Actor-Network Theory

The increasing spread of constructivist approaches in the field of social sciences related to space in the last two decades of the 20th century also meant a shift in focus away from space 1 in favor of space 3 (Kazig & Weichhart, 2009). Alternatives to this focus, which is strongly oriented towards space 3, have already been presented with phenomenology (Sect. 6.6) and action-theoretical and *Thirdspace* approaches (Sects. 6.7 and 6.9). In the following, two theoretical approaches will be presented that have explicitly committed themselves to overcoming a dualistic world view: assemblage theory and actor-network theory.

The goal of assemblage theory (assemblage = structure) is to bring materialities to stronger social consideration (de Landa, 2006) by, with reference to the French authors Gilles Deleuze and Félix Guattari, integrating materiality into a basically constructivist thought structure in order to avoid essentialist or geodeterministic interpretations (Mattissek & Wiertz, 2014). It is directed at the question of how matter becomes socially relevant, namely when it is discursively negotiated (Mattissek & Wiertz, 2014; Miggelbrink, 2014; Müller & Schurr, 2016). It therefore does not ask the essentialist question of what space 1 'essentially' or 'in essence' is, but rather what effects space 1 has on space 3 and how it is negotiated here (van Wezemael & Loepfe, 2009). Accordingly, this approach can also be referred to as the 'contextualization of contexts' (Brenner et al., 2011), whereby the local contexts of the relationships between space 1 and space 3 are examined in the context of general relations between society and space (Brenner et al., 2011). Matter is thus understood as a consequence of discursive negotiations (Mattissek & Wiertz, 2014). Thus, the assemblage theory can be understood as an approach that "addresses social ensembles on the basis of the processes that generate them. It conceptualizes processes of creation and transformation of social ensembles, which are referred to as assemblages, and provides an approach to the analysis of generative processes" (van Wezemael & Loepfe, 2009, p. 108). For example, one can examine how spatial planning constructs space 1 (for example, by mapping vegetation, land use, etc. on the basis of an implicit positivist understanding of science), how different planning (and politics) is negotiated, which space 1-effective claims can be materialized and which not (for example: nature conservation or commercial area; see also Färber, 2014), whereby the assemblage theory is seen as (another) theoretical framework for the investigation of power processes between space 1 and society (and thus also space 3) (McFarlane, 2011). While phenomenological space research is focused on showing the merging of space 1 and space 2, the assemblage theory focuses on merging processes between space 1 and 3 (Fig. 6.17).

The "actor-network theory" (abbreviated ANT) breaks more radically with the thinking in the separate categories of "material world", "individual" and "social world" than phenomenology and the assemblage theory. Its aim is to break up these distinctions, which are common in science but also in everyday use, administration and economy, with the help of the network concept, with a focus on the dichotomies of society and nature and society and technology (Murdoch, 1998; Schulz-Schaeffer, 2000). Based on the work of the French sociologists Michel Callon (1986, 1990) and Bruno Latour (1996, 1997) as well as the British scientist John Law (Law, 1999; Law & Hassard, 1999; summarized: Kneer, 2009), the ANT consistently goes beyond the classical understanding of social theory, because it is about the connections between the "natural" and the "man-made". Thus, social, technical and natural units and factors are not treated as explanans but as explananda by the "actor-network theory" (Schulz-Schaeffer, 2000, p. 188). The understanding of ANT replaces these separations with a network of relationships. These can take on material or immaterial form. So also acting human and non-human objects become (equal) 'actants', with their relationships to each other in the network being quite variable. Every 'thing' becomes the result of networked relationships, space becomes a relational network in which 'things' (e.g.

Fig. 6.17 Visualization of the space focus through the assemblage theory. Space 2 plays a more mediating role in the relationships between space 1 and 3 (own representation)

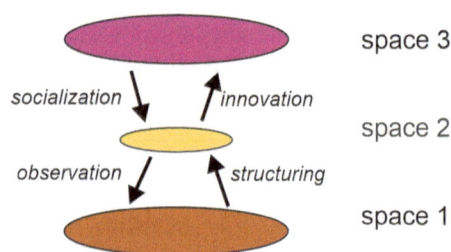

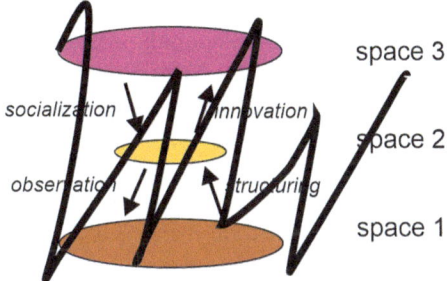

Fig. 6.18 The actor-network theory cannot be represented in the schema chosen here, it rejects the idea of levels in favour of networks. Consequently, the visualization of theories based on Popper's three-world theory reaches its limits here (own representation)

trees, houses, people) concretize. ANT dissolves spaces 1, 2 and 3 into each other and replaces them with a completely new network structure of initially equal actants. However, the principle of equality of the actants, which are also arranged relationally to each other in space, is undermined by power relations, because certain actants can force others to do or not to do something against their 'will': "Power is interwoven in networks of relations and has different forms of expression—dominant power, resistance to power" (Bosco, 2014, p. 158). Accordingly, an ANT-based form of geography deals with the differentiated, always changing relationships of power, networks and places, which are constituted by different actants—which in turn is reflected in material and non-material spatial relations (Bosco, 2015; Höhne & Umlauf, 2014).

The ANT integrates science into the network of actors, creating a self-observation problem. This is also—to a weaker extent—characteristic of constructivism, for example in the form of observing society, in which those driving science are themselves involved. However, if these can abstract from their involvement here, they are themselves entangled in the networks of actors in the ANT. The ANT is also confronted with another criticism: the creation of its own terminology, which makes it more difficult to achieve 'spontaneous connectivity', a criticism also levelled at other complex theory buildings, such as autopoietic system theory. The terminology of the ANT is a result of formulating its own research programme as well as facilitating interdisciplinary work, since all researchers must acquire a new terminology that is located beyond specific disciplinary logics (cf. Färber, 2014; Schulz-Schaeffer, 2000; Fig. 6.18).

6.11 Everything Back to Synthesis? Neopragmatic Understandings of Space

Another strategy for coping with complex issues is chosen by the neo-pragmatic theory; it does not dissolve the concepts of space 1, 2 and 3, and does not exclude other theories, such as ANT, but rather integrates different theoretical perspectives and methods appropriately to the object, in order to also generate knowledge trans-fer in and out of practice. It is therefore not only interdisciplinary, but also trans-disciplinary (Chilla et al., 2015; Eckardt, 2014; Kühne, 2018d, 2019b; Kühne & Jenal, 2020a; Warms & Schroeder, 1999). "Neo-pragmatic approaches" are based on pragmatic traditions, such as those of philosophers such as William James, Charles S. Peirce and John Dewey (see Sect. 3.5.1) and have significantly influ-enced the "Chicago School"; however, they extend it by a theoretical dimension.

Pragmatism is—as shown—focused on the effects of action, in that meanings and truths should determine action, and not moral principles or grand theories. This makes usefulness in specific, concrete contexts the criterion, not consistency with principles (Joas, 1988; Schubert et al., 2010; Steiner, 2014a, b). Accordingly, in pragmatism, 'truth', 'theory', 'practice', etc. are not thought of as separate, but form "a unity mediated by the experience process" (Steiner 2014a, p. 258). In phi-losophy, the neopragmatic approach is particularly associated with Richard Rorty (1982, 1991), as well as with Hilary Putnam (1995). The neopragmatic approach builds on postmodern thought and accordingly rejects notions of universal truth and unshakeable objectivity. Instead, it recognizes pluralistic world views and con-tingency, and is normatively oriented towards result-open, democratic negotiation processes (see: Hildebrand, 2003, 2005; Rorty, 1989).

The neopragmatic goes beyond the pragmatic approach with its strong focus on action (as it has been developed in recent years; Kersting, 2012; Steiner, 2009, 2014a; Weichhart, 2005) by integrating the meta-perspective. While modern sci-ence was concerned with its dichotomous world view to establish theory unity, to understand only one theory as the relevant access to the world, which was then operationalized as far as possible with an empirical access, the neopragmatic view pleads for a stronger orientation towards the understanding of complex objects (such as space, region or landscape; Kühne, 2018d). When dealing with such com-plex objects, even partially conflicting perspectives, such as different construc-tivist with positivist, can be combined in the neopragmatic view (Eckardt, 2014; Fine, 2000; Kühne, 2018d), this also offers a perspective for the integration of natural science, social science and cultural science approaches to a complex object (Steiner, 2014a). Space 1, for example, can be framed with a positivist approach, the reference of space 2 to space 1 phenomenologically, the socialization of cer-tain ideas of space social constructivist and the struggle for interpretive sover-eignty (in space 3) discourse-theoretical, if the complex object requires this with a complex question (for example, the question of the genesis, the economic and social importance of a settlement).

Neopragmatic research is characterized by a combination of triangulations (Denzin, 1970): Constitutively by a theory triangulation (as described above, different theories are used in research), derived from this: Method triangulation (by the application of different methods that result from the different theoretical perspectives). In addition, data triangulations (data from different sources) and researcher triangulations (several researchers deal with the aforementioned trian-gulations together; see also Flick, 2011; Kuckartz, 2014; Fig. 6.19) are required.

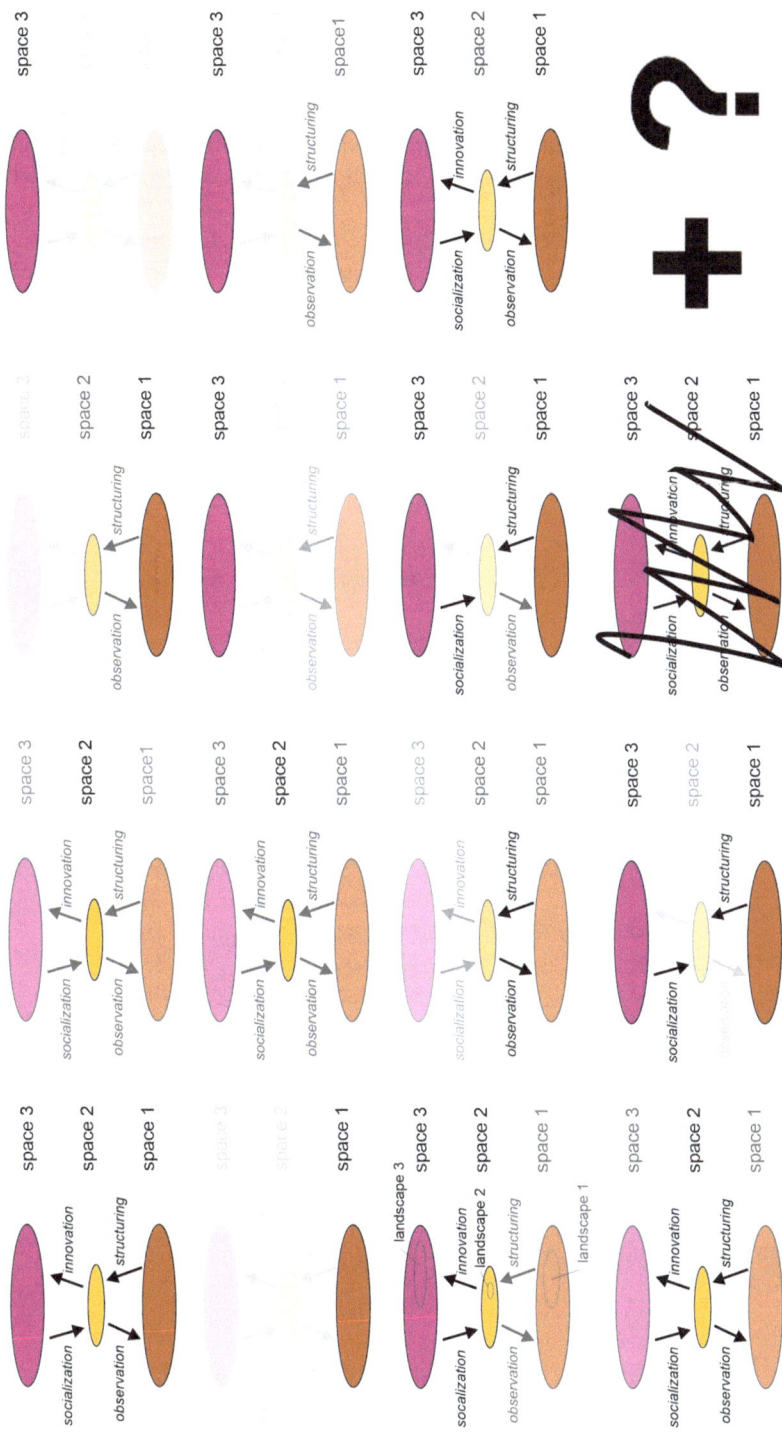

Fig. 6.19 Visualization of a neopragmatic space reference. Since different theoretical approaches to space are possible from this perspective, the most diverse space-related theories can be used (potentially and individually justified), even those that have not yet been invented (own representation)

This creates a more differentiated understanding of a complex object, which would only allow a partial capture by means of a single perspective (theoretical as well as empirical), from which impulses can also be derived from and for research in civil society, administration or politics. In particular, neopragmatic approaches also have a great potential for knowledge, which are characterized by a certain exploratory character, that is, where there is still a barely secure state of knowledge (Kühne, 2019b).

So everything back to the beginning? Does the neo-pragmatic perspective bring back a holistic approach to the investigation of a space container? Yes and no. Yes, because it again focuses on complex, multi-dimensional (here: space 1, 2 and 3 related) questions. No, because it integrates theories that do not work explicitly container-spatially. In addition, the space container is not understood as an objectively given, but as a construct. And no, because the chosen theoretical and empirical approaches as well as the selection of data and the disciplinary origins of the researchers have to be explained and justified.

Textbox 32: Interim Conclusion to: Understandings of Space and Theories of Space

The space concepts presented illustrate the diversity of social space constructions and relationships. A synthesis of the space concepts presented can be made based on Pierre Bourdieu's concept: The social space encompasses and orders the social positions of people. These people have certain ideas about space, which are socially largely determined, but must be newly established through individual actualization (Werlen). On the level of societies, different functional sub-systems also differentiate themselves. Through the spatially-related action of these sub-systems, the appropriated physical space is inscribed in the physical space as the basis of the material world (Läpple). These relational inscriptions often reduce the influence of the physical space on people (Stichweh). In the course of functional differentiation, society (at least) has produced a social sub-system that deals with the scientific construction of space by conceiving space (at least) as a physical object, as a social idea of space and as an everyday world of action (Soja). These relations on the different levels of spatial imagination are based on specific power relations, which manifest themselves in the social definition of ideas of space, but also in inscriptions in physical spaces. So "minorities" (Paris, 2005), such as economically disadvantaged people, immigrants, women, etc., have less chance to inscribe their needs in these physical spaces than those in power, simply because they are often underrepresented in decision-making bodies.

Further Reading
Egner 2010: Compact introduction to geographical theory formation.
Läpple 1992: Now a classic article on social science concepts of space.
Aitken and Valentine (eds.) 2014: Collection which introduces the recent discussion on space concepts and practices in the Anglo-Saxon world in detail, enriched with autobiographical contributions from geographers who are currently attributed greater importance to the subject.
Oßenbrügge & Vogelpohl (eds.) 2014: Introductory overview of (especially critical) theoretical perspectives of urban and spatial research.

Original Literature
Bourdieu 1991: A very instructive article on the entanglements of society with material spaces.
Werlen 1997: A book that changed social geography in Germany sustainably by bringing the actions of subjects into focus.
Soja 1996: One of the classic works of the *Los Angeles School of Urbanism* on the development of postmodern spatial relations.
Wylie 2005: A very impressive essay on phenomenological approaches to spaces.

References

Altvater, E. (1987). *Sachzwang Weltmarkt. Verschuldungskrise, blockierte Industrialisierung und ökologische Gefährdung – der Fall Brasilien.* VSA.
Aschenbrand, E. (2017). *Die Landschaft des Tourismus. Wie Landschaft von Reiseveranstaltern inszeniert und von Touristen konsumiert wird.* Springer VS.
Bathelt, H., & Glückler, J. (2003). Toward a relational economic geography. *Journal of Economic Geography, 3*(2), 117–144. https://doi.org/10.1093/jeg/3.2.117.
Berger, P. L., & Luckmann, T. (1966). *The social construction of reality. A treatise in the sociology of knowledge.* Anchor Books.
Berleant, A. (1997). *Living in the landscape. Toward an aesthetics of environment.* University Press of Kansas.
Berr, K., & Kühne, O. (2020). *„Und das ungeheure Bild der Landschaft …".* The Genesis of Landscape Understanding in the German-speaking Regions. Springer VS.
Blackbourn, D. (2007). *Die Eroberung der Natur. Eine Geschichte der deutschen Landschaft.* Random House.
Boggs, J. S., & Rantisi, N. M. (2003). The ‚relational turn' in economic geography. *Journal of Economic Geography, 3*(2), 109–116.
Bojadžiev, T. (2003). *Die Nacht im Mittelalter.* Königshausen & Neumann.
Bollnow, O. F. (1963). *Mensch und Raum.* Kohlhammer.
Boschma, R. A., & Frenken, K. (2006). Why is economic geography not an evolutionary science? Towards an evolutionary economic geography. *Journal of Economic Geography, 6*(3), 273–302.
Bosco, F. J. (2014). Actor-network theory, networks, and relational geographies. In S. C. Aitken & G. Valentine (Hrsg.), *Approaches to human geography. Philosophies, theories, people and practices* (2. Aufl., S. 150–162). SAGE.

Bosco, F. J. (2015). Actor-network theory, networks, and relational geographies. In S. C. Aitken & G. Valentine (Hrsg.), *Approaches to human geography. Philosophies, theories, people and practices* (2. Aufl., S. 150–162). SAGE.

Bourdieu, P. (1979 [frz. Original 1972]). *Entwurf einer Theorie der Praxis auf der ethnologischen Grundlage der kabylischen Gesellschaft.* Suhrkamp.

Bourdieu, P. (1985). *Sozialer Raum und „Klassen". Leçon sur la leçon; 2 Vorlesungen.* Suhrkamp.

Bourdieu, P. (1991). Physischer, sozialer und angeeigneter physischer Raum. In M. Wentz (Hrsg.), *Stadt-Räume* (S. 25–34). Campus.

Bourdieu, P. (2005 [1983a]). Ökonomisches Kapital – Kulturelles Kapital – Soziales Kapital. In P. Bourdieu (Hrsg.), *Die verborgenen Mechanismen der Macht* (S. 49–80). VSA.

Bourdieu, P., & Wacquant, L. (1996). Die Ziele der reflexiven Soziologie. Chicago-Seminar 1987. In H. Beister (Hrsg.), *Reflexive Anthropologie* (1. Aufl., S. 95–249). Suhrkamp.

Brenner, N., Madden, D. J., & Wachsmuth, D. (2011). Assemblage urbanism and the challenges of critical urban theory. *City, 15*(2), 225–240.

Bruns, D. (2016). Kulturell diverse Raumaneignung. In F. Weber & O. Kühne (Hrsg.), *Fraktale Metropolen. Stadtentwicklung zwischen Devianz, Polarisierung und Hybridisierung* (S. 231–240). Springer VS.

Bruns, D., & Kühne, O. (2015). Gesellschaftliche Transformation und die Entwicklung von Landschaft. Eine Betrachtung aus der Perspektive der sozialkonstruktivistischen Landschaftstheorie. In O. Kühne, K. Gawroński, & J. Hernik (Hrsg.), *Transformation und Landschaft. Die Folgen sozialer Wandlungsprozesse auf Landschaft* (S. 17–34). Springer VS.

Bruns, D., & Münderlein, D. (2019). Interkulturelle Konstruktion. In O. Kühne, F. Weber, K. Berr, & C. Jenal (Hrsg.), *Handbuch Landschaft* (S. 313–319). Springer VS.

Bunge, M. (1984). *Das Leib-Seele-Problem. Ein psychobiologischer Versuch* (Die Einheit der Gesellschaftswissenschaften, Bd. 37). J. C. B. Mohr (Paul Siebeck).

Burckhardt, L. (2006). Landschaftsentwicklung und Gesellschaftsstruktur (1977). In M. Ritter & M. Schmitz (Hrsg.), *Warum ist Landschaft schön? Die Spaziergangswissenschaft* (S. 19–32). Martin Schmitz.

Callon, M. (1986). The sociology of an actor-network: The case of the electric vehicle. In M. Callon, J. Law, & A. Rip (Hrsg.), *Mapping the dynamics of science and technology* (S. 19–34). Palgrave Macmillan.

Callon, M. (1990). Techno-economic networks and irreversibility. *The Sociological Review, 38*(1), 132–161.

Chemero, A. (2003). An outline of a theory of affordances. *Ecological Psychology, 15*(2), 181–195. https://doi.org/10.1207/S15326969ECO1502_5.

Chilla, T., Kühne, O., Weber, F., & Weber, F. (2015). „Neopragmatische" Argumente zur Vereinbarkeit von konzeptioneller Diskussion und Praxis der Regionalentwicklung. In O. Kühne & F. Weber (Hrsg.), *Bausteine der Regionalentwicklung* (S. 13–24). Springer VS.

Cosgrove, D. (1984). *Social formation and symbolic landscape.* University of Wisconsin Press.

Cosgrove, D. (1993). *The Palladian landscape. Geographical change and its cultural representations in sixteenth-century Italy.* Pennsylvania State University Press.

Dahrendorf, R. (1979). *Lebenschancen. Anläufe zur sozialen und politischen Theorie* (Suhrkamp-Taschenbuch, Bd. 559). Suhrkamp.

Dahrendorf, R. (2007). *Auf der Suche nach einer neuen Ordnung. Vorlesungen zur Politik der Freiheit im 21. Jahrhundert* (Krupp-Vorlesungen zu Politik und Geschichte am Kulturwissenschaftlichen Institut im Wissenschaftszentrum Nordrhein-Westfalen, 4. Aufl., Bd. 3). C. H. Beck.

Dangschat, J. (2014). Stadt und Raum in der Soziologie. In J. Oßenbrügge & A. Vogelpohl (Hrsg.), *Theorien in der Raum- und Stadtforschung. Einführungen* (S. 57–67). Westfälisches Dampfboot.

Dear, M. (2005). The Los Angeles school of Urbanism. In B. J. L. Berry & J. O. Wheeler (Hrsg.), *Urban geography in America, 1950–2000. Paradigms and personalities* (S. 327–347). Routledge.

Dear, M. J. (2000). *The postmodern urban condition.* Wiley-Blackwell.

Deffner, V., & Haferburg, C. (2014). Pierre Bourdieu: Habitus und Habitat als Verhältnis von Subjekt, Sozialem und Macht. In J. Oßenbrügge & A. Vogelpohl (Hrsg.), *Theorien in der Raum- und Stadtforschung. Einführungen* (S. 328–347). Westfälisches Dampfboot.

DeMarrais, E., Gosden, C., & Renfrew, C. (Hrsg.). (2004). *Rethinking materiality. The engagement of mind with the material world.* McDonald Institute for Archaeological Research.

Denzin, N. K. (1970). *The research act. A theoretical introduction to sociological methods.* Aldine Publishing Company.

Dirksmeier, P. (2007). Mit Bourdieu gegen Bourdieu empirisch denken. Habitusanalyse mittels reflexiver Fotografie. *ACME: An International Journal for Critical Geographies, 6*(1), 73–97.

Dirksmeier, P. (2015). *Urbanität als Habitus. Zur Sozialgeographie städtischen Lebens auf dem Land.* transcript.

Drexler, D. (2013). Die Wahrnehmung der Landschaft – ein Blick auf das englische, französische und ungarische Landschaftsverständnis. In D. Bruns & O. Kühne (Hrsg.), *Landschaften: Theorie, Praxis und internationale Bezüge. Impulse zum Landschaftsbegriff mit seinen ästhetischen, ökonomischen, sozialen und philosophischen Bezügen mit dem Ziel, die Verbindung von Theorie und Planungspraxis zu stärken* (S. 37–54). Oceano.

Dünne, J., & Günzel, S. (Hrsg.). (2006). *Raumtheorie. Grundlagentexte aus Philosophie und Kulturwissenschaften* (1. Aufl.). Suhrkamp.

Eckardt, F. (2014). *Stadtforschung. Gegenstand und Methoden.* Springer VS.

Egner, H. (2008). *Gesellschaft, Mensch, Umwelt – beobachtet. Ein Beitrag zur Theorie der Geographie* (Erdkundliches Wissen, Bd. 145). Steiner.

Egner, H. (2010). *Theoretische Geographie.* WBG.

Eisel, U. (2009). *Landschaft und Gesellschaft. Räumliches Denken im Visier* (Raumproduktionen: Theorie und gesellschaftliche Praxis, Bd. 5). Westfälisches Dampfboot.

Elias, N., & Schröter, M. (Hrsg.). (1994). *Über die Zeit. Arbeiten zur Wissenssoziologie II* (Suhrkamp-Taschenbuch Wissenschaft, 5. Aufl., Bd. 756). Suhrkamp.

Ellin, N. (1999). *Postmodern Urbanism.* Princeton Architectural Press.

Escher, A., & Petermann, S. (Hrsg.). (2016). *Raum und Ort* (Basistexte Geographie, Bd. 1). Franz Steiner.

Färber, A. (2014). Potenziale freisetzen: Akteur-Netzwerk-Theorie und Assemblageforschung in der interdisziplinären kritischen Stadtforschung. *sub\urban zeitschrift für kritische stadtforschung, 2*(1), 95–103.

Fine, A. (2000). Der Blickpunkt von niemand besonderen. In M. Sandbothe (Hrsg.), *Die Renaissance des Pragmatismus. Aktuelle Verflechtungen zwischen analytischer und kontinentaler Philosophie* (S. 59–77). Velbrück.

Flick, U. (2007). *Qualitative Sozialforschung. Eine Einführung.* Rowohlt.

Flick, U. (2011). *Triangulation.* Springer Fachmedien.

Fliedner, D. (2015). *Sozialgeographie.* De Gruyter.

Fontaine, D. (2017). Ästhetik simulierter Welten am Beispiel Disneylands. In O. Kühne, H. Megerle, & F. Weber (Hrsg.), *Landschaftsästhetik und Landschaftswandel* (RaumFragen: Stadt – Region – Landschaft, S. 105–120). Springer VS.

Foucault, M. (1991). Andere Räume. In M. Wentz (Hrsg.), *Stadt-Räume* (S. 65–72). Campus.

Fröhlich, H. (2003). *Learning from Los Angeles – Zur Rolle von Los Angeles in der Diskussion um die postmoderne Stadt* (Beiträge zur Stadt- und Regionalplanung, Bd. 5). Selbstverlag.

Fuchs-Heinritz, W., & König, A. (2005). *Pierre Bourdieu. Eine Einführung.* UVK.

Funken, C., & Löw, M. (2007). Ego-Shooters Container. Raumkonstruktionen im elektronischen Netz. In R. Maresch & N. Werber (Hrsg.), *Raum, Wissen, Macht* (Suhrkamp Taschenbuch Wissenschaft, Bd. 1603, S. 69–91). Suhrkamp. 7. Nachdr.

Gibson, J. J. (1979). *The ecological approach to visual perception.* Houghton Mifflin.

Giddens, A. (1984). *The constitution of society. Outline of the theory of structuration.* University of California Press.

Glasze, G. (2013). *Politische Räume. Die diskursive Konstitution eines „geokulturellen Raums" – die Frankophonie.* transcript.

Glasze, G., & Mattissek, A. (2009). Die Hegemonie- und Diskurstheorie von Laclau und Mouffe. In G. Glasze & A. Mattissek (Hrsg.), *Handbuch Diskurs und Raum. Theorien und Methoden für die Humangeographie sowie die sozial- und kulturwissenschaftliche Raumforschung* (S. 153–179). transcript.

Greider, T., & Garkovich, L. (1994). Landscapes: The social construction of nature and the environment. *Rural Sociology, 59*(1), 1–24. https://doi.org/10.1111/j.1549-0831.1994.tb00519.x.

Hägerstrand, T. (1970). What about people in regional science? *Papers in Regional Science, 24*(1), 7–24. https://doi.org/10.1111/j.1435-5597.1970.tb01464.x.

Hard, G. (Hrsg.). (2002). *Landschaft und Raum. Aufsätze zur Theorie der Geographie* (Osnabrücker Studien zur Geographie, Bd. 22). Universitätsverlag Rasch.

Hard, G. (2003). Studium in einer diffusen Disziplin. In G. Hard (Hrsg.), *Dimensionen geographischen Denkens. Aufsätze zur Theorie der Geographie* (Osnabrücker Studien zur Geographie, Bd. 23, S. 173–230). V & R Unipress.

Hard, G. (2008). Der Spatial Turn, von der Geographie her beobachtet. In J. Döring & T. Thielmann (Hrsg.), *Spatial Turn. Das Raumparadigma in den Kultur- und Sozialwissenschaften* (S. 263–316). transcript.

Hasse, J. (2012). *Atmosphären der Stadt. Aufgespürte Räume.* Jovis.

Heidegger, M. (2005 [1927]). *Die Grundprobleme der Phänomenologie.* Klostermann.

Helmut Klueter. (1986). *Raum als Element sozialer Kommunikation.* Selbstverlag Geographisches Institut Universität.

Herrmann, H. (2010). Raumbegriffe und Forschungen zum Raum – eine Einleitung. In H. Herrmann (Hrsg.), *RaumErleben. Zur Wahrnehmung des Raumes in Wissenschaft und Praxis* (S. 7–29). Budrich.

Hildebrand, D. L. (2003). The neopragmatist turn. *Southwest Philosophy Review, 19*(1), 79–88.

Hildebrand, D. L. (2005). Pragmatism, neopragmatism, and public administration. *Administration & Society, 37*(3), 345–359.

Höhne, S., & Umlauf, R. (2014). Die Akteur-Netzwerk Theorie. Zur Vernetzung und Entgrenzung des Sozialen. In J. Oßenbrügge & A. Vogelpohl (Hrsg.), *Theorien in der Raum- und Stadtforschung. Einführungen* (S. 195–214). Westfälisches Dampfboot.

Hoskins, W. G. (2005 [1955]). *The making of the English landscape.* The Folio Society.

Hubbard, P., Bartley, B., Fuller, D., & Kitchin, R. (2005). *Thinking geographically. Space, theory and contemporary human geography.* continuum.

Husserl, E. (1913). *Ideen zu einer reinen Phänomenologie und phänomenologischen Philosopie. Erster Buch: Allgemeine Einführung in die reine Phänomenologie.* Niemeyer.

Ingold, T. (2002). *The perception of the environment. Essays on livelihood, dwelling and skill.* Routledge.

Ipsen, D. (2002). Raum als Landschaft. In D. Ipsen & D. Läpple (Hrsg.), *Soziologie des Raumes: Räume der Gesellschaft – soziologische Perspektiven* (S. 86–111). Fernuniversität.

Jenal, C., & Berr, K. (2019). Landschaft als Konflikt. Wenn erlernte Deutungsmuster mit neuen Sichtweisen konkurrieren. *Stadt+Grün, 68*(12), 18–23.

Joas, H. (1988). Symbolischer Interaktionismus. Von der Philosophie des Pragmatismus zu einer soziologischen Forschungstradition. *Kölner Zeitschrift für Soziologie und Sozialpsychologie, 40*, 417–446.

Jørgensen, M., & Phillips, L. (2002). *Discourse analysis as theory and method.* SAGE.

Kant, I. (1959 [1781]). *Kritik der reinen Vernunft.* Felix Meiner.

Kazig, R. (2007). Atmosphären – Konzept für einen nicht repräsentationellen Zugang zum Raum. In C. Berndt & R. Pütz (Hrsg.), *Kulturelle Geographien. Zur Beschäftigung mit Raum und Ort nach dem Cultural Turn* (S. 167–187). transcript.

Kazig, R. (2013). Landschaft mit allen Sinnen – Zum Wert des Atmosphärenbegriffs für die Landschaftsforschung. In D. Bruns & O. Kühne (Hrsg.), *Landschaften: Theorie, Praxis und internationale Bezüge. Impulse zum Landschaftsbegriff mit seinen ästhetischen, ökonomischen, sozialen und philosophischen Bezügen mit dem Ziel, die Verbindung von Theorie und Planungspraxis zu stärken* (S. 221–232). Oceano.

Kazig, R. (2019). Atmosphären und Landschaft. In O. Kühne, F. Weber, K. Berr, & C. Jenal (Hrsg.), *Handbuch Landschaft* (S. 453–460). Springer VS.

Kazig, R., & Weichhart, P. (2009). Die Neuthematisierung der materiellen Welt in der Humangeographie. *Berichte zur deutschen Landeskunde, 83*(2), 109–128.

Kersting, P. (2012). Geomorphologie, Pragmatismus und integrative Ansätze in der Geographie. *Berichte zur deutschen Landeskunde, 86*(1), 49–65.

Kirchberg, V. (2015). Das Museum als öffentlicher Raum in der Stadt. In J. Baur (Hrsg.), *Museumsanalyse. Methoden und Konturen eines neuen Forschungsfeldes* (S. 231–266). transcript.

Kneer, G. (2004). Differenzierung bei Luhmann und Bourdieu. Ein Theorievergleich. In A. Nassehi & G. Nollmann (Hrsg.), *Bourdieu und Luhmann. Ein Theorienvergleich* (S. 25–56). Suhrkamp.

Kneer, G. (2009). Akteur-Netzwerk-Theorie. In G. Kneer & M. Schroer (Hrsg.), *Handbuch Soziologische Theorien* (S. 19–39). VS Verlag.

Koriako, D. (2005). Was sind und wozu dienen reine Anschauungen? Kritische Fragen und Anmerkungen zu Kants Raumtheorie. *Kant-Studien, 96*(1), 20–40.

Krahmer, A. (2017). Edward W. Soja: Thirdspace. In F. Eckardt (Hrsg.), *Schlüsselwerke der Stadtforschung* (S. 47–68). Springer VS.

Kramer, C. (2012). „Alles hat seine Zeit" – die „Time Geography" im Licht des „Material Turn". In N. Weixlbaumer (Hrsg.), *Anthologie zur Sozialgeographie* (Abhandlungen zur Geographie und Regionalforschung, Bd. 16, S. 83–105). Institut für Geographie und Regionalforschung.

Kubsch, R. (2007). *Die Postmoderne. Abschied von der Eindeutigkeit.* Hänssler.

Kuckartz, U. (2014). *Mixed Methods. Methodologie, Forschungsdesigns und Analyseverfahren.* Springer VS.

Kühne, O. (2006). *Landschaft in der Postmoderne. Das Beispiel des Saarlandes.* DUV.

Kühne, O. (2008). Kritische Geographie der Machtbeziehungen – konzeptionelle Überlegungen auf der Grundlage der Soziologie Pierre Bourdieus. *geographische revue, 10*(2), 40–50.

Kühne, O. (2012). *Stadt – Landschaft – Hybridität. Ästhetische Bezüge im postmodernen Los Angeles mit seinen modernen Persistenzen.* Springer VS.

Kühne, O. (2015). The streets of Los Angeles: Power and the infrastructure landscape. *Landscape Research, 40*(2), 139–153. https://doi.org/10.1080/01426397.2013.788691.

Kühne, O. (2018a [2020 erschienen]). Die Landschaften 1, 2 und 3 und ihr Wandel. Perspektiven für die Landschaftsforschung in der Geographie – 50 Jahre nach Kiel. *Berichte. Geographie und Landeskunde, 92* (3–4), 217–231.

Kühne, O. (2018b). *Landschaft und Wandel. Zur Veränderlichkeit von Wahrnehmungen.* Springer VS.

Kühne, O. (2018c). Die Landschaften 1, 2 und 3 und ihr Wandel. Perspektiven für die Landschaftsforschung in der Geographie – 50 Jahre nach Kiel. *Berichte. Geographie und Landeskunde, 3–4,* 217–231.

Kühne, O. (2018d). *Landschaftstheorie und Landschaftspraxis. Eine Einführung aus sozialkonstruktivistischer Perspektive* (2., akt. u. überarb. Aufl.). Springer VS.

Kühne, O. (2019a). *Landscape theories. A brief introduction.* Springer VS.

Kühne, O. (2019b). Sich abzeichnende theoretische Perspektiven für die Landschaftsforschung: Neopragmatismus, Akteur-Netzwerk-Theorie und Assemblage-Theorie. In O. Kühne, F. Weber, K. Berr, & C. Jenal (Hrsg.), *Handbuch Landschaft* (S. 153–162). Springer VS.

Kühne, O. (2019c). Die Sozialisation von Landschaft. In O. Kühne, F. Weber, K. Berr, & C. Jenal (Hrsg.), *Handbuch Landschaft* (S. 301–312). Springer VS.

Kühne, O. (2020). Landscape conflicts. A theoretical approach based on the three worlds theory of Karl Popper and the conflict theory of Ralf Dahrendorf, illustrated by the example of the energy system transformation in Germany. *Sustainability, 12*(17), 1–20. https://doi. org/10.3390/su12176772.

Kühne, O., & Jenal, C. (2020a). *Baton Rouge – The multivillage metropolis. A neopragmatic landscape biographical approach on spatial pastiches, hybridization, and differentiation.* Springer VS.

Kühne, O., & Jenal, C. (2020b). The threefold landscape dynamics – Basic considerations, conflicts and potentials of virtual landscape research. In D. Edler, C. Jenal, & O. Kühne (Hrsg.), *Modern approaches to the visualization of landscapes* (S. 389–402). Springer VS.

Kühne, O., & Weber, F. (2015). Der Energienetzausbau in Internetvideos – eine quantitativ ausgerichtete diskurstheoretisch orientierte Analyse. In S. Kost & A. Schönwald (Hrsg.), *Landschaftswandel – Wandel von Machtstrukturen* (S. 113–126). Springer VS.

Kühne, O., & Weber, F. (2018 [online first 2017]). Conflicts and negotiation processes in the course of power grid extension in Germany. *Landscape Research 43*(4), 529–541. https://doi. org/10.1080/01426397.2017.1300639.

Laclau, E., & Mouffe, C. (1985). *Hegemony and socialist strategy. Towards a radical democratic politics.* Verso.

Lamnek, S. (2010). *Qualitative Sozialforschung* (5., überarb. Aufl.). Beltz.

de Landa, M. (2006). *A new philosophy of society. Assemblage theory and social complexity.* continuum.

Langer, A., & Wrana, D. (2013). Diskursforschung und Diskursanalyse. In B. Friebertshäuser, A. Langer, & A. Prengel (Hrsg.), *Handbuch. Qualitative Forschungsmethoden in der Erziehungswissenschaft* (S. 335–349). Beltz.

Läpple, D. (1991). Gesellschaftszentriertes Raumkonzept. In M. Wentz (Hrsg.), *Stadt-Räume* (S. 35–46). Campus.

Läpple, D. (1992). Essay über den Raum. Für ein gesellschaftswissenschaftliches Raumkonzept. In H. Häußermann, D. Ipsen, R. Krämer-Badoni, D. Läpple, M. Rodenstein, & W. Siebel (Hrsg.), *Stadt und Raum. Soziologische Analysen* (2. Aufl., S. 157–207). Centaurus.

Latour, B. (1996). *Petite réflexion sur le culte moderne des dieux Faitiches.* Synthélabo groupe.

Latour, B. (1997). The trouble with actor-network theory. *Soziale Welt, 47,* 369–381.

Law, J. (1999). After ANT: Complexity, naming and topology. *The Sociological Review, 47*(S1), 1–14.

Law, J., & Hassard, J. (Hrsg.). (1999). *Actor network theory and after* (Sociological review Monographs). Blackwell Publishers.

Lefèbvre, H. (1974). La production de l'espace. *L'Homme et la société, 31–32*(1), 15–32.

Leibenath, M., & Otto, A. (2012). Diskursive Konstituierung von Kulturlandschaft am Beispiel politischer Windenergiediskurse in Deutschland. *Raumforschung und Raumordnung, 70*(2), 119–131. https://doi.org/10.1007/s13147-012-0148-0.

Lippuner, R. (2008). Raumbilder der Gesellschaft. Zur Räumlichkeit des Sozialen in der Systemtheorie. In J. Döring & T. Thielmann (Hrsg.), *Spatial Turn. Das Raumparadigma in den Kultur- und Sozialwissenschaften* (S. 341–363). transcript.

Lippuner, R. (2011). Gesellschaft, Umwelt und Technik: Zur Problemstellung einer »Ökologie sozialer Systeme«. *Soziale Systeme. Zeitschrift für soziologische Theorie, 17*(2), 308–335.

Löw, M. (2001). *Raumsoziologie.* Suhrkamp.

Löw, M., Steets, S., & Stoetzer, S. (2008). *Einführung in die Stadt- und Raumsoziologie* (2., akt. Aufl.). Budrich.

Luhmann, N. (1984). *Soziale Systeme. Grundriß einer allgemeinen Theorie.* Suhrkamp.

Luhmann, N. (1995). *Die Kunst der Gesellschaft.* Suhrkamp.

Luhmann, N. (2001 [1997]). *Die Gesellschaft der Gesellschaft.* Suhrkamp.

Lyotard, J.-F. (1979). *La condition postmoderne. Rapport sur le savoir.* Les Éditions de Minuit.

Maischatz, K. (2010). Eine Einführung in das Sozialkapital-Konzept anhand der zentralen Vertreter. In A. Fischer (Hrsg.), *Die soziale Dimension von Nachhaltigkeit – Beziehungsgeflecht zwischen Nachhaltigkeit und Benachteiligtenförderung. Berufliche Bildung und zukünftige*

Entwicklung (Leuphana-Schriften zur Berufs- und Wirtschaftspädagogik, Bd. 3, S. 31–54). Schneider-Verlag Hohengehren.

Massey, D. (2006). Keine Entlastung für das Lokale. In H. Berking (Hrsg.), *Die Macht des Lokalen in einer Welt ohne Grenzen* (S. 25–31). Campus.

Mattissek, A. (2008). *Die neoliberale Stadt. Diskursive Repräsentationen im Stadtmarketing deutscher Großstädte.* transcript.

Mattissek, A. (2010). Stadtmarketing in der neoliberalen Stadt. Potentiale von Gouvernementalitäts- und Diskursanalyse für die Untersuchung aktueller Prozesse der Stadtentwicklung. In J. Angermüller & S. van Dyk (Hrsg.), *Diskursanalyse meets Gouvernementalitätsforschung. Perspektiven auf das Verhältnis von Subjekt, Sprache, Macht und Wissen* (S. 129–154). Campus.

Mattissek, A., & Wiertz, T. (2014). Materialität und Macht im Spiegel der Assemblage-Theorie: Erkundungen am Beispiel der Waldpolitik in Thailand. *Geographica Helvetica, 69*(3), 157–169.

McFarlane, C. (2011). Assemblage and critical urbanism. *City, 15*(2), 204–224.

Merleau-Ponty, M. (1945). *Phénoménologie de la perception* (Bibliothèque des idées). Gallimard.

Meyer, F., & Miggelbrink, J. (2018). „Der Konjuktiv ist das Problem". Zirkularität, Performativität und Reifikation in der geographischen Forschung. In F. Meyer, J. Miggelbrink, & K. Beurskens (Hrsg.), *Ins Feld und zurück – Praktische Probleme qualitativer Forschung in der Sozialgeographie* (S. 17–23). Springer Spektrum.

Michler, T., Aschenbrand, E., & Leibl, F. (2019). Gestört, aber grün: 30 Jahre Forschung zu Landschaftskonflikten im Nationalpark Bayerischer Wald. In K. Berr & C. Jenal (Hrsg.), *Landschaftskonflikte* (S. 291–311). Springer VS.

Miggelbrink, J. (2014). Diskurs, Machttechnik, Assemblage. Neue Impulse für eine regionalgeographische Forschung. *Geographische Zeitschrift, 102*(1), 25–40.

Moran, D. (2000). *Introduction to phenomenology.* Routledge.

Moran, D. (2002). *Introduction to phenomenology.* Routledge.

Müller, G. (1977). Zur Geschichte des Wortes Landschaft. In A. Hartlieb von Wallthor & H. Quirin (Hrsg.), *„Landschaft" als interdisziplinäres Forschungsproblem. Vorträge und Diskussionen des Kolloquiums am 7./8. November 1975 in Münster* (S. 3–13). Aschendorff.

Müller, M., & Schurr, C. (2016). Assemblage thinking and actor-network theory: Conjunctions, disjunctions, cross-fertilisations. *Transactions of the Institute of British Geographers, 41*(3), 217–229.

Murdoch, J. (1998). The spaces of actor-network theory. *Geoforum, 29*(4), 357–374.

Niemann, H.-J. (2019). Karl Poppers Spätwerk und seine ‚Welt 3'. In G. Franco (Hrsg.), *Handbuch Karl Popper* (Living reference work, S. 1–18). Springer Reference Geisteswissenschaften.

Nissen, U. (1998). *Kindheit, Geschlecht und Raum. Sozialisationstheoretische Zusammenhänge geschlechtsspezifischer Raumaneignung.* Beltz.

Olwig, K. R. (2002). *Landscape, nature, and the body politic. From Britain's Renaissance to America's new world.* University of Wisconsin Press.

Paasi, A. (1998). Boundaries as social processes: Territoriality in the world of flows. *Geopolitics, 3*(1), 69–88. https://doi.org/10.1080/14650049808407608.

Paasi, A. (1999). The changing pedagogies of space: Representation of the other in finnish school geography textbooks. In A. Buttimer, S. Brunn, & U. Wardenga (Hrsg.), *Text and image. Social construction of regional knowledges* (Beiträge zur Regionalen Geographie, Bd. 49, S. 226–237). Institut für Länderkunde.

Paris, R. (2005). *Normale Macht. Soziologische Essays.* UVK.

Parsons, T. (1951). *The social system.* Free Press.

Parsons, T. (1991 [1951]). *The social system.* Routledge.

Popper, K. R. (1973). *Objektive Erkenntnis. Ein evolutionärer Entwurf.* Hoffmann und Campe.

Popper, K. R. (1979). Three worlds. Tanner lecture, Michigan, April 7, 1978. *Michigan Quarterly Review* (1), 141–167. https://tannerlectures.utah.edu/_documents/a-to-z/p/popper80.pdf. Zugegriffen: 12. Mai 2020.

Popper, K. R. (2018 [1984]). *Alle Menschen sind Philosophen*. Piper (Herausgegeben von Heidi Bohnet und Klaus Stadler).

Popper, K. R. (2019 [1987]). *Auf der Suche nach einer besseren Welt. Vorträge und Aufsätze aus dreißig Jahren*. Piper.

Popper, K. R., & Eccles, J. C. (1977). *Das Ich und sein Gehirn*. Piper.

Pott, A. (2007). *Orte des Tourismus. Eine raum- und gesellschaftstheoretische Untersuchung*. transcript.

Putnam, H. (1995). *Pragmatism: An open question*. Blackwell Publishers.

Rau, S. (2017). *Räume. Konzepte, Wahrnehmungen, Nutzungen* (Historische Einführungen, 2., akt. Aufl., Bd. 14). Campus.

Rebay-Salisbury, K. (2013). Phänomenologie und Landschaft: der menschliche Körper in Bewegung. In R. Karl & J. Leskovar (Hrsg.), *Interpretierte Eisenzeiten. Fallstudien, Methoden, Theorie: Tagungsbeträge der 5. Linzer Gespräche zur interpretativen Eisenzeitarchäologie* (Studien zur Kulturgeschichte von Oberösterreich, Folge 37, S. 61–70). Oberösterreichisches Landesmuseum.

Redepenning, M., & Wilhelm, J. (2014). Raumforschung mit luhmannscher Systemtheorie. In J. Oßenbrügge & A. Vogelpohl (Hrsg.), *Theorien in der Raum- und Stadtforschung. Einführungen* (S. 310–327). Westfälisches Dampfboot.

Risse-Kappen, T. (1995). Reden ist nicht billig. Zur Debatte um Kommunikation und Rationalität. *Zeitschrift für Internationale Beziehungen, 2*(1), 171–184.

Ronneberger, K., & Vogelpohl, A. (2014). Henri Lefebvre: Die Produktion des Raumes und die Urbanisierung der Gesellschaft. In J. Oßenbrügge & A. Vogelpohl (Hrsg.), *Theorien in der Raum- und Stadtforschung. Einführungen* (S. 251–270). Westfälisches Dampfboot.

Rorty, R. (1982). *Consequences of pragmatism. Essays: 1972–1980*. University of Minnesota Press.

Rorty, R. (1989). *Kontingenz, Ironie und Solidarität*. Suhrkamp.

Rorty, R. (1991). *Objectivity, relativism, and truth*. Cambridge University Press.

Rosa, H. (2013). *Beschleunigung und Entfremdung. Entwurf einer Kritischen Theorie spätmoderner Zeitlichkeit*. Suhrkamp.

Schamp, E. W. (2012). Evolutionäre Wirtschaftsgeographie. *Zeitschrift für Wirtschaftsgeographie, 56*(1–2), 121–128.

Schenk, W. (2017). Landschaft. In L. Kühnhardt & T. Mayer (Hrsg.), *Bonner Enzyklopädie der Globalität* (Bd. 1, 2, S. 671–684). Springer VS.

Schneider, U. (2004). *Die Macht der Karten. Eine Geschichte der Kartographie vom Mittelalter bis heute*. Primus.

Schroer, M. (2006). *Räume, Orte, Grenzen. Auf dem Weg zu einer Soziologie des Raums*. Suhrkamp.

Schubert, H.-J., Joas, H., & Wenzel, H. (2010). *Pragmatismus zur Einführung. Kreativität, Handlung, Deduktion, Induktion, Abduktion, Chicago School, Sozialreform, symbolische Interaktion* (Zur Einführung, Bd. 382). Junius.

Schulz-Schaeffer, I. (2000). Akteur-Netzwerk-Theorie: Zur Koevolution von Gesellschaft, Natur und Technik. In J. Weyer & J. Abel (Hrsg.), *Soziale Netzwerke. Konzepte und Methoden der sozialwissenschaftlichen Netzwerkforschung* (Lehr- und Handbücher der Soziologie, S. 187–210). Oldenbourg.

Schütz, A. (1960 [1932]). *Der sinnhafte Aufbau der sozialen Welt. Eine Einleitung in die Verstehende Soziologie* (2. Aufl). Julius Springer. (Originalarbeit erschienen 1932).

Schütz, A. (1971a [1962]). *Gesammelte Aufsätze 1. Das Problem der Wirklichkeit*. Martinus Nijhoff.

Schütz, A. (1971b). *Gesammelte Aufsätze 3. Studien zur phänomenologischen Philosophie*. Martinus Nijhoff.

Sloterdijk, P. (1987). *Kopernikanische Mobilmachung und ptolemäische Abrüstung. Ästhetischer Versuch* (Edition Suhrkamp). Suhrkamp.

Smith, N. (1984). *Uneven development. Nature, capital and the production of space.* Blackwell.

Soja, E. W. (1996). *Thirdspace. Journeys to Los Angeles and other real-and-imagined places.* Blackwell.

Soja, E. W. (2003). Thirdspace – Die Erweiterung des Geographischen Blicks. In H. Gebhardt, P. Reuber, & G. Wolkersdorfer (Hrsg.), *Kulturgeographie. Aktuelle Ansätze und Entwicklungen* (Spektrum Lehrbuch, S. 269–288). Spektrum Akademischer.

Soja, E. W. (2014). *My Los Angeles. From urban restructuring to regional urbanization.* University of California Press.

Sokolowski, R. (2000). *Introduction to phenomenology.* Cambridge University Press.

Sørensen, M. L. S., & Rebay-Salisbury, K. (Hrsg.). (2012). *Embodied knowledge. Perspectives on believe and technology.* Oxbow Books.

Sorokin, P. (1959). *Social and cultural mobility.* Free Press of Glencoe.

Steiner, C. (2009). Materie oder Geist? Überlegungen zur Überwindung dualistischer Erkenntniskonzepte aus der Perspektive einer Pragmatischen Geographie. *Berichte zur deutschen Landeskunde, 83*(2), 129–142.

Steiner, C. (2014a). *Pragmatismus – Umwelt – Raum. Potenziale des Pragmatismus für eine transdisziplinäre Geographie der Mitwelt* (Erdkundliches Wissen, Bd. 155). Franz Steiner.

Steiner, C. (2014b). Von Interaktion zu Transaktion – Konsequenzen eines pragmatischen Mensch-Umwelt-Verständnisses für eine Geographie der Mitwelt. *Geographica Helvetica, 69*(3), 171–181.

Stemmer, B. (2016). *Kooperative Landschaftsbewertung in der räumlichen Planung. Sozialkonstruktivistische Analyse der Landschaftswahrnehmung der Öffentlichkeit.* Springer VS.

Stichweh, R. (1998). Raum, Region und Stadt in der Systemtheorie. *Soziale Systeme, 4*(2), 341–358.

Stichweh, R. (2003). Raum und moderne Gesellschaft. Aspekte der sozialen Kontrolle des Raumes. In T. Krämer-Badoni (Hrsg.), *Die Gesellschaft und ihr Raum. Raum als Gegenstand der Soziologie* (Stadt, Raum und Gesellschaft, Bd. 21, S. 93–102). Leske + Budrich.

Sturm, G. (2000). *Wege zum Raum. Methodologische Annäherungen an ein Basiskonzept raumbezogener Wissenschaften.* VS Verlag.

Thibaud, J.-P. (2003). Die sinnliche Umwelt von Städten. Zum Verständnis urbaner Atmosphären. In M. Hauskeller (Hrsg.), *Die Kunst der Wahrnehmung. Beiträge zu einer Philosophie der sinnlichen Erkenntnis* (S. 280–297). SFG-Servicecenter Fachverlage.

Tilley, C. (1997). *A phenomenology of landscape. Places, paths and monuments* (Explorations in anthropology). Berg.

Tilley, C. (2005). Phenomenological archaeology. In C. Renfrew & P. Bahn (Hrsg.), *Archaeology. The key concepts* (Routledge key guides, S. 151–155). Routledge.

Tuan, Y.-F. (1989a). *Space and place. The perspective of experience* (5. Aufl.). University of Minnesota Press.

Tuan, Y.-F. (1989b). Surface phenomena and aesthetic experience. *Annals of the Association of American Geographers, 79*(2), 233–241. https://doi.org/10.1111/j.1467-8306.1989.tb00260.x.

Vester, H.-G. (1993). *Soziologie der Postmoderne.* Quintessenz.

Voelzkow, H. (2000). Korporatismus in Deutschland: Chancen, Risiken und Perspektiven. In E. Holtmann (Hrsg.), *Zwischen Wettbewerbs- und Verhandlungsdemokratie. Analysen zum Regierungssystem der Bundesrepublik Deutschland* (S. 185–212). Westdeutscher.

Vorwerg, C. (2013). *Raumrelationen in Wahrnehmung und Sprache. Kategorisierungsprozesse bei der Benennung visueller Richtungsrelationen* (Studien zur Kognitionswissenschaft). Deutscher Universitätsverlag.

Wardenga, U. (2002). Alte und neue Raumkonzepte für den Geographieunterricht. *Geographie heute, 23*(200), 8–11.

Warms, C. A., & Schroeder, C. A. (1999). Bridging the gulf between science and action: The „new fuzzies" of neopragmatism. *Advances in Nursing Science, 22*(2), 1–10.

Weber, F. (2017). Widerstände im Zuge des Stromnetzausbaus – eine diskurstheoretische Analyse der Argumentationsmuster von Bürgerinitiativen in Anschluss an Laclau und Mouffe. *Berichte. Geographie und Landeskunde, 91*(2), 139–154.

Weber, F. (2018). Ein diskurstheoretischer Zugriff auf ‚Landschaft' und ‚Stadtlandhybride' – Annäherungen zwischen Makro- und Mikroperspektive. In S. Hennecke, H. Kegler, K. Klaczynski, & D. Münderlein (Hrsg.), *Diedrich Bruns wird gelehrt haben. Eine Festschrift* (S. 122–131). Kassel University Press.

Weber, F. (2019a). Diskurstheoretische Landschaftsforschung. In O. Kühne, F. Weber, K. Berr, & C. Jenal (Hrsg.), *Handbuch Landschaft* (S. 105–117). Springer VS.

Weber, F. (2019b). Der Stromnetzausbau in Deutschland – Eine Konturierung des Konfliktes in Anschluss an Chantal Mouffe und Ralf Dahrendorf. In K. Berr & C. Jenal (Hrsg.), *Landschaftskonflikte* (S. 423–437). Springer VS.

Weichhart, P. (1993). How does the person fit into the human ecological triangle? From dualism to duality: The transactional worldview. In D. Steiner & M. Nauser (Hrsg.), *Human ecology. Fragments of anti-fragmentary views of the world* (S. 103–124). Routledge.

Weichhart, P. (1999). Die Räume zwischen den Welten und die Welt der Räume. In P. Meusburger (Hrsg.), *Handlungszentrierte Sozialgeographie. Benno Werlens Entwurf in kritischer Diskussion* (Erdkundliches Wissen, Bd. 130, S. 67–94). Steiner.

Weichhart, P. (2005). Auf der Suche nach der „dritten Säule". Gibt es Wege von der Rhetorik zur Pragmatik? In D. Müller-Mahn & U. Wardenga (Hrsg.), *Möglichkeiten und Grenzen integrativer Forschungsansätze in Physischer Geographie und Humangeographie* (forum ifl, Bd. 2, S. 109–136). Selbstverlag Leibniz-Institut für Länderkunde e. V.

Weichhart, P. (2008). *Entwicklungslinien der Sozialgeographie. Von Hans Bobek bis Benno Werlen* (Sozialgeographie kompakt, Bd. 1). Franz Steiner.

Weichhart, P. (2018a). *Entwicklungslinien der Sozialgeographie. Von Hans Bobek bis Benno Werlen* (Sozialgeographie kompakt, 2., vollst. überarb. u. erw. Aufl., Bd. 1). Franz Steiner.

Weichhart, P. (2018b [2020 erschienen]). Die Landschaft der Landschaften. *Berichte. Geographie und Landeskunde, 92*(3–4), 203–216.

Werlen, B. (1995). Landschafts- und Länderkunde in der Spät-Moderne. In U. Wardenga (Hrsg.), *Kontinuität und Diskontinuität der deutschen Geographie in Umbruchphasen. Studien zur Geschichte der Geographie* (Münstersche geographische Arbeiten, Bd. 39, S. 161–176). Inst. für Geographie.

Werlen, B. (1997). *Sozialgeographie alltäglicher Regionalisierungen* (Globalisierung, Region und Regionalisierung, Bd. 2, Erdkundliches Wissen Schriftenreihe für Forschung und Praxis, Bd. 119). Steiner.

Werlen, B. (2000). *Sozialgeographie. Eine Einführung*. Haupt.

van Wezemael, J., & Loepfe, M. (2009). Veränderte Prozesse der Entscheidungsfindung in der Raumentwicklung. *Geographica Helvetica, 64*(2), 106–118.

Wolf, A. (2020). Landschaftskonflikte im Zuge der Energiewende: Die Windenergieanlagen von Wadgassen (Saarland). In R. Duttmann, O. Kühne, & F. Weber (Hrsg.), *Landschaft als Prozess (in diesem Band)*. Springer VS.

Wylie, J. (2005). A single day's walking: Narrating self and landscape on the South West Coast Path. *Transactions of the Institute of British Geographers, 30*(2), 234–247. https://doi.org/10.1111/j.1475-5661.2005.00163.x.

Wylie, J. (2019). Landscape and phenomenology. In P. Howard, I. Thompson, E. Waterton, & M. Atha (Hrsg.), *The Routledge companion to landscape studies* (2. Aufl., S. 127–138). Routledge.

Developmental Lines and Breaks in Geography—Outline of a History of the Discipline

The development of the 'expert-like special knowledge' in geography was shaped on the one hand by paradigm shifts or also by paradigm parallels, but on the other hand not globally uniform, but rather in national traditions, the different prioritization of the subject in different states, but also by recourse to everyday language specificities. Only in recent decades has the increasing importance of an international (English-speaking) geography become apparent. The following will now outline the development of German-speaking geography, whereby—as a result of the current importance of Anglo-Saxon geography—also references to Anglo-Saxon geography will be made (for an overview: Fig. 7.1).

Geography has long been based on the synthetic description and interpretation of spaces (in the sense of Space 1). Early attempts to derive imprints of peoples from climatic conditions can already be found in ancient Greece (Schultz, 2005). This geo-deterministic view shaped scientific geography until the beginning of the 20th century. While the topographies of the 18th century were still largely descriptive, the "country studies" developed in the 19th century, which were based on three approaches (Born, 1980; Wardenga, 2001): Firstly, the description of hitherto un- or little-researched parts of the earth's surface (for example by Alexander von Humboldt); secondly, the systematic evaluation of existing literature (such as by Carl Ritter) and thirdly, the cartographic recording of the earth's surface (such as by August Petermann). Ferdinand von Richthofen standardized the data collection with his "Guide for Research Travellers" (von Richthofen, 1886). He was endeavouring to standardise the data collection by means of standardised observation questions. Alfred Hettner developed this approach into his "country-specific schema" with the help of reflected methodological considerations (cf. Hettner, 1927; in more detail: Schlottmann & Wintzer, 2019). This approach—as Peter Weichhart (in: Schurr & Weichhart, 2020, p. 54) points out—was by no means a "cookbook instructions for writing country studies", but a fully-fledged content theory. The "logical" or "natural" sequence of the chapters of a country study

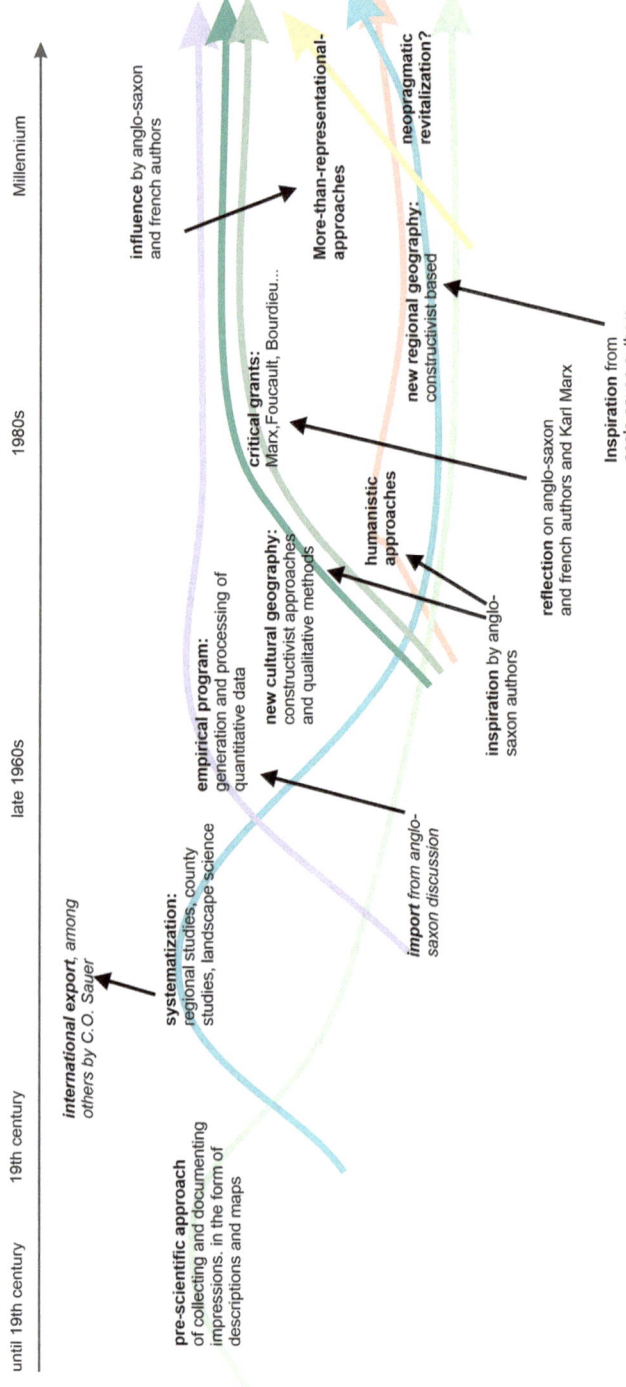

Fig. 7.1 Main research programs of (Western) German geography and their international relations. The arrows indicate the relative importance and are not to be understood absolutely (own representation)

would correspond, for example, to the causal relationships between the physical structure and the cultural conditions. This way of doing geography was also shaped by school geography, which at that time had a great influence on the development of the subject: Schlottmann and Wintzer (2019) see the didactic principle of school geography in teaching pupils the basic features of the earth in the form of a living and design space for humans on the basis of factual knowledge.

For in addition to the trinity of observation, literature evaluation and cartographic recording, "classical geography" knew another trinity, consisting of "geography", "regional geography" and "landscape geography". "Geography" dealt with foreign countries, "regional geography" with one's own country, with the same methodological approach. "Landscape geography" in turn dealt with the research of (sub)regional units, which were "delimited from themselves", that is, "essences" of landscapes were to be captured. The investigation of the "countries" and "landscapes" differed methodologically to a certain extent (so the delimitation problem was hardly present in country and regional geography), but the understanding of "world as a natural and cultural space" (Glasze, 2015, p. 23) was common to the three approaches to space. Research of this tradition pursued the goal of capturing the "essence" of landscapes and countries, which arose from the mutual shaping of nature and culture (see Sect. 3.2; in detail: Eisel, 1982; Hard, 2002; Körner & Eisel, 2003; Piechocki, 2010). This traditional geographical research combined an essentialist core in order to capture the "essence" of landscapes and countries, for which purpose (also) empirical methods, which can be attributed to a positivist tradition, were used. The classical geographical concept of landscape arose from two traditions of thought (Hard, 1977, p. 15): "(1) the 'physiognomic' tradition of the versatile interested traveller, combined with a 'naive' world view and 'landscape eye', and (2) the 'regionalist' tradition of 'thinking in earth spaces' and earth space divisions".

The strongly object-related and culture- and nature-influenced understanding of landscapes in German-speaking geography was anchored in North American geography by Carl O. Sauer (1889–1975). In his classic definition, Carl O. Sauer (1925, p. 46) writes: "The cultural landscape is fashioned from a natural landscape by a cultural group. Culture is the agent, the natural is the medium, the cultural landscape is the result". The founder (in the 1920s) of the "Berkeley School", one of the leading schools of geographical research in the USA in the following decades, understood cultural landscape as a real object, which he interpreted as a "superorganism" (in the terminology used by us as an inscription of Space 3 in Space 1). For him, culture had a decisive influence on the physical space, thus setting itself apart from the previously dominant geodeterminism (which assumed that natural factors determined culture). This tradition of "cultural geography" focused strongly on the investigation of cultural inscriptions in "the landscape", their regional delimitation (in comparison to other landscapes), with a focus on specific cultural phenomena (such as customs; Mathewson, 2009; Wagner & Mikesell, 1962), here one can observe a gain in meaning of Space 3. Sauer and his followers developed an understanding of their role as experts who were strongly oriented towards the maxim of the low involvement of scientists: An expert an be calles a person, standing out from the respective "phenomena, the more, the better it is to investigate them objectively" (Wylie, 2007, p. 41).

After the Second World War, there was a stronger methodological differentiation of data collection and processing, but the search for the "essence" of a section of the earth's surface remained present (a typical example can be found in Brill, 1963). The synthetic form of geographical research continued to dominate the "hypen geographies" (in German: "Bindestrichgeographien", because in German the designation of the specialization with the word geography can be provided with a hyphen; e.g. "Stadt-Geographie"), because the "philosophical conception of geography dating back to the interwar period, that country studies are the core of geographical science" (Wardenga, 2001, p. 20) remained unchanged. In addition to the methodological expansion of country studies, which were staged as the "crowning achievement of geography" (e.g. Bobek, 1957; Schmithüsen, 1959) and the "essential core of the discipline of geography" (Guelke, 1977), the interest in an "appealing presentation" began. A further theoretical or methodological underpinning of geography, in particular of country, regional or landscape studies, which was particularly concerned with the questions of the progressing social modernization, the emerging globalization and the scientific-theoretical considerations, remained largely undeveloped (Wardenga, 2020). The result was a stagnation of the world view of this "traditional geography", which continued to be based on a "well-ordered mosaic of spatially segmented natural and social units" (Blotevogel, 1996, p. 13), in which "both the increasingly important spatial interlocking connections and the conflictual nature of space formation were structurally ignored" (Blotevogel, 1996, p. 13).

Both in the Anglo-Saxon world and in Germany, this paradigm was criticized, although the consequences differed completely. The criticism related to the holistic-organistic landscape and cultural understanding, which was, for example, insensitive to individual (spatial) development (e.g. Duncan, 1980), but also to the 'object fetishism' (Duncan, 1990) of classical cultural geography. In the Anglo-Saxon world, the 'new cultural geography' (Cosgrove, 1984, 1989; Cosgrove & Jackson, 1987; Duncan 1990) emerged from the criticism of the Berkeley School. Although the *new cultural geography* also dealt with historical contexts (however, contextualized and theoretically framed), social aspects were also considered and urban spaces were also included in the investigations (Focus shift to Space 3). There was an interest "in the contingent nature of culture, in dominant ideologies and in forms of resistance to them" (Cosgrove & Jackson, 1987, p. 95). With the *new cultural geography* there is a transition from an essentialist attitude (in connection with empirical access) to a social constructivist-dominated view. For Cosgrove (1984), in his famous definition, landscape is not only the world we see, but also a way of seeing the world. So in the *new cultural geography* 'landscape' always remains present, which was and is understood as a 'text' or as a everyday world construction (in the sense of Landscape 3; Kühne, 2019).

The development in German geography was different. Here, the criticism of the paradigm of traditional geography was cult-mined at the Kiel Geographer's Day in 1969. This was—in particular by the student body, but also by teacher associations (Wardenga, 2020)—criticized as empirically unverifiable, methodologically hardly justifiable and easily ideologizable as well as socially irrelevant. With regard to its reference to material objects, it was accused of "too simple[r] realism" (Kaufmann, 2005, p. 102). The holistic traditional geography was, framed by some retreat skirmishes (more detailed: see Aschauer, 2001; Gebhardt, 2016; Oßenbrügge, 2014; Wardenga, 1996, 2001), replaced by a mainly positivist-empirical paradigm, which, based on Anglo-American models, was oriented towards space models (Bartels, 1968; Kitchin, 2015; for the Anglo-Saxon language area: Harvey, 1969). In addition to Bartels, Hard (1970a, b) shaped the criticism of the "traditional paradigm", in which, "anticipating the later 'linguistic turn'", (Gebhardt, 2019, p. 291) explained the terminology of the landscape science practiced so far as too imprecise for a scientific examination of spatial phenomena.

While in physical geography 'landscape', understood as a material object, remained as a research object and was developed there in connection with ecosystem approaches to geo- or landscape ecology (e.g. Eisel, 2009; Kirchhoff & Trepl, 2009; Leser, 1991, 2019), it was in the "mainstream in anthropogeography it was not very career-enhancing [...], to speak of landscape" (Schenk, 2006, p. 17). An exception was Historical Geography, here 'landscape', in a classical sense, remained present even after 1969. This was concerned with the material (partly immaterial) heritage of landscape, understood as a 'historically grown cultural landscape' (Fehn, 1976; Schenk et al., 1997; Schenk, 2001, 2006, 2011). This historical focus on landscape heritage was in turn intensified in Anglo-Saxon geography since the 00s (Harvey & Wilkinson, 2019). In Germany, however, as Gebhardt (2019, p. 291) states, "human geography has almost completely abandoned a field that used to be part of its core competence", namely object-oriented landscape research (see Textbox 33).

Textbox 33: Mainstream and Relationships of Authority in Science

The mainstream of a science comprises an understanding shared by most representatives of the field about the content, methods, theoretical approaches, etc. of their own discipline, as the reduction of its content to a few programmes (Schwarz, 2011). How a mainstream (within and outside of the sciences) arises and is maintained is in turn a question with which (science) sociology is concerned (see also Sects. 5.3 and 5.4.2). The mainstream is also secured by the fear of failure, in particular by authority relationships. These are based on a double recognition process (Popitz, 1992, p. 29): "On the recognition of the superiority of others as the norm-setting, decisive and on the striving to be recognized by these decision-makers, to receive signs of approval". The recognition relationships are at the same time asymmetrical and reciprocal: "the recognition of the superiority of other persons, the attribution of prestige, and the fixation of our striving for recognition on such superior persons or groups. We want to be recognized by those whom we particularly recognize" (Popitz, 1992). The power of academic students, on the other hand, is severely restricted, but overlaid by self-interest: It consists in the desire to "have students with good positions" (Bourdieu, 1992). These principles of recognition work to a high degree conformist and disciplining, they are based on authority, the "recognition of the *values* they represent" (Sofsky & Paris, 1994, p. 29) and the stabilization of the *mainstream* defined by those of higher status. In securing the *mainstream*, the so-called "Matthew effect" is of central importance, which states that "scientists with high reputation or even those who work at a well-known and respected institution receive attention and recognition that goes beyond their actual achievements, dissociates itself from their direct assessment and becomes autonomous" (Weingart, 2015, p. 23). The mainstream is discursive (see Sect. 6.3) is secured by hegemonizing it, and in turn secures for those who follow it publications, third-party funding and public attention—in short, reputation, as long as the mainstream does not change, triggered for example by a paradigm shift (see Chap. 4; Figs. 7.1, 7.2 and 7.3). ◀

The "quantitative revolution" brought with it, in addition to the abandonment of the search for an "essence" and the spread of increasingly specialized methods of data collection and processing (increasingly Geographic Information Systems), also a positivist world view (Arnreiter & Weichhart, 1998). The category of investigation became generally "space", on a medium-scale level, in particular in German-speaking geography, the concept of the "region" made a career. Here it was assumed that it could be grasped in a more positivist way than the "essentially

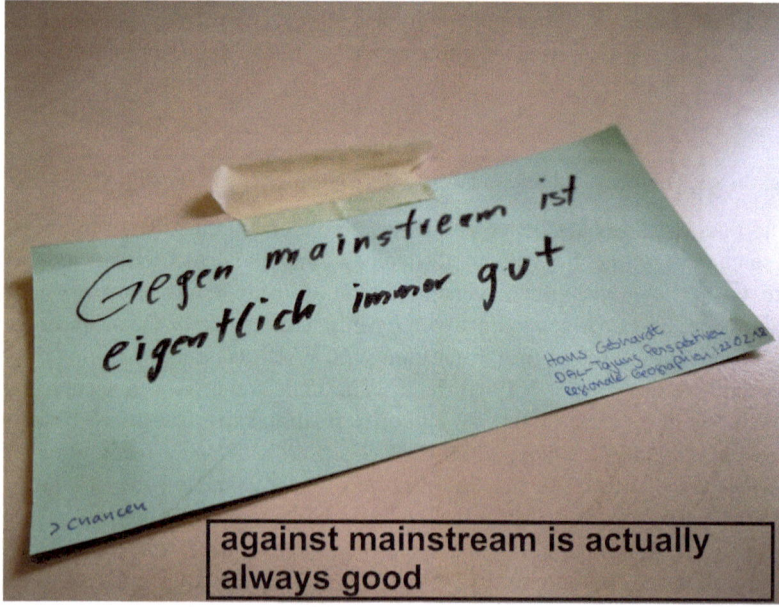

Fig. 7.2 The cartographic documentation of the critical position of the geographer Hans Gebhardt (at a meeting of the German Academy of Regional Studies) with regard to the mainstream makes it clear that scientific progress is primarily made outside or, at best, at the margins of the mainstream (with thanks to Hans Gebhardt for permission to reproduce)

The manuscript as submitted:

... what is left after appraisal and revision can swing in either direction ...

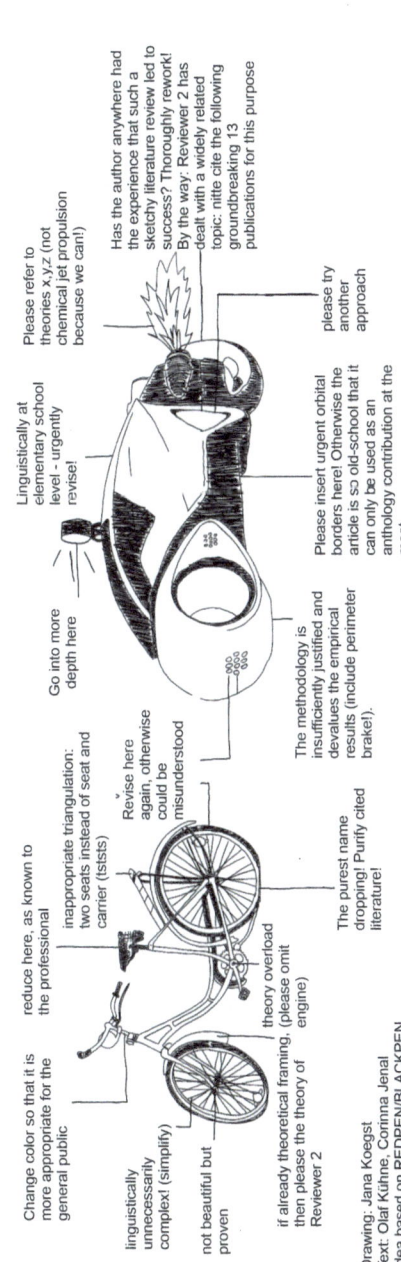

Basically:
Paper tries to be too innovative, so it's not connectable to the mainstream

reduce here, as known to the professional

inappropriate triangulation: two seats instead of seat and carrier (tststs)

Revise here again, otherwise could be misunderstood

Go into more depth here

Linguistically at elementary school level - urgently revise!

Please refer to theories x,y,z (not chemical jet propulsion because we can!)

Basically:
the paper falls short of its potential, in what it goes beyond the mainstream is not made clear

Has the author anywhere had the experience that such a sketchy literature review led to success? Thoroughly rework! By the way: Reviewer 2 has dealt with a widely related topic: nitte cite the following groundbreaking 13 publications for this purpose

please try another approach

Please insert urgent orbital borders here! Otherwise the article is so old-school that it can only be used as an anthology contribution at the most

The methodology is insufficiently justified and devalues the empirical results (include perimeter brake!).

The purest name dropping! Purify cited literature!

theory overload

if already theoretical framing, (please omit engine) then please the theory of Reviewer 2

not beautiful but proven

linguistically unnecessarily complex! (simplify)

Change color so that it is more appropriate for the general public

Drawing: Jana Koegst
Text: Olaf Kühne, Corinna Jenal
Idea based on REDPEN/BLACKPEN

Fig. 7.3 Caricature of possible results of peer reviews: The originally intended instrument of scientific quality assurance often has a mainstream-conforming effect, sometimes has unintended side effects (drawing: Jana Koegst; texts: Olaf Kühne/Corinna Jenal; idea: Redpen/Blackpen)

burdened concept of landscape" (Chilla et al., 2016). The discussion about the two concepts remained much more open for "landscape" in Anglo-Saxon geography (Howe, 2011; Mathewson, 2011). The term "landscape" is still much more present here.

The very different presence of the word "landscape" in German-speaking physical and human geography after 1969 can be understood as an indicator of a more general development: With the farewell to a "holistic geography", physical geography and human geography moved away from each other (cf. Werlen, 2003). The importance of the 'dash-geographies' decreased to the disadvantage of the synthetic overview of spaces (now referred to as 'regional geographies'), although these did not disappear in the 1970s and 1980s, with the integration of physical and human geographical content decreasing (Wardenga, 2001). The addition of 'facts' (often with reminiscences of the Hettner scheme) replaced integration and the construction of a 'being'. Even if a 'problem-oriented regional geography' was formulated in the 1980s, which considered spatially limited sections thematically (an early example: Stewig, 1979), Aschauer (2001, 2002) and Wardenga (2001) noted that regional work focused more on representation than on an independent research perspective. Internationally, the 'new regional geography' emerged. Based on the work of Gilbert (1988; supplemented in particular by Thrift 1991), various perspectives, Marxist and humanist approaches (see Textbox 34) and practice theories were applied to the topic of region (Entrikin, 1996; see Gebhardt et al., 2004; Paasi, 2002) and also connections between the different scale levels came into focus (observations (Paasi, 2009). Regions were no longer understood as 'chambers' delimited from each other, but as dynamically interwoven and connected spatial constructs with local and supra-regional units. As early as the 1990s, criticism of purely constructivist approaches to 'region' was too one-sided, as they did not sufficiently take into account materialities, for example (Holmén, 1995; Murphy, 1991). Here, two paradigmatic reformulations can be seen exemplarily in the 'regional geography': First, the criticism of the quantitative paradigm by the increase in importance of constructivist approaches and, as a result, a stronger empirical focus on qualitative methods since the 1980s, which is reflected, for example, in the development of discourse-theoretical space research (Sect. 6.4), social constructivist landscape research (Sect. 6.5) and the turn to an action-theoretical geography, which, however, already extends into the following category (Sect. 6.7). Secondly, the criticism of the constructivist paradigm, which attributed to this a lack of reference to Space 1. From this criticism, in addition to the return to phenomenological approaches (Sect. 6.6) to space, also arise newer approaches such as the assemblage theory and the actor-network theory (both in Sect. 6.10) or the development of a neopragmatic approach (Sect. 6.11). These approaches in turn make it possible to integrate physical and human geography (see Gebhardt, 2016, 2019)—from this perspective, the consideration of the topic 'landscape'—beyond a search for a speculative 'essence' or an additive recording of 'facts'—appears very up-to-date, for example in the question of which changes at the level of objects (Landscape 1), which individual feelings and evaluations

(Landscape 2) on the basis of which interpretation and evaluation patterns (Landscape 3) trigger (see, among many others: Gebhardt, 2019; Kühne, 2018, 2019; Kühne et al., 2018). At the same time, an increasing focus of geographical interest has been directed towards media content in general and visualisations in particular, not only in the form of conveying research results, but also as an object of research (Döring & Thielmann, 2009; Edler, 2020; Gebhardt, 2019; Jenal, 2020; Schlottmann & Miggelbrink, 2015). With these orientations, German-speaking geography has clearly joined international currents.

Textbox 34: Humanistic Geographies

Humanistic geographies are different approaches that were developed in the English-speaking world in the 1960s, in contrast to the positivist view that was dominant there at the time. They have a theoretical proximity to phenomenology and thus place the individual in his or her needs, but also in his or her ability to shape spaces, but also in the advocacy for people with reduced social participation opportunities in the center of interest. The widespread methodological understanding approach (qualitative methods, often participatory observation) can be seen as an expression of empathy for people, cultural peculiarities and the specifics of places (see Textbox 30; Buttimer & Seamon, 1980; Rodaway, 2014; Seamon, 2014; Tuan, 1976). ◀

One exception remained: While in Anglo-Saxon (specifically North American) geography a strong Marxist expression emerged, the resonance of Marxist ideas in West German geography remained subdued, only in the 1990s did the 'critical geography' (Gebhardt, 2016; Belina, 2009) strengthen, without using the concept of landscape. This was different in the Anglo-Saxon world: Neo-Marxist approaches were sometimes combined with the 'new cultural geography' here (e.g. with Cosgrove & Daniels, 1988). The neo-Marxist-oriented geography (e.g. Harvey, 1996) appropriated the concept of landscape in order to, for example, criticise unequal opportunities for the manifestation of one's own interests in Landscape 1 as well as the development of existing conditions perpetuating landscape-aesthetic ideas (Landscape 3) (Daniels, 1989; Warren, 1994), which contribute to the elimination of economic inequalities (such as the unequal distribution of property in space) by means of aestheticisation (Bermingham, 1989; Cosgrove, 1998; see Cox, 2014 for a summary). In recent years, the critical perspective has gained importance in German-speaking geography, which is not only reflected in the occupation of professorships by Critical Geographers, but also in an increase in publications and the establishment of specifically critical-geographical publication organs (such as the internet journal 'sub/urban. zeitschrift für kritische stadtforschung' or the series 'Raumproduktionen' of the Westfälisches Dampfboot publishing house). This has also internationalised German-speaking geography. This development can be interpreted in different ways: From the 'object of polarised space relations', from the expansion of critical perspectives, based on a

Marxist approach, for example to perspectives based on authors such as Bourdieu or Foucault or (this will be discussed again later) feminist perspectives (see, for example, Kemper & Wiegand, 2014) and/or sociological from the experience of the uncertainty of scientific working conditions (fixed-term contracts for young academics, often referred to as 'precarization'), whereby the social experience co-determines the scientific perspective (in the sense of Mannheim). Contrary to a 'classical' understanding of science (mode 1), which tries to separate science and politics, but also an understanding which critically reflects the entanglements of research with economy and politics (as well as the public) (mode 2), Critical Geography claims to not only name the identified deficiencies, but to actively contribute to their elimination (a frequent topic: gentrification), thus blurring the boundary between science and activism (Samers et al., 2014). In the context of a power-critical engagement with the production of spaces, a feminist perspective on space has also established itself, which deals with the question of how gender relations are solidified in, through and via spaces (Bauriedl & Schurr, 2014). The focus is on how (gender-specific) identities and relationships (re-)produce social inequalities (Marquardt, 2015). Gender is primarily understood in the sense of *gender*, as a socially (re)produced gender (only secondarily as a biological gender in the sense of *sex;* Butler, 2011). The question of how specifically 'male' thinking has been inscribed there and 'female' perspectives marginalized is connected with the feminist perspective on the development of the discipline (Marquardt, 2015; see also: Gilbert, 2008; Mott & Roberts, 2014). With the latter approach, feminist geography develops a meta-perspective on (here) geographical theory formation—in its gender-specific dependence.

Textbox 35: Interim Conclusion on: Development Lines and -Breaks in Geography—Outline of a Disciplinary History

The history of geography can be read as the history of a professional version, in the form of the trinity of country, regional and landscape studies, its canonization, in the form of the schematic attention to Spaces 1, the crisis and a largely new orientation as well as subsequent diversification. The crisis arose essentially from a thematic and methodological/theoretical closure towards social and scientific developments. The paradigmatic realignment took place in (West) Germany with greater vehemence than in the Anglo-Saxon world. Since the 1990s, German-speaking geography has been oriented more internationally. Today, geography is a "multiparadigmatic science" (Schurr & Weichhart, 2020): Depending on the research object and the research question, positivist approaches predominate, which primarily rely on quantitative methods, a multitude of constructivist approaches, which have an affinity for qualitative methods, have been added in the last two decades "more-than-representational" approaches and in addition there are also "residual stocks" of classical regional studies. Finally, critical approaches (in Germany) have gained in importance.

Further Reading

Schlottmann and Wintzer 2019: A very comprehensive and finely structured history of ideas of geographical thinking.

Wardenga 2020: A sociological analysis of the events at and around the Geographers' Conference in Kiel in 1969, which shows that the paradigm shift around Kiel had a long prehistory.

Schurr and Weichhart 2020: A dialogue on the developments in geography since the Geographers' Conference in Kiel in 1969 and the current challenges facing the discipline and the profession.

Hubbard 2002: An introduction to the development, manifestations and topics of Anglo-American geography.

Original Literature

Hard 1970b: An article (among many by the author) which is still worth reading today for its importance to the history of the discipline.

Cosgrove 1984: One of the fundamental works for both new cultural geography and for social and cultural research on landscape.

Paasi 2002: An article (among many by the author) in which he contributes to a new (theoretical) foundation of regional geography.

References

Arnreiter, G., & Weichhart, P. (1998). Rivalisierende Paradigmen im Fach Geographie. In G. Schurz & P. Weingartner (Hrsg.), *Koexistenz rivalisierender Paradigmen. Eine post-kuhnsche Bestandsaufnahme zur Struktur gegenwärtiger Wissenschaft* (S. 53–85). Westdeutscher.

Aschauer, W. (2001). *Landeskunde als adressatenorientierte Form der Darstellung. Ein Plädoyer mit Teilen einer Landeskunde des Landesteils Schleswig* (Forschungen zur deutschen Landeskunde, Bd. 249). Deutsche Akademie für Landeskunde.

Aschauer, W. (2002). Zwischen Theorie und Praxis. Anmerkungen zur Konzeption von Landeskunde. *Berichte zur deutschen Landeskunde, 76*(4), 253–271.

Bartels, D. (1968). *Zur wissenschaftstheoretischen Grundlegung einer Geographie des Menschen* (Erdkundliches Wissen, Bd. 19). Franz Steiner.

Bauriedl, S., & Schurr, C. (2014). Zusammenprall der Identitäten. Soziale und kulturelle Differenz in Städten aus Sicht der feministischen Forschung. In J. Oßenbrügge & A. Vogelpohl (Hrsg.), *Theorien in der Raum- und Stadtforschung. Einführungen* (S. 136–255). Westfälisches Dampfboot.

Belina, B. (2009). Theorie, Kritik und Relevanz in der deutschsprachigen sozialwissenschaftlichen Geographie 40 Jahre nach Kiel. *Rundbrief Geographie, 221*, 18–20.

Bermingham, A. (1989). *Landscape and ideology. The English rustic tradition 1740–1860.* University of California Press.

Blotevogel, H. H. (1996). Aufgaben und Probleme der Regionalen Geographie heute. Überlegungen zur Theorie der Landes- und Länderkunde anläßlich des Gründungskonzepts des Instituts für Länderkunde, Leipzig. *Berichte zur deutschen Landeskunde, 70*(1), 11–40.

Bobek, H. (1957). Gedanken über das logische System der Geographie. *Mitteilungen der Geographischen Gesellschaft Wien, 99*(2), 122–157.

Born, M. (1980). *Geographische Landeskunde des Saarlandes.* Saarbrücker Druckerei und Verlag (Aus dem Nachlass herausgegeben von Renate Born und Helmut Frühauf).

Bourdieu, P. (1992). *Homo academicus* (Suhrkamp-Taschenbuch Wissenschaft, Bd. 1002). Suhrkamp (französische Originalausgabe 1984).

Brill, D. (1963). *Baton Rouge, LA. Aufstieg, Funktionen und Gestalt einer jungen Großstadt des neuen Industriegebietes am unteren Mississippi* (Schriften des Geographischen Instituts der Universität Kiel, Bd. 21,2). Selbstverlag des Geographischen Instituts der Universität Kiel.

Butler, J. (2011). *Gender trouble. Feminism and the subversion of identity*. Routledge.

Buttimer, A., & Seamon, D. (Hrsg.). (1980). *The human experience of space and place*. Croom Helm.

Chilla, T., Kühne, O., & Neufeld, M. (2016). *Regionalentwicklung* (UTB, Bd. 4566). Ulmer.

Cosgrove, D. (1984). *Social formation and symbolic landscape*. University of Wisconsin Press.

Cosgrove, D. (1989). A terrain of metaphor: Cultural geography 1988–89. *Progress in Human Geography, 13*(4), 566–575. https://doi.org/10.1177/030913258901300406.

Cosgrove, D. (1998). *Social formation and symbolic landscape*. University of Wisconsin Press.

Cosgrove, D., & Daniels, S. (Hrsg.). (1988). *The Iconography of landscape. Essays on the symbolic representation, design and use of past environments* (Cambridge Studies in Historical Geography, Bd. 9). Cambridge University Press.

Cosgrove, D., & Jackson, P. (1987). New directions in cultural geography. *Area, 19*(2), 95–101.

Cox, K. R. (2014). *Making human geography*. Guilford Press.

Daniels, S. (1989). Marxism, culture, and the duplicity of landscape. In R. Peet & N. Thrift (Hrsg.), *New models in geography Vol. 2. The political-economy perspective* (S. 196–220). Unwin Hyman.

Döring, J., & Thielmann, T. (Hrsg.). (2009). *Mediengeographie. Theorie – Analyse – Diskussion*. transcript.

Duncan, J. S. (1980). The superorganic in American cultural geography. *Annals of the Association of American Geographers, 70*(2), 181–198.

Duncan, J. S. (1990). *The city as text: The politics of landscape interpretation in the Kandyan Kingdom*. Cambridge University Press.

Edler, D. (2020). Where spatial visualization meets landscape research and „Pinballology": Examples of landscape construction in Pinball Games. *KN – Journal of Cartography and Geographic Information*. https://doi.org/10.1007/s42489-020-00044-1.

Eisel, U. (1982). Die schöne Landschaft als kritische Utopie oder als konservatives Relikt. Über die Kristallisation gegnerischer politischer Philosophien im Symbol „Landschaft". *Soziale Welt, 33*(2), 157–168.

Eisel, U. (2009). *Landschaft und Gesellschaft. Räumliches Denken im Visier* (Raumproduktionen: Theorie und gesellschaftliche Praxis, Bd. 5). Westfälisches Dampfboot.

Entrikin, J. N. (1996). Place and region 2. *Progress in Human Geography, 20*(2), 215–221. https://doi.org/10.1177/030913259602000206.

Fehn, K. (1976). Historische Geographie. Eigenständige Wissenschaft und Teilwissenschaft der Geographie. *Mitteilungen der Geogrphischen Gesellschaft München, 61*, 35–51.

Gebhardt, H. (2016). Entwicklungspfade und Perspektiven der Humangeographie im deutschsprachigen Raum – einige Leitlinien. In J. Aistleitner, M. Coy, & J. Stötter (Hrsg.), *Die Welt verstehen – eine geographische Herausforderung. Eine Festschrift der Geographie Innsbruck für Axel Borsdorf* (Innsbrucker geographische Studien, Bd. 40, S. 43–59). Geographie Innsbruck Selbstverlag.

Gebhardt, H. (2019). Landeskunde und Landschaft – eine kritische Betrachtung. In O. Kühne, F. Weber, K. Berr, & C. Jenal (Hrsg.), *Handbuch Landschaft* (S. 289–298). Springer VS.

Gebhardt, H., Reuber, P., & Wolkersdorfer, G. (2004). Konzepte und Konstruktionsweisen regionaler Geographien im Wandel der Zeit. *Berichte zur deutschen Landeskunde, 78*(3), 293–312.

Gilbert, A. (1988). The new regional geography in English and French-speaking countries. *Progress in Human Geography, 12*(2), 208–228. https://doi.org/10.1177/030913258801200203.

Gilbert, A.-F. (2008). Feministische Geographien: Ein Streifzug in die Zukunft. In P. Moss & K. Falconer Al-Hindi (Hrsg.), *Feminisms in geography. Rethinking space, place, and knowledges* (S. 96–113). Rowman & Littlefield.

Glasze, G. (2015). Identitäten und Räume als politisch: Die Perspektive der Diskurs- und Hegemonietheorie. *Europa Regional, 21*(1–2), 23–34.

Guelke, L. (1977). Regional geography. *The Professional Geographer, 29*(1), 1–7. https://doi. org/10.1111/j.0033-0124.1977.00001.x.

Hard, G. (1970a). „Was ist eine Landschaft?". Über Etymologie als Denkform in der geographischen Literatur. In D. Bartels (Hrsg.), *Wirtschafts- und Sozialgeographie* (Neue wissenschaftliche Bibliothek, Bd. 35, S. 66–84). Kiepenheuer & Witsch.

Hard, G. (1970b). *Die „Landschaft" der Sprache und die „Landschaft" der Geographen. Semantische und forschungslogische Studien*. Ferdinand Dümmlers.

Hard, G. (1977). Zu den Landschaftsbegriffen der Geographie. In A. Hartlieb von Wallthor & H. Quirin (Hrsg.), *„Landschaft" als interdisziplinäres Forschungsproblem. Vorträge und Diskussionen des Kolloquiums am 7./8. November 1975 in Münster* (S. 13–24). Aschendorff.

Hard, G. (2002). Zu Begriff und Geschichte von „Natur" und „Landschaft" in der Geographie des 19. und 20. Jahrhunderts [1983 erstveröffentlicht]. In G. Hard (Hrsg.), *Landschaft und Raum. Aufsätze zur Theorie der Geographie* (Osnabrücker Studien zur Geographie, Bd. 22, S. 171–210). Universitätsverlag Rasch.

Harvey, D. (1969). *Explanation in geography*. Arnold.

Harvey, D. (1996). *Justice, nature and the geography of difference*. Blackwell.

Harvey, D., & Wilkinson, T.J. (2019). Landscape and heritage: emerging landscapes of heritage. In P. Howard, I. Thompson, E. Waterton, & M. Atha (Hrsg.), *The Routledge companion to landscape studies* (2. Aufl., S. 176–191). Routledge.

Hettner, A. (1927). *Die Geographie. Ihre Geschichte, ihr Wesen und ihre Methoden*. Hirt.

Holmén, H. (1995). What's new and what's regional in the ‚new regional geography'? *Geografiska Annaler: Series B, Human Geography, 77*(1), 47–63. https://doi.org/10.1080/04353 684.1995.11879680.

Howe, N. (2011). Landscape versus region. *The Wiley-Blackwell Companion to Human Geography, 16*, 114–129. https://doi.org/10.1002/9781444395839.ch7.

Hubbard, P. (2002). *Thinking geographically. Space, theory and contemporary human geography*. continuum.

Jenal, C. (2020). Visualizations of ‚landscape' in protest movements. On exclusive and inclusive patterns of vision and interpretation using the example of resistance to the expansion of the electricity grid in Germany. In D. Edler, C. Jenal, & O. Kühne (Hrsg.), *Modern approaches to the visualization of landscapes* (S. 427–445). Springer VS.

Kaufmann, S. (2005). *Soziologie der Landschaft*. VS Verlag.

Kemper, J., & Wiegand, F. (2014). Marxistische Stadtforschung. In J. Oßenbrügge & A. Vogelpohl (Hrsg.), *Theorien in der Raum- und Stadtforschung. Einführungen* (S. 215–233). Westfälisches Dampfboot.

Kirchhoff, T., & Trepl, L. (2009). Landschaft, Wildnis, Ökosystem: zur kulturbedingten Vieldeutigkeit ästhetischer, moralischer und theoretischer Naturauffassungen. Einleitender Überblick. In T. Kirchhoff & L. Trepl (Hrsg.), *Vieldeutige Natur. Landschaft, Wildnis und Ökosystem als kulturgeschichtliche Phänomene* (Sozialtheorie, S. 13–68). transcript.

Kitchin, R. (2015). Positivist geography. In S. C. Aitken & G. Valentine (Hrsg.), *Approaches to human geography. Philosophies, theories, people and practices* (2. Aufl., S. 23–34). SAGE.

Körner, S., & Eisel, U. (2003). Naturschutz als kulturelle Aufgabe – theoretische Rekonstruktrion und Anregungen für eine inhaltliche Erweiterung. In S. Körner, A. Nagel, & U. Eisel (Hrsg.), *Naturschutzbegründungen* (S. 5–49). Selbstverlag.

Kühne, O. (2018). *Landschaft und Wandel. Zur Veränderlichkeit von Wahrnehmungen*. Springer VS.

Kühne, O. (2019). *Landscape theories. A brief introduction*. Springer VS.

Kühne, O., Weber, F., & Jenal, C. (2018). *Neue Landschaftsgeographie. Ein Überblick (Essentials)*. Springer VS.

Leser, H. (1991). *Landschaftsökologie. Ansatz, Modelle, Methodik, Anwendung* (UTB, 3., völlig neubearb. Aufl., Bd. 521). Ulmer.

Leser, H. (2019). Landschaftsökologie. In O. Kühne, F. Weber, K. Berr, & C. Jenal (Hrsg.), *Handbuch Landschaft* (S. 181–191). Springer VS.

Marquardt, N. (2015). *Feministische Geographie*. https://gender-glossar.de/f/item/50-feministische-geographie. Zugegriffen: 27. Nov. 2020.

Mathewson, K. (2009). Carl Sauer and his crititcs. In W. M. Denevan & K. Mathewson (Hrsg.), *Carl Sauer on culture and landscape. Readings and commentaries* (S. 9–28). Louisiana State University Press.

Mathewson, K. (2011). Landscape versus region. *The Wiley-Blackwell Companion to Human Geography, 16*, 130.

Mott, C., & Roberts, S. M. (2014). *Not everyone has (the) balls. Urban exploration and the persistence of Masculinist geography.* Wiley.

Murphy, A. B. (1991). Regions as social constructs: The gap between theory and practice. *Progress in Human Geography, 15*(1), 23–35. https://doi.org/10.1177/030913259101500102.

Oßenbrügge, J. (2014). Zur Theoriediskussion in der Geographie und geographischen Stadtforschung. In J. Oßenbrügge & A. Vogelpohl (Hrsg.), *Theorien in der Raum- und Stadtforschung. Einführungen* (S. 24–33). Westfälisches Dampfboot.

Paasi, A. (2002). Place and region: Regional worlds and words. *Progress in Human Geography, 26*(6), 802–811. https://doi.org/10.1191/0309132502ph404pr.

Paasi, A. (2009). Regional geography I. In R. Kitchin & N. Thrift (Hrsg.), *International encyclopedia of human geography* (Bd. 9, S. 214–227). Elsevier.

Piechocki, R. (2010). *Landschaft – Heimat – Wildnis. Schutz der Natur – aber welcher und warum?* Beck.

Popitz, H. (1992). *Phänomene der Macht* (2., stark erw. Aufl.). Mohr Siebeck.

von Richthofen, F. (1886). *Führer für Forschungsreisende. Anleitung zu Beobachtungen über Gegenstände der physischen Geographie und Geologie.* Oppenheim.

Rodaway, P. (2014). Humanism and people-centered methods. In S. C. Aitken & G. Valentine (Hrsg.), *Approaches to human geography. Philosophies, theories, people and practices* (2. Aufl., S. 334–343). SAGE.

Samers, M., Bigger, P., & Belcher, O. (2014). To build another world: Activism in the light of Marxist geographical thought. In S. C. Aitken & G. Valentine (Hrsg.), *Approaches to human geography. Philosophies, theories, people and practices* (2. Aufl., S. 344–360). SAGE.

Sauer, C. O. (1925). *The morphology of landscape* (Universitiy of California publications in geography). Universitiy of California Press.

Schenk, W. (2001). Kulturlandschaft in Zeiten verschärfter Nutzungskonkurrenz: Genese, Akteure, Szenarien. In Akademie für Raumforschung und Landesplanung (Hrsg.), *Die Zukunft der Kulturlandschaft zwischen Verlust, Bewahrung und Gestaltung* (Forschungs- und Sitzungsberichte, Bd. 215, S. 30–44). Selbstverlag.

Schenk, W. (2006). Der Terminus „gewachsene Kulturlandschaft" im Kontext öffentlicher und raumwissenschaftlicher Diskurse zu „Landschaft" und „Kulturlandschaft". In U. Matthiesen, R. Danielzyk, S. Heiland, & S. Tzschaschel (Hrsg.), *Kulturlandschaften als Herausforderung für die Raumplanung. Verständnisse – Erfahrungen – Perspektiven* (Forschungs- und Sitzungsberichte, Bd. 228, S. 9–21). Selbstverlag.

Schenk, W. (2011). *Historische Geographie* (Geowissen kompakt). WBG.

Schenk, W., Fehn, K., & Denecke, D. (Hrsg.). (1997). *Kulturlandschaftspflege. Beiträge der Geographie zur räumlichen Planung.* Borntraeger.

Schlottmann, A., & Miggelbrink, J. (Hrsg.). (2015). *Visuelle Geographien. Zur Produktion, Aneignung und Vermittlung von RaumBildern.* transcript.

Schlottmann, A., & Wintzer, J. (2019). *Weltbildwechsel. Ideengeschichten geographischen Denkens und Handelns* (utb Geographie, 1. Aufl.). Haupt.

Schmithüsen, J. (1959). Das System der geographischen Wissenschaft. *Berichte zur deutschen Landeskunde, 23*, 1–14.

Schultz, H.-D. (2005). Zwischen fordernder Natur und freiem Willen: Das Politische an der „klassischen" deutschen Geographie. *Erdkunde, 59*(1), 1–21.

Schurr, C., & Weichhart, P. (2020). From Margin to Center? Theoretische Aufbrüche in der Geographie seit Kiel 1969. *Geographica Helvetica, 75*(2), 53–67.

Schwarz, H. (2011). Über Tabuthemen in der Wissenschaft, Programmförderung und Mainstreamforschung. *Forschung und Lehre, 18*(5), 354–355.

Seamon, D. (2014). Lived emplacementand the locality of being: A return to humanistic geography? In S. C. Aitken & G. Valentine (Hrsg.), *Approaches to human geography. Philosophies, theories, people and practices* (2. Aufl., S. 35–48). SAGE.

Sofsky, W., & Paris, R. (1994). *Figurationen sozialer Macht. Autorität, Stellvertretung, Koalition* (Suhrkamp Taschenbuch Wissenschaft, Bd. 1135). Suhrkamp.

Stewig, R. (Hrsg.). (1979). *Probleme der Länderkunde* (Wege der Forschung, Bd. 391). Wissenschaftliche Buchgesellschaft.

Tuan, Y.-F. (1976). Humanistic geography. *Annals of the Association of American Geographers, 66*(2), 266–276. https://doi.org/10.1111/j.1467-8306.1976.tb01089.x.

Wagner, P. L., & Mikesell, M. W. (Hrsg.). (1962). *Readings in cultural geography.* University of Chicago Press.

Wardenga, U. (1996). Von der Landeskunde zur „Landeskunde". In G. Heinritz, G. Sandner, & R. Wießner (Hrsg.), *Der Weg der deutschen Geographie. Rückblick und Ausblick* (50. Deutscher Geographentag, Bd. 4, S. 132–141). Franz Steiner.

Wardenga, U. (2001). Theorie und Praxis der länderkundlichen Forschung und Darstellung in Deutschland. In F.-D. Grimm & U. Wardenga (Hrsg.), *Zur Entwicklung des länderkundlichen Ansatzes* (Beiträge zur Regionalen Geographie, Bd. 53, S. 9–35). Selbstverlag.

Wardenga, U. (2020). Vergangene Zukünfte – oder: Die Verhandlung neuer Möglichkeitsräume in der Geographie. Futures past – Or: The negotiation of new spaces of possibility in geography. *Geographische Zeitschrift, 108*(1), 4–22. https://doi.org/10.25162/gz-2019-0009.

Warren, S. (1994). Disneyfication of the metropolis: Popular resistance in Seattle. *Journal of Urban Affairs, 16*(2), 89–107. https://doi.org/10.1111/j.1467-9906.1994.tb00319.x.

Weingart, P. (2015). *Wissenschaftssoziologie.* transcript.

Werlen, B. (2003). Kulturgeographie und kulturtheoretische Wende. In H. Gebhardt, P. Reuber, & G. Wolkersdorfer (Hrsg.), *Kulturgeographie. Aktuelle Ansätze und Entwicklungen* (Spektrum Lehrbuch, S. 251–268). Spektrum Akademischer.

Wylie, J. (2007). *Landscape.* Routledge.

Conclusion

8

As we have shown in this book, people in everyday life as well as in the sciences try to orient themselves in the "world" with linguistic means such as words, concepts, sentences, judgments, conclusions and theoretical generalizations, to recognize and understand something and to grasp the "objects" of everyday or scientific "world" in everyday life as in the sciences. Whoever wants to do science should therefore be informed about the logical, linguistic, semantic and pragmatic basics of scientific theory and the sciences. These include in particular the basics of concept formation, definition theory, statement logic, the correct structure of arguments and typical conclusion forms in the sciences, as well as typical logical, semantic or linguistic errors, misunderstandings, false conclusions and forms of incorrect argumentation. This information also includes the knowledge of the differences and similarities of the different "science-sciences" as well as a knowledge of how difficult it is to determine science and scientificity more precisely as well as the decisive concept of "truth" for science. Also the historical and systematic passage through the decisive philosophical positions as well as the discussion of the decisive problems of the theory of science make the richness of current discussions clear, the knowledge of which is not to be underestimated for an informed reception and independent discussion. It also became clear that and how an appropriate understanding of science in theory and practice depends on the consideration of its historical and social contexts.

Science can therefore be understood as a dynamic system of knowledge relationships with different degrees of probability and certainty. Science is therefore a process, not a state. Scientific knowledge is always contextualized, with other scientific knowledge, with prescientific knowledge, in social relationships, in particular with economy and politics. Science also means diversity of perspectives. If this is lost, it means a loss of interpretation and explanation possibilities for the inner and outer world. Science does not only mean a process of genesis, re-measurement, rejection, supplement and further development in relation to knowledge, but

© The Author(s), under exclusive license to Springer Fachmedien Wiesbaden
GmbH, part of Springer Nature 2022
O. Kühne and K. Berr, *Science, Space, Society*,
https://doi.org/10.1007/978-3-658-39140-9_8

also its social contextualisation changes. While it used to be a strongly self-referential system of production of basic knowledge (Mode 1), today it has opened up strongly towards application-orientation and integration into economic and political processes (Mode 2). A development which is ambivalently assessed, on the one hand it has gained in social relevance, on the other hand it has become more dependent on social (in particular political and economic) influences. For example, the delicate question remains of which basic assumptions sciences cannot do without despite all contextualisation and thus also external determination, whether there is such a thing as a "hard core" of theories, paradigms or sciences which is to be preserved against all social and historical developments if possible. And the fact of the diversity of differentiated sciences, paradigms, theories and methods poses the challenge for science to ask for practicable and appropriate forms of dealing with these pluralisms. This pluralisation is promoted by a paradox which consists in the fact that new knowledge does not lead to less non-knowledge, but on the contrary to the awareness of ever *more* non-knowledge. Pictorially speaking, in the metaphor of the Enlightenment, the light of knowledge illuminates the dark infinite space of possible knowledge and throws a light on what we do not yet know, but only become aware of through the new knowledge. Put differently: Behind every door of new knowledge, scientists stand in front of further doors of the not-yet-known. This is not meant to belittle what has been achieved, but modesty would be a good virtue for all scientists—and this insight can also be played back against exaggerated claims of society on the sciences of society.

The increased contextualization is also reflected in the development of the discipline of geography: Geography has developed from a discipline of description of countries and landscapes (Space 1) to a multi-paradigmatic science that is internationally networked and inter- and transdisciplinary contextualized. This contextualization can be evaluated ambivalently, as an opening or as a loss of the core of the subject (Schurr & Weichhart, 2020). So it is no longer clear what is meant by "space". Speaking of developments in "the space" appears in view of the variety of space concepts and theories only in a few clearly delineated contexts to be possible (the examples are rather to be found in physical, than in human geography). If different theories and concepts are available, there is not only the need to examine which framework is appropriate for the topic to be dealt with, the decision must also be justified and the justification documented. This can also be used as an indicator for the distinction between (in particular social and cultural) scientific knowledge and non- or partially scientific knowledge: The knowledge of theoretical approaches to a research object, their weighing and justified selection and finally the reflection of one's own decisions in the light of alternative theories, one's own and other empirical results.

Currently, three basic currents can be found in the sciences (i.e. humanities, cultural, social and natural sciences) which selectively unite different theoretical, world view and methodological approaches (text box 36). These are not necessarily clearly separated units. Rather, there are transitions and "turbulences" between these basic currents and also scientific actors change between the perspectives, depending on the questions, but also during the course of their scientific biography.

Textbox 36: Three Current Basic Currents of the Sciences

1. The critical perspectives: These are more or less intense on (neo) Marxist thinking. Other theory traditions (such as constructivism) are taken up when they are (neo) Marxist framed. The critical perspectives are represented in particular in social and cultural contexts. Critical perspectives have a tendency to change society through politics, which leads to a tendency to de-differentiation of science and politics. Marxism is accused of a teleological essentialism (namely to know how society must develop quasi-naturally), the neo-Marxism (which renounces this teleological orientation) an undirected criticism of existing conditions, without showing specific alternatives (critics of this approach see something like social romanticism here). It is also criticized that science is used as an accomplice of political interests.

2. Socio-ecological-transformative perspectives: These perspectives are based on very different world views, which can range from conservative to neo-Marxist. They are strongly oriented towards making societies more sustainable. Here there is a tendency towards the undifferentiation of science and society as a whole, namely to align society with scientific findings. This relies heavily on syntheses from natural and social sciences research. Here a teleological attitude is criticized, namely knowing today how future societies will live (want to) and thus excluding the possibility of alternative development. Critics see here approaches of social romanticism, but especially eco-romanticism (which is often found in an idealization of pre-modern social conditions). In addition, the merging of science and society is subject to criticism, so that it cannot establish a detached observational perspective on society in relation to society.

3. Continuation of the tradition of the Enlightenment: These perspectives continue the tradition of "classical science": They separate science from politics in particular and the rest of (non-scientific) society in general. In terms of ideology, they are nevertheless close to liberal thinking, from a variety of competing theories and methods, the most suitable one may be found. Frequent scientific perspectives are positivistic, critical rationalistic, but also constructivist approaches, depending on which scientific question is to be analyzed. This tradition is predominantly found in natural sciences and humanities, but can also be found in social and cultural sciences. Criticism of this perspective is that knowledge production is primarily a self-purpose, so this type of science has little social binding, thus threatening to become socially irrelevant. It is also criticized that here scientific knowledge is given a higher position than other social knowledge, which, inter alia, leads to hierarchical communication.

Geography draws on natural, social, as well as cultural and humanities traditions as a science. All three of these basic currents can be found in it. ◄

But not only theoretical approaches to and from science have become more complex through the differentiation of basic currents and the development of competing paradigms, but also doing science has become more complex with the
transition from Mode 1 to Mode 2 science. From a more or less self-regulating
system, the logic of which was to produce "true" knowledge, Mode 2 has become
a Pastiche confronted with different science-external claims, which requires scientists to be skilled in coordination, project management, networking, public
relations and much more, as the caricature in Fig. 8.1 represents. And yes: the
last word of this book. Pastiche. This means "not simply Entdifferenzierung",
here the dissolution of the difference between science and non-science, it "presupposes Differenzbildung in order to then lead to Hybridkreuzungen, Rekombi-
nationen, Reintegrationen" (Vester, 1993, p. 29; more on the concept of Pastiche
see e.g. Hoesterey, 2001). Science therefore enters into alliances, connections up
to symbioses of different duration with systems previously separated from it such
as politics, economy or even public, scientists become social media stars, politics
determines the development of universities by target agreements (with), companies finance research projects, civil society organizations influence what comes
into the focus of scientific attention. And here too: the future is open.

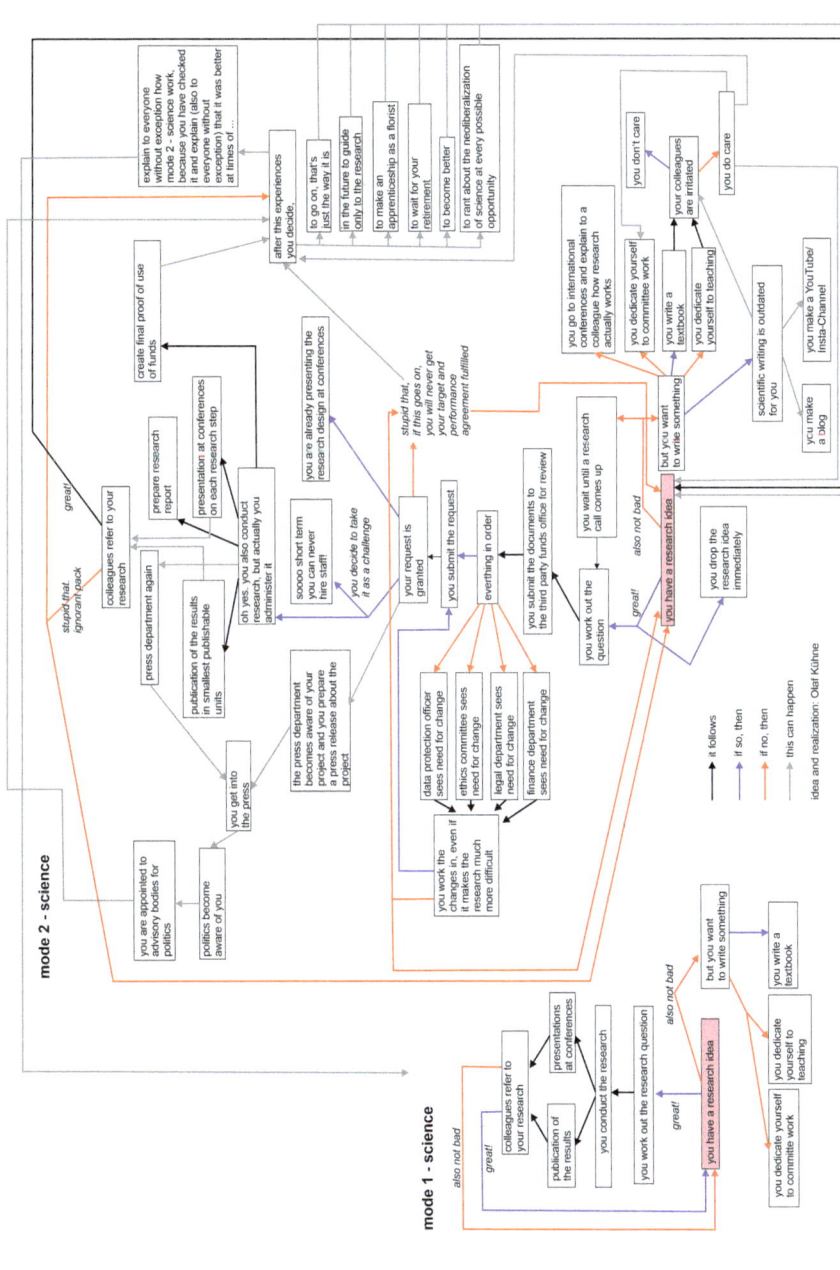

Fig. 8.1 The new complexity of Mode-2 sciences poses 'challenges' that can lead those affected to caricature … (own representation)

References

Hoesterey, I. (2001). *Pastiche. Cultural memory in art, film, literature*. Indiana University Press.

Schurr, C., & Weichhart, P. (2020). From Margin to Center? Theoretische Aufbrüche in der Geographie seit Kiel 1969. *Geographica Helvetica, 75*(2), 53–67.

Vester, H.-G. (1993). *Soziologie der Postmoderne*. Quintessenz.

References

Abel, G. (2005). Die Kunst des Neuen. Kreativität als Problem der Philosophie. In G. Abel (Hrsg.), *Kreativität. XX. Deutscher Kongress für Philosophie, 26.–30. September 2005 in Berlin* (Sektionsbeiträge, Bd. 2, S. 1–21). Universitätsverlag der TU Berlin.

Abel, G. (2007). Kreativität – Worin besteht sie und was macht sie so wertvoll? In G. Abel, J. Conzett, U. P. Jauch, & P. M. Da Rocha (Hrsg.), *Grenzüberschreitungen im Entwurf* (Architekturvorträge der ETH Zürich, Bd. 5, S. 10–43). gta.

Agnoli, J. (1968). *Die Transformation der Demokratie*. Europäische Verlaganstalt.

Aitken, S. C., & Valentine, G. (Hrsg.). (2014). *Approaches to human geography. Philosophies, theories, people and practices* (2. Aufl.). SAGE.

Aitken, S. C., & Valentine, G. (Hrsg.). (2015). *Approaches to human geography. Philosophies, theories, people and practices* (2. Aufl.). SAGE.

Albert, G. (2005). *Hermeneutischer Positivismus und dialektischer Essentialismus Vilfredo Paretos*. VS Verlag.

Altvater, E. (1987). *Sachzwang Weltmarkt. Verschuldungskrise, blockierte Industrialisierung und ökologische Gefährdung – der Fall Brasilien*. VSA.

Andersson, G. B. J. (1988). *Kritik und Wissenschaftsgeschichte. Kuhns, Lakatos' und Feyerabends Kritik des kritischen Rationalismus* (Die Einheit der Gesellschaftswissenschaften, Bd. 54). Mohr.

von Aquin, T. (2013 [1256–1259]). *Über die Wahrheit. Quaestiones disputatae de veritate*. Marix (In der Übersetzung von Eith Stein).

Aristoteles. (1991 [348–345 v. u. Z.]). *Metaphysik. Schriften zur ersten Philosophie*. Reclam.

Aristoteles. (2001). *Die Nikomachische Ethik. Griechisch-deutsch* (Sammlung Tusculum). Artemis & Winkler (Übersetzt von Olof Gigon, neu heraugegeben von Rainer Nickel).

Aristoteles. (2004 [367–344 v. u. Z.]). *Topik*. Reclam.

Aristoteles. (2009). *Politik. Neuausgabe*. Rowohlt (Nach der Übersetzung von Franz Susemihl mit Einleitung, Bibliographie und zusätzlichen Anmerkungen von Wolfgang Kullmann).

Arnreiter, G., & Weichhart, P. (1998). Rivalisierende Paradigmen im Fach Geographie. In G. Schurz & P. Weingartner (Hrsg.), *Koexistenz rivalisierender Paradigmen. Eine post-kuhnsche Bestandsaufnahme zur Struktur gegenwärtiger Wissenschaft* (S. 53–85). Westdeutscher.

Aschauer, W. (2001). *Landeskunde als adressatenorientierte Form der Darstellung. Ein Plädoyer mit Teilen einer Landeskunde des Landesteils Schleswig* (Forschungen zur deutschen Landeskunde, Bd. 249). Deutsche Akademie für Landeskunde.

Aschauer, W. (2002). Zwischen Theorie und Praxis. Anmerkungen zur Konzeption von Landeskunde. *Berichte zur deutschen Landeskunde, 76*(4), 253–271.

Aschenbrand, E. (2017). *Die Landschaft des Tourismus. Wie Landschaft von Reiseveranstaltern inszeniert und von Touristen konsumiert wird*. Springer VS.

Bacon, F. (1990 [1620]). *Neues Organon. Herausgegeben und mit einer Einleitung von Woflgang Krohn* (Teilband 1 und Teilband 2). Wissenschaftliche Buchgesellschaft.

Bacon, F. (2006 [1605]). *Über die Würde und die Förderung der Wissenschaften: London 1605/1623. Aus dem Englischen übertragen von Jutta Schlösser. Herausgegeben und mit einem Anhang versehen von Hermann Klenner* (Haufe-Schriftenreihe zur rechtswissenschaftlichen Grundlagenforschung, Bd. 19). Haufe.

Balsiger, P. W. (2005). *Transdisziplinarität. Systematisch-vergleichende Untersuchung disziplinenübergreifender Wissenschaftspraxis.* Fink.

Balzert, H., Schäfer, C., Schröder, M., & Kern, U. (2010). *Wissenschaftliches Arbeiten. Wissenschaft, Quellen, Artefakte, Organisation, Präsentation* (Soft skills). Herdecke W3L.

Barnes, B. (2013 [1974]). *Scientific knowledge and sociological theory.* Routledge.

Bärsch, C.-E. (1981). Sozialismus. In J. H. Schoeps, J. H. Knoll, & C.-E. Bärsch (Hrsg.), *Konservativismus, Liberalismus, Sozialismus. Einführung, Texte, Bibliographien* (Uni-Taschenbücher Politologie, Neuere Geschichte, Soziologie, Bd. 1032, S. 140–249). Fink.

Bartelborth, T. (1996). Der Schluß auf die beste Erklärung. In C. Hubig (Hrsg.), *Cognitio humana – Dynamik des Wissens und der Werte* (S. 552–559). Akademie.

Bartels, D. (1968). *Zur wissenschaftstheoretischen Grundlegung einer Geographie des Menschen* (Erdkundliches Wissen, Bd. 19). Franz Steiner.

Bathelt, H., & Glückler, J. (2003). Toward a relational economic geography. *Journal of Economic Geography, 3*(2), 117–144. https://doi.org/10.1093/jeg/3.2.117.

Bauberger, S. (2016). *Wissenschaftstheorie. Eine Einführung.* Kohlhammer.

Bauer, L., & Wall-Strasser, S. (2016). Liberalismus/Neoliberalismus. *Politik und Zeitgeschehen, 4.* https://www.voegb.at/cs/Satellite?blobcol=urldata&blobheadername1=content-type&blob headername2=contentdisposition&blobheadervalue1=application%2Fpdf&blobheadervalue 2=inline%3B+filename%3D%22PZG-04_Liberalismus%252FNeoliberalismus.pdf%22&blo bkey=id&blobnocache=false&blobtable=MungoBlobs&blobwhere=1342594922248&ssbi nary=true&site=S08.

Bauriedl, S., & Schurr, C. (2014). Zusammenprall der Identitäten. Soziale und kulturelle Differenz in Städten aus Sicht der feministischen Forschung. In J. Oßenbrügge & A. Vogelpohl (Hrsg.), *Theorien in der Raum- und Stadtforschung. Einführungen* (S. 136–255). Westfälisches Dampfboot.

Beck, U. (1986). *Risikogesellschaft. Auf dem Weg in eine andere Moderne* (Edition Suhrkamp, Bd. 1365). Suhrkamp.

Beck, U. (2007). *Weltrisikogesellschaft. Auf der Suche nach der verlorenen Sicherheit* (Edition Zweite Moderne). Suhrkamp.

Becker, W. (2013). *Macht ohne Maß und kein Ende? Katholizismus, Kapitalismus (Imperialismus) und Kommunismus.* Engelsdorfer.

Beckermann, A. (2014). *Einführung in die Logik* (De Gruyter Studium, 4., durchgesehene. Aufl.). De Gruyter.

Beckmann, J. P. (1997). Rationalismus I. In G. Müller (Hrsg.), *Theologische Realenzyklopädie.* (Theologische Realenzyklopädie, Bd. 28, S. 161–170). De Gruyter.

Belina, B. (2009a). Kriminalitätskartierung – Produkt und Mittel neoliberalen Regierens, oder: Wenn falsche Abstraktionen durch die Macht der Karte praktisch wahr gemacht werden. *Geographische Zeitschrift, 97*(4), 192–212.

Belina, B. (2009b). Theorie, Kritik und Relevanz in der deutschsprachigen sozialwissenschaftlichen Geographie 40 Jahre nach Kiel. *Rundbrief Geographie, 221,* 18–20.

Bender, G. (2004). Modus 2 – Wissenserzeugung in globalen Netzwerken? In U. Matthiesen (Hrsg.), *Stadtregion und Wissen. Analysen und Plädoyers für eine wissensbasierte Stadtpolitik* (S. 87–96). VS Verlag.

Berger, P. L., & Luckmann, T. (1966). *The social construction of reality. A treatise in the sociology of knowledge.* Anchor Books.

Berleant, A. (1997). *Living in the landscape. Toward an aesthetics of environment.* University Press of Kansas.

Berlin, I. (1995 [1969]). *Freiheit. Vier Versuche.* Fischer.

Bermingham, A. (1989). *Landscape and ideology. The English rustic tradition 1740–1860.* University of California Press.

Berr, K. (Hrsg.). (2018). *Transdisziplinäre Landschaftsforschung. Grundlagen und Perspektiven.* Springer VS.

Berr, K. (2019). Konflikt und Ethik. In K. Berr & C. Jenal (Hrsg.), *Landschaftskonflikte* (S. 109–129). Springer VS.

Berr, K. (2020). Vom Wahren, Schönen und Guten. Philosophische Zugänge zu Landschaftsprozessen. In R. Duttmann, O. Kühne, & F. Weber (Hrsg.), *Landschaft als Prozess.* Springer VS.

Berr, K., & Kühne, O. (2019). Moral und Ethik von Landschaft. In O. Kühne, F. Weber, K. Berr, & C. Jenal (Hrsg.), *Handbuch Landschaft* (S. 351–365). Springer VS.

Berr, K., & Kühne, O. (2020). *„Und das ungeheure Bild der Landschaft …".* The Genesis of Landscape Understanding in the German-speaking Regions. Springer VS.

Berr, K., Jenal, C., Kühne, O., & Weber, F. (2019). *Landschaftsgovernance. Ein Überblick zu Theorie und Praxis.* Springer VS.

Birnbacher, D. (2006). *Natürlichkeit.* De Gruyter.

Blackbourn, D. (2007). *Die Eroberung der Natur. Eine Geschichte der deutschen Landschaft.* Random House.

Bloor, D. (1982). Durkheim and Mauss revisited: Classification and the sociology of knowledge. *Studies in History and Philosophy of Science London, 13*(4), 267–297.

Bloor, D. (1991 [1976]). *Knowledge and social imagery.* University of Chicago Press.

Blotevogel, H. H. (1996). Aufgaben und Probleme der Regionalen Geographie heute. Überlegungen zur Theorie der Landes- und Länderkunde anläßlich des Gründungskonzepts des Instituts für Länderkunde, Leipzig. *Berichte zur deutschen Landeskunde, 70*(1), 11–40.

BNatSchG. (2009 [1976]). *Gesetz über Naturschutz und Landschaftspflege (Bundesnaturschutzgesetz).* https://www.gesetze-im-internet.de/bnatschg_2009. Zugegriffen: 17. Mai 2018.

Bobek, H. (1957). Gedanken über das logische System der Geographie. *Mitteilungen der Geographischen Gesellschaft Wien, 99*(2), 122–157.

Boggs, J. S., & Rantisi, N. M. (2003). The ‚relational turn' in economic geography. *Journal of Economic Geography, 3*(2), 109–116.

Bojadžiev, T. (2003). *Die Nacht im Mittelalter.* Königshausen & Neumann.

Bollnow, O. F. (1963). *Mensch und Raum.* Kohlhammer.

Bonney, R., Cooper, C. B., Dickinson, J., Kelling, S., Phillips, T., Rosenberg, K. V., & Shirk, J. (2009). Citizen science: A developing tool for expanding science knowledge and scientific literacy. *BioScience, 59*(11), 977–984. https://doi.org/10.1525/bio.2009.59.11.9.

Bormann, K. (1987). *Platon.* Alber.

Born, M. (1980). *Geographische Landeskunde des Saarlandes.* Saarbrücker Druckerei und Verlag (Aus dem Nachlass herausgegeben von Renate Born und Helmut Frühauf).

Borsche, T. (1996). Einleitung. Sprachphilosophische Überlegungen zu einer Geschichte der Sprachphilosophie. In T. Borsche (Hrsg.), *Klassiker der Sprachphilosophie. Von Platon bis Noam Chomsky* (S. 7–13). Beck.

Böschen, S. (2005). Reflexive Wissenspolitik. Formierung und Strukturierung von Gestaltungsöffentlichkeit. In A. Bogner & H. Torgersen (Hrsg.), *Wozu Experten? Ambivalenzen der Beziehung von Wissenschaft und Politik* (S. 241–265). VS Verlag.

Boschma, R. A., & Frenken, K. (2006). Why is economic geography not an evolutionary science? Towards an evolutionary economic geography. *Journal of Economic Geography, 6*(3), 273–302.

Bosco, F. J. (2014). Actor-network theory, networks, and relational geographies. In S. C. Aitken & G. Valentine (Hrsg.), *Approaches to human geography. Philosophies, theories, people and practices* (2. Aufl., S. 150–162). SAGE.

Bosco, F. J. (2015). Actor-network theory, networks, and relational geographies. In S. C. Aitken & G. Valentine (Hrsg.), *Approaches to human geography. Philosophies, theories, people and practices* (2. Aufl., S. 150–162). SAGE.

Bourdieu, P. (1979 [frz. Original 1972]). *Entwurf einer Theorie der Praxis auf der ethnologis-chen Grundlage der kabylischen Gesellschaft*. Suhrkamp.

Bourdieu, P. (1982). *Leçon sur la leçon*. Les Éditions de Minuit.

Bourdieu, P. (1985). *Sozialer Raum und „Klassen". Leçon sur la leçon; 2 Vorlesungen*. Suhrkamp.

Bourdieu, P. (1987 [1979]). *Die feinen Unterschiede. Kritik der gesellschaftlichen Urteilskraft* (Suhrkamp Taschenbuch Wissenschaft, Bd. 658). Suhrkamp.

Bourdieu, P. (1991). Physischer, sozialer und angeeigneter physischer Raum. In M. Wentz (Hrsg.), *Stadt-Räume* (S. 25–34). Campus.

Bourdieu, P. (1992). *Homo academicus* (Suhrkamp-Taschenbuch Wissenschaft, Bd. 1002). Suhrkamp (französische Originalausgabe 1984).

Bourdieu, P. (1996). Die Praxis der reflexiven Anthropologie. Einleitung zum Seminar an der École des hates études en sciences sociales. Paris, Oktober 1987. In P. Bourdieu & L. Wacquant (Hrsg.), *Reflexive Anthropologie* (S. 251–294). Suhrkamp.

Bourdieu, P. (1998). *Der Einzige und sein Eigenheim*. VSA.

Bourdieu, P. (2000). *Les structures sociales de l'économie*. Seuil.

Bourdieu, P. (2001). *Meditationen: zur Kritik der scholastischen Vernunft*. Suhrkamp.

Bourdieu, P. (2005a [1983a]). Ökonomisches Kapital – Kulturelles Kapital – Soziales Kapital. In P. Bourdieu (Hrsg.), *Die verborgenen Mechanismen der Macht* (S. 49–80). VSA.

Bourdieu, P. (2005b [1983b]). Politik, Bildung und Sprache [1997]. In P. Bourdieu (Hrsg.), *Die verborgenen Mechanismen der Macht. Schriften zu Politik und Kultur 1* (S. 13–30). VSA.

Bourdieu, P., & Wacquant, L. (1996). Die Ziele der reflexiven Soziologie. Chicago-Seminar 1987. In H. Beister (Hrsg.), *Reflexive Anthropologie* (1. Aufl., S. 95–249). Suhrkamp.

Bredemeier, K. (2005). *Schwarze Rhetorik. Macht und Magie der Sprache*. Goldmann.

Brenner, N., Madden, D. J., & Wachsmuth, D. (2011). Assemblage urbanism and the challenges of critical urban theory. *City, 15*(2), 225–240.

Brill, D. (1963). *Baton Rouge, LA. Aufstieg, Funktionen und Gestalt einer jungen Großstadt des neuen Industriegebietes am unteren Mississippi* (Schriften des Geographischen Instituts der Universität Kiel, Bd. 21,2). Selbstverlag des Geographischen Instituts der Universität Kiel.

Broad, C. D. (1952). *Ethics and the history of philosophy. Selected essays*. Routledge.

Bruns, D. (2016). Kulturell diverse Raumaneignung. In F. Weber & O. Kühne (Hrsg.), *Fraktale Metropolen. Stadtentwicklung zwischen Devianz, Polarisierung und Hybridisierung* (S. 231–240). Springer VS.

Bruns, D., & Kühne, O. (2015). Gesellschaftliche Transformation und die Entwicklung von Landschaft. Eine Betrachtung aus der Perspektive der sozialkonstruktivistischen Landschaftstheorie. In O. Kühne, K. Gawroński, & J. Hernik (Hrsg.), *Transformation und Landschaft. Die Folgen sozialer Wandlungsprozesse auf Landschaft* (S. 17–34). Springer VS.

Bruns, D., & Münderlein, D. (2019). Interkulturelle Konstruktion. In O. Kühne, F. Weber, K. Berr, & C. Jenal (Hrsg.), *Handbuch Landschaft* (S. 313–319). Springer VS.

Bucher, T. G. (1998). *Einführung in die angewandte Logik*. De Gruyter.

Bunge, M. (1984). *Das Leib-Seele-Problem. Ein psychobiologischer Versuch* (Die Einheit der Gesellschaftswissenschaften, Bd. 37). J. C. B. Mohr (Paul Siebeck).

Burckhardt, L. (2006). Landschaftsentwicklung und Gesellschaftsstruktur (1977). In M. Ritter & M. Schmitz (Hrsg.), *Warum ist Landschaft schön? Die Spaziergangswissenschaft* (S. 19–32). Martin Schmitz.

Burnet, J. (1962). *Platonis Opera recognovit brevique adnotatione critica instruxit Ioannes Burnet. Tomus II*. Clarendon.

Burr, V. (2005). *Social constructivism*. Routledge.

Butler, J. (2001). *Psyche der Macht. Das Subjekt der Unterwerfung*. Suhrkamp.

Butler, J. (2011). *Gender trouble. Feminism and the subversion of identity*. Routledge.

Buttimer, A., & Seamon, D. (Hrsg.). (1980). *The human experience of space and place*. Croom Helm.

Callon, M. (1984). Some elements of a sociology of translation: Domestication of the scallops and the fishermen of St Brieuc Bay. *The Sociological Review, 32*(1), 196–233.

Callon, M. (1986). The sociology of an actor-network: The case of the electric vehicle. In M. Callon, J. Law, & A. Rip (Hrsg.), *Mapping the dynamics of science and technology* (S. 19–34). Palgrave Macmillan.

Callon, M. (1990). Techno-economic networks and irreversibility. *The Sociological Review, 38*(1), 132–161.

Carnap, R. (1931). Die physikalische Sprache als Universalsprache der Wissenschaft. *Erkenntnis, 2*(1), 432–465. https://doi.org/10.1007/BF02028172.

Carnap, R. (1962). *Logical foundations of probability*. University of Chicago Press.

Carnap, R. (1966 [1928]). *Scheinprobleme in der Philosophie. Das Fremdpsychische und der Realismusstreit*. Suhrkamp.

Carrier, M. (2005). Verwertungsdruck und Erkenntnisgewinn: Philosophische Reflexion angewandter Forschung. *Information Philosophie, 3*, 7–19.

Carrier, M. (2007). Erkenntnisgewinn und Nutzenmehrung. Eine verwickelte Beziehung. In M. Carrier, W. Krohn, & P. Weingart (Hrsg.), *Nachrichten aus der Wissensgesellschaft. Analysen zur Veränderung der Wissenschaft* (S. 93–110). Velbrück-Wissenschaft.

Carrier, M. (2017). *Wissenschaftstheorie zur Einführung* (Zur Einführung, 4., überarb. Aufl., Bd. 353). Junius.

Carrier, M., Krohn, W., & Weingart, P. (Hrsg.). (2007). *Nachrichten aus der Wissensgesellschaft. Analysen zur Veränderung der Wissenschaft*. Velbrück-Wissenschaft.

Chalmers, A. F. (1999). *Grenzen der Wissenschaft*. Springer.

Chalmers, A. F. (2006 [1996]). *Wege der Wissenschaft. Einführung in die Wissenschaftstheorie*. Springer.

Chemero, A. (2003). An outline of a theory of affordances. *Ecological Psychology, 15*(2), 181–195. https://doi.org/10.1207/S15326969ECO1502_5.

Chilla, T., Kühne, O., Weber, F., & Weber, F. (2015). „Neopragmatische" Argumente zur Vereinbarkeit von konzeptioneller Diskussion und Praxis der Regionalentwicklung. In O. Kühne & F. Weber (Hrsg.), *Bausteine der Regionalentwicklung* (S. 13–24). Springer VS.

Chilla, T., Kühne, O., & Neufeld, M. (2016). *Regionalentwicklung* (UTB, Bd. 4566). Ulmer.

Cosgrove, D. (1984). *Social formation and symbolic landscape*. University of Wisconsin Press.

Cosgrove, D. (1989a). Geography is everywhere: Culture and symbolism in human landscapes. In D. Gregory & R. Walford (Hrsg.), *Horizons in human geography* (S. 118–135). Macmillan Press LTD.

Cosgrove, D. (1989b). A terrain of metaphor: Cultural geography 1988–89. *Progress in Human Geography, 13*(4), 566–575. https://doi.org/10.1177/030913258901300406.

Cosgrove, D. (1993). *The Palladian landscape. Geographical change and its cultural representations in sixteenth-century Italy*. Pennsylvania State University Press.

Cosgrove, D. (1998). *Social formation and symbolic landscape*. University of Wisconsin Press.

Cosgrove, D., & Daniels, S. (Hrsg.). (1988). *The Iconography of landscape. Essays on the symbolic representation, design and use of past environments* (Cambridge Studies in Historical Geography, Bd. 9). Cambridge University Press.

Cosgrove, D., & Jackson, P. (1987). New directions in cultural geography. *Area, 19*(2), 95–101.

Cox, K. R. (2014). *Making human geography*. Guilford Press.

Cresswell, T. (2012). *Geographic thought. A critical introduction*. Wiley-Blackwell.

Dahrendorf, R. (1957). *Soziale Klassen und Klassenkonflikt in der industriellen Gesellschaft*. Enke.

Dahrendorf, R. (1971). *Die Idee des Gerechten im Denken von Karl Marx*. Verlag für Literatur und Zeitgeschehen.

Dahrendorf, R. (1972). *Konflikt und Freiheit. Auf dem Weg zur Dienstklassengesellschaft*. Piper.

Dahrendorf, R. (1979). *Lebenschancen. Anläufe zur sozialen und politischen Theorie* (Suhrkamp-Taschenbuch, Bd. 559). Suhrkamp.

Dahrendorf, R. (1980). *Die neue Freiheit. Überleben und Gerechtigkeit in einer veränderten Welt.* Suhrkamp.

Dahrendorf, R. (1983). *Die Chancen der Krise. Über die Zukunft des Liberalismus.* Deutsche Verlags-Anstalt.

Dahrendorf, R. (1992). *Der moderne soziale Konflikt. Essay zur Politik der Freiheit.* Deutsche Verlags-Anstalt.

Dahrendorf, R. (2007a). *Auf der Suche nach einer neuen Ordnung. Vorlesungen zur Politik der Freiheit im 21. Jahrhundert* (Krupp-Vorlesungen zu Politik und Geschichte am Kulturwissenschaftlichen Institut im Wissenschaftszentrum Nordrhein-Westfalen, 4. Aufl., Bd. 3). C. H. Beck.

Dahrendorf, R. (2007b). Freiheit – eine Definition. In U. Ackermann (Hrsg.), *Welche Freiheit. Plädoyers für eine offene Gesellschaft* (S. 26–39). Matthes & Seitz.

Dahrendorf, R. (2008). *Versuchungen der Unfreiheit. Die Intellektuellen in Zeiten der Prüfung* (Beck'sche Reihe, Bd. 1875). C. H. Beck.

Dangschat, J. (2014). Stadt und Raum in der Soziologie. In J. Oßenbrügge & A. Vogelpohl (Hrsg.), *Theorien in der Raum- und Stadtforschung. Einführungen* (S. 57–67). Westfälisches Dampfboot.

Daniels, S. (1989). Marxism, culture, and the duplicity of landscape. In R. Peet & N. Thrift (Hrsg.), *New models in geography Vol. 2. The political-economy perspective* (S. 196–220). Unwin Hyman.

Davis, B. D. (1978). The moralistic fallacy. *Nature, 272,* 390. https://doi.org/10.1038/272390a0.

Dear, M. (2005). The Los Angeles School of Urbanism. In B. J. L. Berry & J. O. Wheeler (Hrsg.), *Urban geography in America, 1950–2000. Paradigms and personalities* (S. 327–347). Routledge.

Dear, M. J. (2000). *The postmodern urban condition.* Wiley-Blackwell.

Deffner, V., & Haferburg, C. (2014). Pierre Bourdieu: Habitus und Habitat als Verhältnis von Subjekt, Sozialem und Macht. In J. Oßenbrügge & A. Vogelpohl (Hrsg.), *Theorien in der Raum- und Stadtforschung. Einführungen* (S. 328–347). Westfälisches Dampfboot.

DeMarrais, E., Gosden, C., & Renfrew, C. (Hrsg.). (2004). *Rethinking materiality. The engagement of mind with the material world.* McDonald Institute for Archaeological Research.

Demmerling, C. (2019). Szientismus. In J. Ritter (Hrsg.), *Historisches Wörterbuch der Philosophie* (Völlig neubearbeitete Ausgabe des „Wörterbuchs der philosophischen Begriffe" von Rudolf Eisler, Sonderausgabe, St – T, Bd. 10, S. 872–876). WBG Academic.

Denzin, N. K. (1970). *The research act. A theoretical introduction to sociological methods.* Aldine Publishing Company.

Descartes, R. (1990 [1637]). *Discours de la méthode. Französisch – Deutsch* (Philosophische Bibliothek, Bd. 261, Unveränderter Nachdruck). Meiner (Von der Methode des richtigen Vernunftgebrauchs und der wissenschaftlichen Forschung. Übersetzt und herausgegeben von Lüder Gäbe).

Descartes, R. (2008 [1637]). *Meditationes de prima philosophia. Lateinisch – Deutsch* (Philosophische Bibliothek, Bd. 597). Meiner.

Detel, W. (2018). *Grundkurs Philosophie* (Erkenntnis- und Wissenschaftstheorie, Bd. 4, 3., vollst. durchgesehene u. erw.). Reclam.

Dewey, J. (2016). *Logik. Die Theorie der Forschung* (Suhrkamp Taschenbuch Wissenschaft, 2. Aufl.). Suhrkamp.

DFG (Deutsche Forschungsgemeinschaft). (Hrsg.). (2013). *Sicherung guter wissenschaftlicher Praxis – Safeguarding Good Scientific Practice. Denkschrift – Memorandum.* https://www. dfg.de/download/pdf/dfg_im_profil/reden_stellungnahmen/download/empfehlung_wiss_ praxis_1310.pdf. Zugegriffen: 07. Juni 2019.

Diaz-Bone, R., & Schubert, K. (1996). *William James zur Einführung.* Junius.

Diels, H. (2004–2005). *Die Fragmente der Vorsokratiker. Griechisch und Deutsch* (Unveränderter Nachdruck der 6. Aufl. 1952). Weidmann (Herausgegeben von Walther Kranz).

Diemer, A. (2019). Geisteswissenschaften. In J. Ritter (Hrsg.), *Historisches Wörterbuch der Philosophie* (Völlig neubearbeitete Ausgabe des „Wörterbuchs der philosophischen Begriffe" von Rudolf Eisler, Sonderausgabe, G – H, Bd. 3, S. 211–215). WBG Academic.

Dilthey, W. (1924). Die Entstehung der Hermeneutik. In G. Misch (Hrsg.), *Gesammelte Schriften* (Die geistige Welt. Einleitung in die Philosophie des Lebens. Erste Hälfte. Abhandlungen zur Grundlegung der Geisteswissenschaften, Bd. V). B.G. Teubner.

Dilthey, W. (1970 [1910]). *Der Aufbau der geschichtlichen Welt in den Geisteswissenschaften* (Suhrkamp-Taschenbuch Wissenschaft, Bd. 354). Suhrkamp.

Dilthey, W. (2017 [1883]). *Einleitung in die Geisteswissenschaften: Versuch einer Grundlegung für das Studium der Gesellschaft und ihrer Geschichte. Neuausgabe mit einer Biographie des Autors.* Hofenberg.

Dirksmeier, P. (2007). Mit Bourdieu gegen Bourdieu empirisch denken. Habitusanalyse mittels reflexiver Fotografie. *ACME: An International Journal for Critical Geographies, 6*(1), 73–97.

Dirksmeier, P. (2015). *Urbanität als Habitus. Zur Sozialgeographie städtischen Lebens auf dem Land.* transcript.

Doeker, G. (1973). Konservatismus in den Vereinigten Staaten von Amerika. *Der Staat, 12,* 369.

Domański, B. (1997). *Industrial control over the socialist town: Benevolence or exploitation?* Praeger Publishers.

Döring, J., & Thielmann, T. (Hrsg.). (2009). *Mediengeographie. Theorie – Analyse – Diskussion.* transcript.

Drexler, D. (2013). Die Wahrnehmung der Landschaft – ein Blick auf das englische, französische und ungarische Landschaftsverständnis. In D. Bruns & O. Kühne (Hrsg.), *Landschaften: Theorie, Praxis und internationale Bezüge. Impulse zum Landschaftsbegriff mit seinen ästhetischen, ökonomischen, sozialen und philosophischen Bezügen mit dem Ziel, die Verbindung von Theorie und Planungspraxis zu stärken* (S. 37–54). Oceano.

Duden (Hrsg.). (2012). *Basiswissenschule Geografie. 7. Klasse bis Abitur.* Dudenverlag.

Duncan, J. S. (1980). The superorganic in American cultural geography. *Annals of the Association of American Geographers, 70*(2), 181–198.

Duncan, J. S. (1990). *The city as text: The politics of landscape interpretation in the Kandyan Kingdom.* Cambridge University Press.

Dünne, J., & Günzel, S. (Hrsg.). (2006). *Raumtheorie. Grundlagentexte aus Philosophie und Kulturwissenschaften* (1. Aufl.). Suhrkamp.

Düwell, M. (2001). Angewandte Ethik – Skizze eines wissenschaftlichen Profils. In A. Holderegger & J.-P. Wills (Hrsg.), *Interdisziplinäre Ethik. Grundlagen, Methoden, Bereiche. Festgabe für Dietmar Mieth zum sechzigsten Geburtstag* (Studien zur theologischen Ethik, Bd. 89, S. 165–184). Universitätsverlag.

Eckardt, F. (2014). *Stadtforschung. Gegenstand und Methoden.* Springer VS.

Edler, D. (2020). Where spatial visualization meets landscape research and „Pinballology": Examples of landscape construction in Pinball Games. *KN – Journal of Cartography and Geographic Information.* https://doi.org/10.1007/s42489-020-00044-1.

Edmüller, A., & Wilhelm, T. (2009). *Manipulationstechniken.* Haufe Lexware.

Egner, H. (2008). *Gesellschaft, Mensch, Umwelt – beobachtet. Ein Beitrag zur Theorie der Geographie* (Erdkundliches Wissen, Bd. 145). Steiner.

Egner, H. (2010). *Theoretische Geographie.* WBG.

Eisel, U. (1982). Die schöne Landschaft als kritische Utopie oder als konservatives Relikt. Über die Kristallisation gegnerischer politischer Philosophien im Symbol „Landschaft". *Soziale Welt, 33*(2), 157–168.

Eisel, U. (1992). Über den Umgang mit dem Unmöglichen. Ein Erfahrungsbericht über Interdisziplinarität im Studiengang Landschaftsplanung – Teil 2. *Gartenamt, 41*(10), 710–719.

Eisel, U. (1997). Unbestimmte Stimmungen und bestimmte Unstimmigkeiten. Über die guten Gründe der deutschen Landschaftsarchitektur für die Abwendung von der Wissenschaft und die schlechten Gründe für ihre intellektuelle Abstinenz – mit Folgerungen für die Ausbildung in diesem Fach. In S. Bernhard & P. Sattler (Hrsg.), *Vor der Tür. Aktuelle Landschaftsarchitekru aus Berlin* (S. 17–33). Callwey. http://www.ueisel.de/fileadmin/dokumente/eisel/Unbestimmte_ Stimmungen/Eisel_Unbestimmte_Stimmungen_fertig.pdf. Zugegriffen: 12. Jan. 2019.

Eisel, U. (2004). Politische Schubladen als theoretische Heuristik. Methodische Aspekte politischer Bedeutungsverschiebungen in Naturbildern. In L. Fischer (Hrsg.), *Projektionsfläche Natur. Zum Zusammenhang von Naturbildern und gesellschaftlichen Verhältnissen* (S. 29–44). Hamburg University Press.

Eisel, U. (2009). *Landschaft und Gesellschaft. Räumliches Denken im Visier* (Raumproduktionen: Theorie und gesellschaftliche Praxis, Bd. 5). Westfälisches Dampfboot.

Elias, F., Franz, A., Murmann, H., & Weiser, U. W. (Hrsg.). (2014). *Praxeologie. Beiträge zur interdisziplinären Reichweite praxistheoretischer Ansätze in den Geistes- und Sozialwissenschaften* (Materiale Textkulturen, Bd. 3). De Gruyter.

Elias, N., & Schröter, M. (Hrsg.). (1994). *Über die Zeit. Arbeiten zur Wissenssoziologie II* (Suhrkamp-Taschenbuch Wissenschaft, 5. Aufl., Bd. 756). Suhrkamp.

Ellin, N. (1999). *Postmodern Urbanism.* Princeton Architectural Press.

Enders, M., & Szaif, J. (Hrsg.). (2006). *Die Geschichte des philosophischen Begriffs der Wahrheit.* De Gruyter.

Endruweit, G. (2015). *Empirische Sozialforschung. Wissenschaftstheoretische Grundlagen* (UTB Sozialwissenschaften, Bd. 4460). UVK.

Engfer, H.-J. (1996). *Empirismus versus Rationalismus? Kritik eines philosophiegeschichtlichen Schemas.* Schöningh.

Entrikin, J. N. (1996). Place and region 2. *Progress in Human Geography, 20*(2), 215–221. https://doi.org/10.1177/030913259602000206.

Eppler, E. (1975). *Ende oder Wende. Von der Machbarkeit des Notwendigen.* Kohlhammer.

Ernst, G. (2014). *Einführung in die Erkenntnistheorie* (5., bibliograph. akt. Aufl.). WBG – Wissenschaftliche Buchgesellschaft.

Escher, A., & Petermann, S. (Hrsg.). (2016). *Raum und Ort* (Basistexte Geographie, Bd. 1). Franz Steiner.

Esser, B. (1998). Das Selbstverständnis einer Nation. Von der ersten Teilung bis zum Ende des Sozialismus. *Geographische Rundschau, 50*(1), 12–16.

Euchner, W., Stegmann, F. J., Langhorst, P., Jähnichen, T., & Friedrich, N. (2015). *Geschichte der sozialen Ideen in Deutschland: Sozialismus – Katholische Soziallehre – Protestantische Sozialethik. Ein Handbuch.* Springer.

Fainstein, S. S. (2010). *The just city.* Cornell University Press.

Falkenburg, B. (2017). Natur. In T. Kirchhoff, N. C. Karafyllis, D. Evers, B. Falkenburg, M. Gerhard, G. Hartung, et al. (Hrsg.), *Naturphilosophie. Ein Lehr- und Studienbuch* (S. 96–102). Mohr Siebeck/UTB GmbH.

Färber, A. (2014). Potenziale freisetzen: Akteur-Netzwerk-Theorie und Assemblageforschung in der interdisziplinären kritischen Stadtforschung. *sub\urban zeitschrift für kritische stadtforschung, 2*(1), 95–103.

Fehn, K. (1976). Historische Geographie. Eigenständige Wissenschaft und Teilwissenschaft der Geographie. *Mitteilungen der Geogrphischen Gesellschaft München, 61*, 35–51.

Feyerabend, P. (1976). *Wider den Methodenzwang. Skizze einer anarchistischen Erkenntnistheorie.* Suhrkamp.

Feyerabend, P. (1983). *Wider den Methodenzwang. Against method* (Suhrkamp-Taschenbuch Wissenschaft, Bd. 597). Suhrkamp.

Feyerabend, P. (2010 [1975]). *Against method: Outline of an anarchist theory of knowledge.* Verso.

Feyerabend, P. K. (Hrsg.). (1978). *Ausgewählte Schriften* (Bd. 1). Vieweg.

Feyerabend, P. K. (1981). Der Pluralismus als ein methodologisches Prinzip. In P. K. Feyerabend (Hrsg.), *Probleme des Empirismus. Schriften zur Theorie der Erklärung, der Quantentheorie und der Wissenschaftsgeschichte. Ausgewählte Schriften II* (Wissenschaftstheorie Wissenschaft und Philosophie, Bd. 2, S. 7–14). Vieweg+Teubner.

Fierla, I. (1999). *Repetytorium z geografii gospodarczej*. Polskie Wydawnictwo Ekonomiczne.

Fine, A. (2000). Der Blickpunkt von niemand besonderen. In M. Sandbothe (Hrsg.), *Die Renaissance des Pragmatismus. Aktuelle Verflechtungen zwischen analytischer und kontinentaler Philosophie* (S. 59–77). Velbrück.

Finke, P. (2014). *Citizen Science. Das unterschätzte Wissen der Laien*. Oekom.

Fischer, R. (2005). Regulierter Rinderwahnsinn. Die Reform der wissenschaftlichen Politikberatung innerhalb der Europäischen Union. In A. Bogner & H. Torgersen (Hrsg.), *Wozu Experten? Ambivalenzen der Beziehung von Wissenschaft und Politik* (S. 109–130). VS Verlag.

Fleck, L. (1980 [1935]). *Entstehung und Entwicklung einer wissenschaftlichen Tatsache. Einführung in die Lehre vom Denkstil und Denkkollektiv* (Wissenschaftsforschung). Suhrkamp (Mit einer Einleitung herausgegebn von Lothar Schäfer und Thomas Schnelle).

Fleck, L. (1983). *Erfahrung und Tatsache. Gesammelte Aufsätze* (Suhrkamp-Taschenbuch Wissenschaft, Bd. 404). Suhrkamp (Mit einer Einleitung herausgegeben von Lothar Schäfer und Thomas Schnelle).

Fleck, L. (2011). *Denkstile und Tatsachen. Gesammelte Schriften und Zeugnisse* (Suhrkamp-Taschenbuch Wissenschaft, Bd. 1953). Suhrkamp (Herausgegeben und kommentiert von Sylwia Werner und Claus Zittel unter Mitarbeit von Frank Stahnisch).

Flick, U. (2007). *Qualitative Sozialforschung. Eine Einführung*. Rowohlt.

Flick, U. (2011). *Triangulation*. Springer Fachmedien.

Flick, U., Kardorff, E. v., & Steinke, I. (Hrsg.). (2007). *Qualitative Forschung. Ein Handbuch*. Rowohlt.

Fliedner, D. (2015). *Sozialgeographie*. De Gruyter.

Fontaine, D. (2017). Ästhetik simulierter Welten am Beispiel Disneylands. In O. Kühne, H. Megerle, & F. Weber (Hrsg.), *Landschaftsästhetik und Landschaftswandel* (RaumFragen: Stadt – Region – Landschaft, S. 105–120). Springer VS.

Foucault, M. (1977). *Überwachen und Strafen. Die Geburt des Gefängnisses* (Suhrkamp-Taschenbuch Wissenschaft, Bd. 184). Suhrkamp.

Foucault, M. (1983 [1976]). *Der Wille zum Wissen. Sexualität und Wahrheit* (Suhrkamp-Taschenbuch Wissenschaft). Suhrkamp taschenbuch wissenschaft.

Foucault, M. (1991). Andere Räume. In M. Wentz (Hrsg.), *Stadt-Räume* (S. 65–72). Campus.

Foucault, M. (1996). *Diskurs und Wahrheit. Berkeley-Vorlesungen 1983*. Merve.

Foucault, M. (2007 [frz. Original 1971]). *Die Ordnung des Diskurses. Mit einem Essay von Ralf Konersmann*. Fischer Taschenbuch.

Foucault, M. (2019 [frz. Original 1975]). *Überwachen und Strafen. Die Geburt des Gefängnisses*. Suhrkamp.

Franck, G. (2007). *Ökonomie der Aufmerksamkeit. Ein Entwurf*. dtv.

Frege, G. (1879). *Begriffsschrift. Eine der arithmetischen nachgebildete Formelsprache des reinen Denkens*. Verlag von Louis Nebert.

Fröhlich, H. (2003). *Learning from Los Angeles – Zur Rolle von Los Angeles in der Diskussion um die postmoderne Stadt* (Beiträge zur Stadt- und Regionalplanung, Bd. 5). Selbstverlag.

Fuchs-Heinritz, W., & König, A. (2005). *Pierre Bourdieu. Eine Einführung*. UVK.

Funken, C., & Löw, M. (2007). Ego-Shooters Container. Raumkonstruktionen im elektronischen Netz. In R. Maresch & N. Werber (Hrsg.), *Raum, Wissen, Macht* (Suhrkamp Taschenbuch Wissenschaft, Bd. 1603, S. 69–91). Suhrkamp. 7. Nachdr.

Funtowicz, S. O., & Ravetz, J. R. (1990). *Uncertainty and quality in science for policy*. Kluwer Academic Publishers.

Gabriel, G. (1993). *Grundprobleme der Erkenntnistheorie. Von Descartes zu Wittgenstein* (3., durchgesehene. Aufl.). Schöningh.

Gabriel, G. (1998). Erkenntnistheorie. In A. Pieper (Hrsg.), *Philosophische Disziplinen. Ein Handbuch* (Reclam-Bibliothek, Bd. 1643, S. 52–71). Reclam.

Gabriel, G. (2004). Propädeutik. In J. Mittelstraß (Hrsg.), *Enzyklopädie Philosophie und Wissenschaftstheorie* (3, P – So, unveränderte Sonderausgabe, S. 361–362). J. B. Metzler.

Gabriel, G. (2005). Orientierung – Unterscheidung – Vergegenwärtigung. Zur Unverzichtbarkeit nicht propositionaler Erenntnis für die Philosophie. In G. Wolters & M. Carrier (Hrsg.), *Homo Sapiens und Homo Faber. Epistemische und technische Rationalität in Antike und Gegenwart. Festschrift für Jürgen Mittelstraß* (S. 323–334). De Gruyter.

Gabriel, G. (2012). Geltung und Genese als Grundlagenproblem. *Erwägen Wissen Ethik, 23*(4), 475–486.

Gadamer, H.-G. (1975). *Wahrheit und Methode. Grundzüge einer philosophischen Hermeneutik.* Mohr.

Gailing, L., & Leibenath, M. (2012). Von der Schwierigkeit, „Landschaft" oder „Kulturlandschaft" allgemeingültig zu definieren. *Raumforschung und Raumordnung, 70*(2), 95–106. https://doi.org/10.1007/s13147-011-0129-8.

Gamm, G. (2009). *Philosophie im Zeitalter der Extreme. Eine Geschichte philosophischen Denkens im 20. Jahrhundert.* Primus.

Gansland, H. R., & Carrier, M. (2004). Positivismus (historisch). In J. Mittelstraß (Hrsg.), *Enzyklopädie Philosophie und Wissenschaftstheorie* (3, P – So, unveränderte Sonderausgabe, S. 301–303). J. B. Metzler.

Gawlick, G. (Hrsg.). (1995). *Geschichte der Philosophie in Text und Darstellung* (Empirismus, Bd. 4). Reclam.

Gebhardt, H. (2016). Entwicklungspfade und Perspektiven der Humangeographie im deutschsprachigen Raum – einige Leitlinien. In J. Aistleitner, M. Coy, & J. Stötter (Hrsg.), *Die Welt verstehen – eine geographische Herausforderung. Eine Festschrift der Geographie Innsbruck für Axel Borsdorf* (Innsbrucker geographische Studien, Bd. 40, S. 43–59). Geographie Innsbruck Selbstverlag.

Gebhardt, H. (2019). Landeskunde und Landschaft – eine kritische Betrachtung. In O. Kühne, F. Weber, K. Berr, & C. Jenal (Hrsg.), *Handbuch Landschaft* (S. 289–298). Springer VS.

Gebhardt, H., Reuber, P., & Wolkersdorfer, G. (2004). Konzepte und Konstruktionsweisen regionaler Geographien im Wandel der Zeit. *Berichte zur deutschen Landeskunde, 78*(3), 293–312.

Gehlen, A. (2016). *Moral und Hypermoral. Eine pluralistische Ethik.* Klostermann.

Gethmann, C. F. (1987). Vom Bewusstsein zum Handeln. Pragmatische Tendenzen in der deutschen Philosophie der ersten Jahrzehnte des 20. Jahrhunderts. In H. Stachowiak (Hrsg.), *Pragmatik. Handbuch pragmatisches Denken* (2. Aufl., S. 202–232). Meiner.

Gethmann, C. F. (1991). Vielheit der Wissenschaften – Einheit der Lebenswelt. In Akademie der Wissenschaften zu Berlin (Hrsg.), *Einheit der Wissenschaften* (S. 349–371). De Gruyter.

Gethmann, C. F. (2004). Universalienstreit. In J. Mittelstraß (Hrsg.), *Enzyklopädie Philosophie und Wissenschaftstheorie* (4, Sp – Z, unveränderte Sonderausgabe, S. 411–412). J. B. Metzler.

Gethmann, C. F. (2005). Ist das Wahre das Ganze? Methodologische Probleme Integrierter Forschung. In G. Wolters & M. Carrier (Hrsg.), *Homo Sapiens und Homo Faber. Epistemische und technische Rationalität in Antike und Gegenwart. Festschrift für Jürgen Mittelstraß* (S. 391–404). De Gruyter.

Gibbons, M., Limoges, C., Nowotny, H., Schwartzmann, S., Scott, P., & Trow, M. (1994). *The new production of knowledge. The dynamics of science and research in contemporary societies.* Sage.

Gibson, J. J. (1979). *The ecological approach to visual perception.* Houghton Mifflin.

Giddens, A. (1984). *The constitution of society. Outline of the theory of structuration.* University of California Press.

Gilbert, A. (1988). The new regional geography in English and French-speaking countries. *Progress in Human Geography, 12*(2), 208–228. https://doi.org/10.1177/030913258801200203.

Gilbert, A.-F. (2008). Feministische Geographien: Ein Streifzug in die Zukunft. In P. Moss & K. Falconer Al-Hindi (Hrsg.), *Feminisms in geography. Rethinking space, place, and knowledges* (S. 96–113). Rowman & Littlefield.

Giroux, H. A. (2015). Public intellectuals against the Neoliberal University. In N. K. Denzin & M. D. Giardina (Hrsg.), *Qualitative inquiry-past, present, and future. A critical reader* (S. 194–221). Left Coast Press.

von Glasersfeld, E. (1988). Einführung in den radikalen Konstruktivismus. In P. Watzlawick (Hrsg.), *Die erfundene Wirklichkeit. Wie wissen wir, was wir zu wissen glauben? Beiträge zum Konstruktivismus* (5. Aufl., S. 16–38). Piper.

von Glasersfeld, E. (1995). *Radical constructivism. A way of knowing and learning* (Studies in mathematics education series, Bd. 6). Falmer Press.

Glasze, G. (2013). *Politische Räume. Die diskursive Konstitution eines „geokulturellen Raums" – die Frankophonie.* transcript.

Glasze, G. (2015). Identitäten und Räume als politisch: Die Perspektive der Diskurs- und Hegemonietheorie. *Europa Regional, 21*(1–2), 23–34.

Glasze, G., & Mattissek, A. (2009). Die Hegemonie- und Diskurstheorie von Laclau und Mouffe. In G. Glasze & A. Mattissek (Hrsg.), *Handbuch Diskurs und Raum. Theorien und Methoden für die Humangeographie sowie die sozial- und kulturwissenschaftliche Raumforschung* (S. 153–179). transcript.

Gloy, K. (2004). *Wahrheitstheorien. Eine Einführung.* Francke.

Gloy, K. (2005 [1995]). *Die Geschichte des wissenschaftlichen Denkens. Das Verständnis der Natur.* Komet.

Goodman, N. (1978). *Ways of worldmaking* (Harvester studies in philosophy, Bd. 5). Harvester Press.

Grau, A. (2017). *Hypermoral. Die neue Lust an der Empörung* (2. Aufl.). Claudius.

Greider, T., & Garkovich, L. (1994). Landscapes: The social construction of nature and the environment. *Rural Sociology, 59*(1), 1–24. https://doi.org/10.1111/j.1549-0831.1994.tb00519.x.

Greiffenhagen, M. (1971). *Das Dilemma des Konservatismus in Deutschland.* Piper.

Guelke, L. (1977). Regional geography. *The Professional Geographer, 29*(1), 1–7. https://doi.org/10.1111/j.0033-0124.1977.00001.x.

Habermas, J. (Hrsg.). (1983). *Moralbewußtsein und kommunikatives Handeln.* Suhrkamp.

Habermas, J. (1994). Die Moderne – ein unvollendetes Projekt. In W. Welsch (Hrsg.), *Wege aus der Moderne. Schlüsseltexte der Postmoderne-Diskussion* (2., durchgesehene. Aufl., S. 177–192). Akademie.

Habermas, J. (1995). *Theorie des kommunikativen Handelns* (Handlungsrationalität und gesellschaftliche Rationalisierung, Bd. I). Suhrkamp.

Habermas, J., & Luhmann, N. (1972). *Theorie der Gesellschaft oder Sozialtechnologie. Was leistet die Systemforschung?* Suhrkamp.

Hägerstrand, T. (1970). What about people in regional science? *Papers in Regional Science, 24*(1), 7–24. https://doi.org/10.1111/j.1435-5597.1970.tb01464.x.

Han, B.-C. (2005). *Was ist Macht?* Reclam.

Hard, G. (1969). Das Wort Landschaft und sein semantischer Hof. Zu Methode und Ergebnis eines linguistischen Tests. *Wirkendes Wort, 19*, 3–14.

Hard, G. (1970a). „Was ist eine Landschaft?". Über Etymologie als Denkform in der geographischen Literatur. In D. Bartels (Hrsg.), *Wirtschafts- und Sozialgeographie* (Neue wissenschaftliche Bibliothek, Bd. 35, S. 66–84). Kiepenheuer & Witsch.

Hard, G. (1970b). *Die „Landschaft" der Sprache und die „Landschaft" der Geographen. Semantische und forschungslogische Studien.* Ferdinand Dümmlers.

Hard, G. (1977). Zu den Landschaftsbegriffen der Geographie. In A. Hartlieb von Wallthor & H. Quirin (Hrsg.), *„Landschaft" als interdisziplinäres Forschungsproblem. Vorträge und Diskussionen des Kolloquiums am 7./8. November 1975 in Münster* (S. 13–24). Aschendorff.

Hard, G. (Hrsg.). (2002a). *Landschaft und Raum. Aufsätze zur Theorie der Geographie* (Osnabrücker Studien zur Geographie, Bd. 22). Universitätsverlag Rasch.

Hard, G. (2002b). Zu Begriff und Geschichte von „Natur" und „Landschaft" in der Geographie des 19. und 20. Jahrhunderts [1983 erstveröffentlicht]. In G. Hard (Hrsg.), *Landschaft und*

Raum. Aufsätze zur Theorie der Geographie (Osnabrücker Studien zur Geographie, Bd. 22, S. 171–210). Universitätsverlag Rasch.

Hard, G. (2003). Studium in einer diffusen Disziplin. In G. Hard (Hrsg.), *Dimensionen geographischen Denkens. Aufsätze zur Theorie der Geographie* (Osnabrücker Studien zur Geographie, Bd. 23, S. 173–230). V & R Unipress.

Hard, G. (2008). Der Spatial Turn, von der Geographie her beobachtet. In J. Döring & T. Thielmann (Hrsg.), *Spatial Turn. Das Raumparadigma in den Kultur- und Sozialwissenschaften* (S. 263–316). transcript.

Hartmann, D. (2020). *Neues System der philosophischen Wissenschaften im Grundriss* (Erkenntnistheorie, Bd. I). Mentis.

Hartmann, D., & Janich, P. (Hrsg.). (1996). *Methodischer Kulturalismus. Zwischen Naturalismus und Postmoderne* (Suhrkamp-Taschenbuch Wissenschaft, Bd. 1272). Suhrkamp.

Harvey, D. (1969). *Explanation in geography*. Arnold.

Harvey, D. (1996). *Justice, nature and the geography of difference*. Blackwell.

Harvey, D. (2005). *A brief history of neoliberalism*. OUP.

Harvey, D. (2008). The right to the city. *New Left Review, 27*(53), 23–40.

Harvey, D. (2009). *Cosmopolitanism and the geographies of freedom* (The Wellek library lectures). Columbia University Press.

Harvey, D., & Wilkinson, T. J. (2019). Landscape and heritage: Emerging landscapes of heritage. In P. Howard, I. Thompson, E. Waterton, & M. Atha (Hrsg.), *The Routledge companion to landscape studies* (2. Aufl., S. 176–191). Routledge.

Hasse, J. (2012). *Atmosphären der Stadt. Aufgespürte Räume*. Jovis.

Hauskeller, M. (2005). *Was ist Kunst? Positionen der Ästhetik von Platon bis Danto* (Beck'sche Reihe, 8. Aufl.). Beck.

Häußling, R. (2018). Institution. In J. Kopp & A. Steinbach (Hrsg.), *Grundbegriffe der Soziologie* (12. Aufl., S. 191–193). Springer Fachmedien Wiesbaden.

Hayek, F. A. v. (1996). *Die Anmassung von Wissen. Neue Freiburger Studien* (Wirtschaftswissenschaftliche und wirtschaftsrechtliche Untersuchungen, Bd. 32). Mohr.

Hegel, G. W. F. (1970). *Enzyklopädie der philosophischen Wissenschaften im Grundrisse 1830. Erster Teil: Die Wissenschaft der Logik. Mit mündlichen Zusätzen* (8 Bände). Suhrkamp.

Hegel, G. W. F. (1980). *Phänomenologie des Geistes* (Gesammelte Werke, Bd. 9). Meiner.

Heidegger, M. (1963). *Holzwege* (4. Aufl.). Klostermann.

Heidegger, M. (1992). *Der Ursprung des Kunstwerkes*. Reclam.

Heidegger, M. (2005 [1927]). *Die Grundprobleme der Phänomenologie*. Klostermann.

Hein, K. (2006). *Hybride Identitäten. Bastelbiografien im Spannungsverhältnis zwischen Lateinamerika und Europa*. transcript.

Heisenberg, W. (1942). *Die Einheit des naturwissenschaftlichen Weltbildes*. Barth.

Held, K. (1991). Husserls neue Einführung in die Philosophie: der Begriff der Lebenswelt. In C. F. Gethmann (Hrsg.), *Lebenswelt und Wissenschaft. Studien zum Verhältnis von Phänomenologie und Wissenschaftstheorie* (Neuzeit und Gegenwart, Bd. 1, S. 79–113). Bouvier.

Hempel, C. G. (1965). *Aspects of scientific explanation and other essays in the philosophy of science*. Free Press.

Hempel, C. G., & Oppenheim, P. (1948). Studies in the logic of explanation. *Philosophy of Science, 15*(2), 135–175.

Herrmann, H. (2010). Raumbegriffe und Forschungen zum Raum – eine Einleitung. In H. Herrmann (Hrsg.), *RaumErleben. Zur Wahrnehmung des Raumes in Wissenschaft und Praxis* (S. 7–29). Budrich.

Herzog, L. (2013). *Freiheit gehört nicht nur den Reichen. Plädoyer für einen zeitgemäßen Liberalismus* (C. H. Beck Paperback, Bd. 6127). C. H. Beck.

Hettner, A. (1927). *Die Geographie. Ihre Geschichte, ihr Wesen und ihre Methoden*. Hirt.

Hildebrand, D. L. (2003). The neopragmatist turn. *Southwest Philosophy Review, 19*(1), 79–88.

Hildebrand, D. L. (2005). Pragmatism, neopragmatism, and public administration. *Administration & Society, 37*(3), 345–359.

Hillebrandt, F. (2000). Disziplinargesellschaft. In G. Kneer, A. Nassehi & M. Schroer (Hrsg.), *Soziologische Gesellschaftsbegriffe. Konzepte moderner Zeitdiagnosen* (UTB für Wissenschaft Uni-Taschenbücher Soziologie, 2. Aufl., Bd. 1961, S. 101–126). Fink.

Hirschberger, J. (1976). *Geschichte der Philosophie* (Altertum und Mittelalter, Bd. 1). Komet.

Hobbes, T. (2017 [1651]). *Leviathan oder Stoff, Form und Gewalt eines kirchlichen und bürgerlichen Staates. Herausgegeben und eingeleitet von Iring Fetscher. Übersetzt von Walter Euchner* (Suhrkamp-Taschenbuch Wissenschaft, Bd. 462, 16. Aufl.). Suhrkamp.

Hoesterey, I. (2001). *Pastiche. Cultural memory in art, film, literature.* Indiana University Press.

Höffe, O. (1981). *Sittlich-politische Diskurse. Philosophische Grundlagen. Politische Ethik. Biomedizinische Ethik* (Suhrkamp-Taschenbuch Wissenschaft, Bd. 380). Suhrkamp.

Höffe, O. (2001). *Kleine Geschichte der Philosophie.* Beck.

Hofmann, P., & Hirschauer, S. (2012). Die konstruktivistische Wende. In S. Maasen, M. Kaiser, M. Reinhart, & B. Sutter (Hrsg.), *Handbuch Wissenschaftssoziologie* (S. 85–99). Springer.

Höhne, S., & Umlauf, R. (2014). Die Akteur-Netzwerk Theorie. Zur Vernetzung und Entgrenzung des Sozialen. In J. Oßenbrügge & A. Vogelpohl (Hrsg.), *Theorien in der Raum- und Stadtforschung. Einführungen* (S. 195–214). Westfälisches Dampfboot.

Hokema, D. (2009). Die Landschaft der Regionalentwicklung: Wie flexibel ist der Landschaftsbegriff? *Raumforschung und Raumordnung, 67*(3), 239–249.

Hokema, D. (2013). *Landschaft im Wandel? Zeitgenössische Landschaftsbegriffe in Wissenschaft, Planung und Alltag.* Springer VS.

Holmén, H. (1995). What's new and what's regional in the 'new regional geography'? *Geografiska Annaler: Series B, Human Geography, 77*(1), 47–63. https://doi.org/10.1080/04353 684.1995.11879680.

Holzinger, M. (2004). *Natur als sozialer Akteur. Realismus und Konstruktivismus in der Wissenschafts- und Gesellschaftstheorie* (Forschung Soziologie, Bd. 197). VS Verlag.

Honneth, A. (1992). *Kampf um Anerkennung. Zur moralischen Grammatik sozialer Konflikte.* Suhrkamp.

Hoskins, W. G. (2005 [1955]). *The making of the English landscape.* The Folio Society.

Howe, N. (2011). Landscape versus region. *The Wiley-Blackwell Companion to Human Geography, 16*, 114–129. https://doi.org/10.1002/9781444395839.ch7.

Hoyningen-Huene, P. (1998). *Formale Logik. Eine philosophische Einführung.* Reclam.

Hradil, S. (1995). Schicht, Schichtung und Mobilität. In H. Korte & B. Schäfers (Hrsg.), *Einführung in Hauptbegriffe der Soziologie* (6. Aufl., S. 145–164). VS Verlag.

Hradil, S. (2018). Milieu, soziales. In J. Kopp & A. Steinbach (Hrsg.), *Grundbegriffe der Soziologie* (12. Aufl., S. 319–322). Springer Fachmedien Wiesbaden.

Hubbard, P. (2002). *Thinking geographically. Space, theory and contemporary human geography.* continuum.

Hubbard, P., Bartley, B., Fuller, D., & Kitchin, R. (2005). *Thinking geographically. Space, theory and contemporary human geography.* continuum.

Hübner, K. (1978). *Kritik der wissenschaftlichen Vernunft.* Alber.

Hügin, U. (1996). *Individuum, Gemeinschaft, Umwelt. Konzeption einer Theorie der Dynamik anthropogener Systeme.* Lang.

Hume, D. (1961 [1748]). *Eine Untersuchung über den menschlichen Verstand. Übersetzt von Raoul Richter* (Philosophische Bibliothek, Bd. 648). Meiner.

Hume, D. (1978). *Ein Traktat über die menschliche Natur. Buch II. Über die Affekte Buch III. Über Moral* (Unveränderter Nachdruck der 1. Aufl. von 1906 (Buch 2 und 3)). Meiner.

Hunt, R. (2016). *Huts, bothies and buildings out-of-doors: An exploration of the practice, heritage and culture of 'out-dwellings' in rural Scotland.* PhD thesis, University of Glasgow.

Husserl, E. (1913). *Ideen zu einer reinen Phänomenologie und phänomenologischen Philosopie. Erster Buch: Allgemeine Einführung in die reine Phänomenologie.* Niemeyer.

Husserl, E. (1954). *Die Krisis der europäischen Wissenschaften und die transzendentale Phänomenologie. Eine Einleitung in die phänomenologische Philosophie* (Husserliana, Bd. 6). Martinus Nijhoff (Herausgegeben von Walter Biemel).

Ingold, T. (2002). *The perception of the environment. Essays on livelihood, dwelling and skill.* Routledge.

Ipsen, D. (1992). Stadt und Land – Metamorphosen einer Beziehung. In H. Häußermann, D. Ipsen, R. Krämer-Badoni, D. Läpple, M. Rodenstein, & W. Siebel (Hrsg.), *Stadt und Raum. Soziologische Analysen* (2. Aufl., S. 117–156). Centaurus.

Ipsen, D. (2002). Raum als Landschaft. In D. Ipsen & D. Läpple (Hrsg.), *Soziologie des Raumes: Räume der Gesellschaft – soziologische Perspektiven* (S. 86–111). Fernuniversität.

Jaehne, G. (1972). *Landwirtschaft und Landwirtschatliche Zusammenarbeit im Rat für gegenseitige Wirtschaftshilfe Comecon.* Duncker & Humblot.

James, W. (1977). *Der Pragmatismus. Ein neuer Name für alte Denkmethoden* (Philosophische Bibliothek, Bd. 297). Meiner.

James, W. (Hrsg.). (2005 [1909]). *Das pluralistische Universum. Vorlesungen über die gegenwärtige Lage der Philosophie.* Wissenschaftliche Buchgesellschaft.

Janich, P. (1996). *Was ist Wahrheit? Eine philosophische Einführung* (Beck'sche Reihe C. H. Beck Wissen, Bd. 2052, Original-Ausgabe). Beck.

Janich, P. (2001). *Logisch-pragmatische Propädeutik. Ein Grundkurs im philosophischen Reflektieren.* Velbrück Wissenschaft.

Janich, P. (2014). *Sprache und Methode. Eine Einführung in philosophische Reflexion.* Francke.

Janich, P. (2015). *Handwerk und Mundwerk. Über das Herstellen von Wissen.* C. H. Beck.

Jaspers, K. (1975). *Was ist Philosophie? Ein Lesebuch.* Piper.

Jenal, C. (2018). Ikonologie des Protests – Der Stromnetzausbau im Darstellungsmodus seiner Kritiker(innen). In O. Kühne & F. Weber (Hrsg.), *Bausteine der Energiewende* (S. 469–487). Springer VS.

Jenal, C. (2020). Visualizations of ‚landscape' in protest movements. On exclusive and inclusive patterns of vision and interpretation using the example of resistance to the expansion of the electricity grid in Germany. In D. Edler, C. Jenal, & O. Kühne (Hrsg.), *Modern approaches to the visualization of landscapes* (S. 427–445). Springer VS.

Jenal, C., & Berr, K. (2019). Landschaft als Konflikt. Wenn erlernte Deutungsmuster mit neuen Sichtweisen konkurrieren. *Stadt+Grün, 68*(12), 18–23.

Jerusalem, W. (1925 [1909]). Soziologie des Erkennens. In W. Jerusalem (Hrsg.), *Gedanken und Denker. Gesammelte Aufsätze* (o. S.). Wilhelm Braumüller.

Joas, H. (1988). Symbolischer Interaktionismus. Von der Philosophie des Pragmatismus zu einer soziologischen Forschungstradition. *Kölner Zeitschrift für Soziologie und Sozialpsychologie, 40*, 417–446.

Jørgensen, M., & Phillips, L. (2002). *Discourse analysis as theory and method.* SAGE.

Kaiser, M., & Maasen, S. (2010). Wissenschaftssoziologie. In G. Kneer & M. Schroer (Hrsg.), *Handbuch Spezielle Soziologien* (S. 685–705). VS Springer.

Kambartel, F. (2004). Positivismus (systematisch). In J. Mittelstraß (Hrsg.), *Enzyklopädie Philosophie und Wissenschaftstheorie* (3, P – So, unveränderte Sonderausgabe, S. 303–304). J. B. Metzler.

Kamlah, W., & Lorenzen, P. (1967). *Logische Propädeutik oder Vorschule des vernünftigen Redens.* Bibliographisches Institut.

Kammler, C. (2014). Einführung: Konzeptualisierungen der Werke Foucaults. In C. Kammler, R. Parr, U. J. Schneider, & E. Reinhardt-Becker (Hrsg.), *Foucault-Handbuch. Leben – Werk – Wirkung* (S. 9–11). J.B. Metzler.

Kant, I. (1959 [1781]). *Kritik der reinen Vernunft.* Felix Meiner.

Kant, I. (1968). Metaphysische Anfangsgründe der Naturwissenschaften. In I. Kant (Hrsg.), *Kants Werke. Akdemische Textausgabe* (Bd. IV, S. 465–566). De Gruyter.

Karmasin, M., & Ribing, R. (2017). *Die Gestaltung wissenschaftlicher Arbeiten. Ein Leitfaden für Facharbeit/VWA, Seminararbeiten, Bachelor-, Master-, Magister- und Diplomarbeiten sowie Dissertationen* (utb Schlüsselkompetenzen, 9., überarb. u. akt. Aufl., Bd. 2774). Facultas Verlags- und Buchhandels AG.

Kaufmann, S. (2005). *Soziologie der Landschaft.* VS Verlag.

Kazig, R. (2007). Atmosphären – Konzept für einen nicht repräsentationellen Zugang zum Raum. In C. Berndt & R. Pütz (Hrsg.), *Kulturelle Geographien. Zur Beschäftigung mit Raum und Ort nach dem Cultural Turn* (S. 167–187). transcript.

Kazig, R. (2013). Landschaft mit allen Sinnen – Zum Wert des Atmosphärenbegriffs für die Landschaftsforschung. In D. Bruns & O. Kühne (Hrsg.), *Landschaften: Theorie, Praxis und internationale Bezüge. Impulse zum Landschaftsbegriff mit seinen ästhetischen, ökonomischen, sozialen und philosophischen Bezügen mit dem Ziel, die Verbindung von Theorie und Planungspraxis zu stärken* (S. 221–232). Oceano.

Kazig, R. (2019). Atmosphären und Landschaft. In O. Kühne, F. Weber, K. Berr, & C. Jenal (Hrsg.), *Handbuch Landschaft* (S. 453–460). Springer VS.

Kazig, R., & Weichhart, P. (2009). Die Neuthematisierung der materiellen Welt in der Humangeographie. *Berichte zur deutschen Landeskunde, 83*(2), 109–128.

Kemper, J., & Wiegand, F. (2014). Marxistische Stadtforschung. In J. Oßenbrügge & A. Vogelpohl (Hrsg.), *Theorien in der Raum- und Stadtforschung. Einführungen* (S. 215–233). Westfälisches Dampfboot.

Kersting, P. (2012). Geomorphologie, Pragmatismus und integrative Ansätze in der Geographie. *Berichte zur deutschen Landeskunde, 86*(1), 49–65.

Kienzle, B. (2010). David Hume – Kausalprinzip und Induktionsproblem. In A. Beckermann & D. Perler (Hrsg.), *Klassiker der Philosophie heute* (S. 352–372). Reclam.

Kirchberg, V. (2015). Das Museum als öffentlicher Raum in der Stadt. In J. Baur (Hrsg.), *Museumsanalyse. Methoden und Konturen eines neuen Forschungsfeldes* (S. 231–266). transcript.

Kirchhoff, T. (2019a). Ökosystemdienstleistungen. In O. Kühne, F. Weber, K. Berr, & C. Jenal (Hrsg.), *Handbuch Landschaft* (S. 807–822). Springer VS.

Kirchhoff, T. (2019b). Politische Weltanschauungen und Landschaft. In O. Kühne, F. Weber, K. Berr, & C. Jenal (Hrsg.), *Handbuch Landschaft* (S. 383–396). Springer VS.

Kirchhoff, T., & Trepl, L. (2009a). Landschaft, Wildnis, Ökosystem: zur kulturbedingten Vieldeutigkeit ästhetischer, moralischer und theoretischer Naturauffassungen. Einleitender Überblick. In T. Kirchhoff & L. Trepl (Hrsg.), *Vieldeutige Natur. Landschaft, Wildnis und Ökosystem als kulturgeschichtliche Phänomene* (Sozialtheorie, S. 13–68). transcript.

Kirchhoff, T., & Trepl, L. (Hrsg.). (2009b). *Vieldeutige Natur. Landschaft, Wildnis und Ökosystem als kulturgeschichtliche Phänomene* (Sozialtheorie). transcript.

Kitchin, R. (2015). Positivist geography. In S. C. Aitken & G. Valentine (Hrsg.), *Approaches to human geography. Philosophies, theories, people and practices* (2. Aufl., S. 23–34). SAGE.

Kleiner, M. S. (Hrsg.). (2001). *Michel Foucault. Eine Einführung in sein Denken*. Campus.

Kloock, D., & Spahr, A. (2007 [1986]). *Medientheorien. Eine Einführung* (UTB). Fink.

Klueter, H. (1986). *Raum als Element sozialer Kommunikation*. Selbstverlag Geographisches Institut Universität.

Kneer, G. (2004). Differenzierung bei Luhmann und Bourdieu. Ein Theorievergleich. In A. Nassehi & G. Nollmann (Hrsg.), *Bourdieu und Luhmann. Ein Theorienvergleich* (S. 25–56). Suhrkamp.

Kneer, G. (2009a). Akteur-Netzwerk-Theorie. In G. Kneer & M. Schroer (Hrsg.), *Handbuch Soziologische Theorien* (S. 19–39). VS Verlag.

Kneer, G. (2009b). Jenseits von Realismus und Antirealismus. Eine Verteidigung des Sozialkonstruktivismus gegenüber seinen postkonstruktivistischen Kritikern. *Zeitschrift für Soziologie, 38*(1), 5–25. https://doi.org/10.1515/zfsoz-2009-0101.

Kneer, G., & Schroer, M. (Hrsg.). (2009). *Handbuch Soziologische Theorien*. VS Verlag.

Knoblauch, H. (2006). *Wissenssoziologie*. UVK/UTB.

Knoblauch, H. (2013). Wissenssoziologie, Wissensgesellschaft und die Transformation der Wissenskommunikation. *Aus Politik und Zeitgeschichte, 63*(18–20), 9–16.

Knoll, J. H. (1981). Liberalismus. In J. H. Schoeps, J. H. Knoll, & C.-E. Bärsch (Hrsg.), *Konservativismus, Liberalismus, Sozialismus. Einführung, Texte, Bibliographien* (Uni-Taschenbücher Politologie, Neuere Geschichte, Soziologie, Bd. 1032, S. 87–139). Fink.

Knorr, K. (1980). Die Fabrikation von Wissen. Versuch zu einem gesellschaftlich relativierten Wissensbegriff. In N. Stehr & V. Meja (Hrsg.), *Wissenssoziologie. Kölner Zeitschrift für Soziologie und Sozialpsychologie* (Bd. 22, S. 226–245). Westdeutscher. [Themenheft].

Knorr-Cetina, K. (2002a). *Die Fabrikation von Erkenntnis. Zur Anthropologie von Wissenschaft.* Suhrkamp.

Knorr-Cetina, K. (2002b). *Wissenskulturen. Ein Vergleich naturwissenschaftlicher Wissensformen* (Suhrkamp-Taschenbuch Wissenschaft, Bd. 1594, Dt. Erstausg, 1. Aufl.). Suhrkamp.

Koriako, D. (2005). Was sind und wozu dienen reine Anschauungen? Kritische Fragen und Anmerkungen zu Kants Raumtheorie. *Kant-Studien, 96*(1), 20–40.

Körner, S., & Eisel, U. (2003). Naturschutz als kulturelle Aufgabe – theoretische Rekonstruktrion und Anregungen für eine inhaltliche Erweiterung. In S. Körner, A. Nagel, & U. Eisel (Hrsg.), *Naturschutzbegründungen* (S. 5–49). Selbstverlag.

Kornmesser, S., & Büttemeyer, W. (2020). *Wissenschaftstheorie. Eine Einführung.* J.B. Metzler.

Kornmesser, S., & Schurz, G. (2014). Die multiparadigmatische Struktur der Wissenschaften: Einleitung und Übersicht. In S. Kornmesser & G. Schurz (Hrsg.), *Die multiparadigmatische Struktur der Wissenschaften* (S. 11–46). Springer VS.

Koselleck, R. (Hrsg.). (1979). *Historische Semantik und Begriffsgeschichte.* Klett-Cotta.

Kötzle, M. (1999). Eigenart und Eigentum. Zur Genese und Struktur konservativer und liberaler Weltbilder. In S. Körner, T. Heger, A. Nagel, & U. Eisel (Hrsg.), *Naturbilder in Naturschutz und Ökologie* (Landschaftsentwicklung und Umweltforschung, Bd. 111, S. 19–36). TU Berlin.

Kraemer, K., & Bittlingmayer, U. H. (2001). Soziale Polarisierung durch Wissen. In P. A. Berger & D. Konietzka (Hrsg.), *Die Erwerbsgesellschaft. Neue Ungleichheiten und Unsicherheiten* (S. 313–329). VS Verlag.

Krahmer, A. (2017). Edward W. Soja: Thirdspace. In F. Eckardt (Hrsg.), *Schlüsselwerke der Stadtforschung* (S. 47–68). Springer VS.

Kramer, C. (2012). „Alles hat seine Zeit" – die „Time Geography" im Licht des „Material Turn". In N. Weixlbaumer (Hrsg.), *Anthologie zur Sozialgeographie* (Abhandlungen zur Geographie und Regionalforschung, Bd. 16, S. 83–105). Institut für Geographie und Regionalforschung.

Krieger, G. (2011). Substanz. In P. Kolmer & A. G. Wildfeuer (Hrsg.), *Neues Handbuch philosophischer Grundbegriffe* (S. 2146–2158). Alber.

Kronauer, M., & Siebel, W. (Hrsg.). (2013). *Polarisierte Städte. Soziale Ungleichheit als Herausforderung für die Stadtpolitik.* Campus.

Kruse, O. (2017). *Kritisches Denken und Argumentieren. Eine Einführung für Studierende* (UTB, Bd. Nr. 4767). UVK Verlagsgesellschaft mbH; UVK/Lucius.

Kubsch, R. (2007). *Die Postmoderne. Abschied von der Eindeutigkeit.* Hänssler.

Kuckartz, U. (2014). *Mixed Methods. Methodologie, Forschungsdesigns und Analyseverfahren.* Springer VS.

Kuhlmann, W. (2011). Begründung. In M. Düwell, C. Hübenthal, & M. H. Werner (Hrsg.), *Handbuch Ethik* (3., akt. Aufl., S. 319–325). J.B. Metzler.

Kuhn, T. S. (1976). *Die Struktur wissenschaftlicher Revolutionen* (Suhrkamp-Taschenbuch Wissenschaft, Bd. 25, zweite revidierte und um das Postskriptum von 1969 ergänzte Aufl.). Suhrkamp.

Kühne, O. (1999). *Die Wetterlagen-, Tages- und Jahreszeitenabhängigkeit der Verteilung von Lufttemperatur, spezifischer Luftfeuchte, Windfeld, Äquivalenttemperatur und anderer bioklimatisch wirksamer Größen im Lokalklima der Stadt Homburg/Saar.* Dissertation zur Erlangung des akademischen Grades eines Doktors der Philosophie, Universität des Saarlandes, Saarbrücken.

Kühne, O. (2006). *Landschaft in der Postmoderne. Das Beispiel des Saarlandes.* DUV.

Kühne, O. (2008a). *Distinktion – Macht – Landschaft. Zur sozialen Definition von Landschaft.* VS Verlag.

Kühne, O. (2008b). Kritische Geographie der Machtbeziehungen – konzeptionelle Überlegungen auf der Grundlage der Soziologie Pierre Bourdieus. *geographische revue, 10*(2), 40–50.

Kühne, O. (2011). Die Konstruktion von Landschaft aus Perspektive des politischen Liberalismus. Zusammenhänge zwischen politischen Theorien und Umgang mit Landschaft. *Naturschutz und Landschaftsplanung, 43*(6), 171–176.

Kühne, O. (2012). *Stadt – Landschaft – Hybridität. Ästhetische Bezüge im postmodernen Los Angeles mit seinen modernen Persistenzen.* Springer VS.

Kühne, O. (2013). *Landschaftstheorie und Landschaftspraxis. Eine Einführung aus sozialkonstruktivistischer Perspektive.* Springer VS.

Kühne, O. (2015a). The streets of Los Angeles: Power and the infrastructure landscape. *Landscape Research, 40*(2), 139–153. https://doi.org/10.1080/01426397.2013.788691.

Kühne, O. (2015b). Weltanschauungen in regionalentwickelndem Handeln – die Beispiele liberaler und konservativer Ideensysteme. In O. Kühne & F. Weber (Hrsg.), *Bausteine der Regionalentwicklung* (S. 55–69). Springer VS.

Kühne, O. (2017). *Zur Aktualität von Ralf Dahrendorf. Einführung in sein Werk* (Aktuelle und klassische Sozial- und Kulturwissenschaftler|innen). Springer VS.

Kühne, O. (2018a [2020 erschienen]). Die Landschaften 1, 2 und 3 und ihr Wandel. Perspektiven für die Landschaftsforschung in der Geographie – 50 Jahre nach Kiel. *Berichte. Geographie und Landeskunde, 92*(3–4), 217–231.

Kühne, O. (2018b). *Landschaft und Wandel. Zur Veränderlichkeit von Wahrnehmungen.* Springer VS.

Kühne, O. (2018c). Die Landschaften 1, 2 und 3 und ihr Wandel. Perspektiven für die Landschaftsforschung in der Geographie – 50 Jahre nach Kiel. *Berichte. Geographie und Landeskunde, 3–4*, 217–231.

Kühne, O. (2018d). *Landschaftstheorie und Landschaftspraxis. Eine Einführung aus sozialkonstruktivistischer Perspektive* (2., akt. u. überarb. Aufl.). Springer VS.

Kühne, O. (2018e). Reboot „Regionale Geographie" – Ansätze einer neopragmatischen Rekonfiguration „horizontaler Geographien". *Berichte. Geographie und Landeskunde, 92*(2), 101–121.

Kühne, O. (2019a). *Landscape theories. A brief introduction.* Springer VS.

Kühne, O. (2019b). Landschaftsverständnisse in ihrer historischen Gebundenheit – zwischen Gegenständlichkeit, Essenz und Konstruktion. In K. Berr & C. Jenal (Hrsg.), *Landschaftskonflikte* (S. 23–36). Springer VS.

Kühne, O. (2019c). Sich abzeichnende theoretische Perspektiven für die Landschaftsforschung: Neopragmatismus, Akteur-Netzwerk-Theorie und Assemblage-Theorie. In O. Kühne, F. Weber, K. Berr, & C. Jenal (Hrsg.), *Handbuch Landschaft* (S. 153–162). Springer VS.

Kühne, O. (2019d). Die Sozialisation von Landschaft. In O. Kühne, F. Weber, K. Berr, & C. Jenal (Hrsg.), *Handbuch Landschaft* (S. 301–312). Springer VS.

Kühne, O. (2020). Landscape conflicts. A theoretical approach based on the three worlds theory of Karl Popper and the conflict theory of Ralf Dahrendorf, illustrated by the example of the energy system transformation in Germany. *Sustainability, 12*(17), 1–20. https://doi.org/10.3390/su12176772.

Kühne, O., & Duttmann, R. (2019). Recent challenges of the ecosystems services approach from an interdisciplinary point of view. *Raumforschung und Raumordnung Spatial Research and Planning.* https://doi.org/10.2478/rara-2019-0055.

Kühne, O., & Jenal, C. (2020a). *Baton Rouge – The multivillage metropolis. A neopragmatic landscape biographical approach on spatial pastiches, hybridization, and differentiation.* Springer VS.

Kühne, O., & Jenal, C. (2020b). The threefold landscape dynamics – Basic considerations, conflicts and potentials of virtual landscape research. In D. Edler, C. Jenal, & O. Kühne (Hrsg.), *Modern approaches to the visualization of landscapes* (S. 389–402). Springer VS.

Kühne, O., & Weber, F. (2015). Der Energienetzausbau in Internetvideos – eine quantitativ ausgerichtete diskurstheoretisch orientierte Analyse. In S. Kost & A. Schönwald (Hrsg.), *Landschaftswandel – Wandel von Machtstrukturen* (S. 113–126). Springer VS.

Kühne, O., & Weber, F. (2018 [online first 2017]). Conflicts and negotiation processes in the course of power grid extension in Germany. *Landscape Research 43* (4), 529–541. https://doi.org/10.1080/01426397.2017.1300639.

Kühne, O., Weber, F., & Jenal, C. (2018). *Neue Landschaftsgeographie. Ein Überblick (Essentials)*. Springer VS.

Kühne, O., Weber, F., & Berr, K. (2019a). The productive potential and limits of landscape conflicts in light of Ralf Dahrendorf's conflict theory. *Società Mutamento Politica, 10* (19), 77–90. https://oajournals.fupress.net/index.php/smp/article/view/10597. Zugegriffen: 22. Juni 2020.

Kühne, O., Weber, F., & Jenal, C. (2019b). Neue Landschaftsgeographie. In O. Kühne, F. Weber, K. Berr, & C. Jenal (Hrsg.), *Handbuch Landschaft* (S. 119–134). Springer VS.

Kühne, O., Berr, K., Schuster, K., & Jenal, C. (2021). *Freiheit und Landschaft. Auf der Suche nach Lebenschancen mit Ralf Dahrendorf*. Springer.

Kunzmann, P., Burkard, F.-P., & Wiedmann, F. (1993). *dtv-Atlas zur Philosophie. Tafeln und Texte*. dtv.

Kurz, G. (2015). *Das Wahre, Schöne, Gute. Aufstieg, Fall und Fortbestehen einer Trias*. Fink.

Kutschera, F. v., & Breitkopf, A. (2007). *Einführung in die moderne Logik* (8. Aufl.). Karl Alber.

Kutschera, U. (2011). *Darwiniana Nova. Verborgene Kunstformen der Natur*. LIT.

Laclau, E., & Mouffe, C. (1985). *Hegemony and socialist strategy. Towards a radical democratic politics*. Verso.

Lakatos, I. (1974a). Falsifikation und die Methodologie wissenschaftlicher Forschungsprogramme. In I. Lakatos & A. Musgrave (Hrsg.), *Kritik und Erkenntnisfortschritt* (Abhandlungen des Internationalen Kolloquiums über die Philosophie der Wissenschaft, Bd. 4, S. 89–189). Vieweg.

Lakatos, I. (1974b). Die Geschichte der Wissenschaften und ihre rationalen Rekonstruktionen. In I. Lakatos & A. Musgrave (Hrsg.), *Kritik und Erkenntnisfortschritt* (Abhandlungen des Internationalen Kolloquiums über die Philosophie der Wissenschaft, Bd. 4, S. 271–312). Vieweg.

Lamnek, S. (2005). *Qualitative Sozialforschung. Lehrbuch* (4., vollst. überarb. Aufl.). Beltz PVU.

Lamnek, S. (2010). *Qualitative Sozialforschung* (5., überarb. Aufl.). Beltz.

de Landa, M. (2006). *A new philosophy of society. Assemblage theory and social complexity*. continuum.

Langer, A., & Wrana, D. (2013). Diskursforschung und Diskursanalyse. In B. Friebertshäuser, A. Langer, & A. Prengel (Hrsg.), *Handbuch. Qualitative Forschungsmethoden in der Erziehungswissenschaft* (S. 335–349). Beltz.

Läpple, D. (1991). Gesellschaftszentriertes Raumkonzept. In M. Wentz (Hrsg.), *Stadt-Räume* (S. 35–46). Campus.

Läpple, D. (1992). Essay über den Raum. Für ein gesellschaftswissenschaftliches Raumkonzept. In H. Häußermann, D. Ipsen, R. Krämer-Badoni, D. Läpple, M. Rodenstein, & W. Siebel (Hrsg.), *Stadt und Raum. Soziologische Analysen* (2. Aufl., S. 157–207). Centaurus.

Latour, B. (1996). *Petite réflexion sur le culte moderne des dieux Faitiches*. Synthélabo groupe.

Latour, B. (1997). The trouble with actor-network theory. *Soziale Welt, 47*, 369–381.

Latour, B. (2002 [1999]). *Die Hoffnung der Pandora. Untersuchungen zur Wirklichkeit der Wissenschaft*. Suhrkamp.

Latour, B., & Woolgar, S. (2013 [1979]). *Laboratory life: The construction of scientific facts*. Princeton University Press.

Laudan, L. (1977). *Progress and its problems. Towards a theory of scientific growth*. University of California Press.

Lautensach, H. (1973). Über die Erfassung und Abgrenzung von Landschaftsräumen [Erstveröffentlichung 1938]. In K. Paffen (Hrsg.), *Das Wesen der Landschaft* (Wege der Forschung, Bd. 39, S. 20–38). WBG.

Law, J. (1999). After ANT: Complexity, naming and topology. *The Sociological Review, 47*(S1), 1–14.

Law, J., & Hassard, J. (Hrsg.). (1999). *Actor network theory and after* (Sociological review Monographs). Blackwell Publishers.

Le Rond d'Alembert, J. (2011 [1751]). *Discours préliminaire à l'Encyclopédie*. https://philosophie.cegeptr.qc.ca/wp-content/documents/Discours-pr%C3%A9liminaire-%C3%A0-lEncyclop%C3%A9die.pdf. Zugegriffen: 25. Nov. 2020.

Leerhoff, H., Rehkämpfer, K., & Wachtenhofer, T. (2010). *Einführung in die Analytische Philosophie*. Wissenschaftliche Buchgesellschaft.

Lefèbvre, H. (1974). La production de l'espace. *L'Homme et la société, 31–32*(1), 15–32.

Leibenath, M. (2017). Ecosystem services and neoliberal governmentality – German style. *Land Use Policy, 64*, 307–316.

Leibenath, M., & Otto, A. (2012). Diskursive Konstituierung von Kulturlandschaft am Beispiel politischer Windenergiediskurse in Deutschland. *Raumforschung und Raumordnung, 70*(2), 119–131. https://doi.org/10.1007/s13147-012-0148-0.

Leibniz, G. W. (1961). *Neue Abhandlungen über den menschlichen Verstand*. Insel.

Leibniz, G. W. (2019 [1714]. *Monadologie. Französisch/Deutsch* (durchgesehene und bibliographisch ergänzte Ausgabe). Reclam.

Lenk, K. (1989). *Deutscher Konservatismus*. Campus.

Leonhard, J. (2001). *Liberalismus. Zur historischen Semantik eines europäischen Deutungsmusters* (Veröffentlichungen des Deutschen Historischen Instituts London/ Publications of the German Historical Institute London). R. Oldenbourg.

Leser, H. (1991). *Landschaftsökologie. Ansatz, Modelle, Methodik, Anwendung* (UTB, 3., völlig neubearb. Aufl., Bd. 521). Ulmer.

Leser, H. (2019). Landschaftsökologie. In O. Kühne, F. Weber, K. Berr, & C. Jenal (Hrsg.), *Handbuch Landschaft* (S. 181–191). Springer VS.

Levidow, L. (2005). Expert-based policy or policy-based expertise? Regulating GM crops in Europe. In A. Bogner & H. Torgersen (Hrsg.), *Wozu Experten? Ambivalenzen der Beziehung von Wissenschaft und Politik* (S. 86–108). VS Verlag.

Lippuner, R. (2008). Raumbilder der Gesellschaft. Zur Räumlichkeit des Sozialen in der Systemtheorie. In J. Döring & T. Thielmann (Hrsg.), *Spatial Turn. Das Raumparadigma in den Kultur- und Sozialwissenschaften* (S. 341–363). transcript.

Lippuner, R. (2011). Gesellschaft, Umwelt und Technik: Zur Problemstellung einer »Ökologie sozialer Systeme«. *Soziale Systeme. Zeitschrift für soziologische Theorie, 17*(2), 308–335.

List, E. (2004). Einleitung. Interdisziplinäre Kulturforschung auf der Suche nach theoretischer Orientierung. In E. List (Hrsg.), *Grundlagen der Kulturwissenschaften. Interdisziplinäre Kulturstudien* (S. 3–12). Francke.

Locke, J. (1981 [1689]). *Versuch über den menschlichen Verstand. 2 Bände* (Philosophische Bibliothek, Bd. 75, 4., durchgesehene Aufl.). Meiner.

Lorenz, K. (2004). Einheitswissenschaft. In J. Mittelstraß (Hrsg.), *Enzyklopädie Philosophie und Wissenschaftstheorie* (1, A – G, unveränderte Sonderausgabe, S. 530). J. B. Metzler.

Lorenzen, P. (1968). *Methodisches Denken* (Theorie, Bd. 2). Suhrkamp.

Losee, J. (1977). *Wissenschaftstheorie. Eine historische Einführung* (Beck'sche Elementarbücher). Beck.

Löw, M. (2001). *Raumsoziologie*. Suhrkamp.

Löw, M. (2008). Wenn Sex zum Image wird. Über die Leistungsfähigkeit vergeschlechtlichter Großstadtbilder. In D. Schott & M. Toyka-Seid (Hrsg.), *Die europäische Stadt und ihre Umwelt* (S. 193–206). Wissenschaftliche Buchgesellschaft.

Löw, M., Steets, S., & Stoetzer, S. (2008). *Einführung in die Stadt- und Raumsoziologie* (2., akt. Aufl.). Budrich.

Luhmann, N. (1984). *Soziale Systeme. Grundriß einer allgemeinen Theorie*. Suhrkamp.

Luhmann, N. (1986). *Ökologische Kommunikation. Kann die moderne Gesellschaft sich auf ökologische Gefährdungen einstellen?* Westdeutscher.

Luhmann, N. (1995). *Die Kunst der Gesellschaft*. Suhrkamp.

Luhmann, N. (1997). *Gesellschaft der Gesellschaft*. Suhrkamp.

Luhmann, N. (2001 [1997]). *Die Gesellschaft der Gesellschaft*. uhrkamp.
Luhmann, N. (2002 [1990]). *Die Wissenschaft der Gesellschaft* (4. Aufl.). Suhrkamp.
Lyotard, J.-F. (1979). *La condition postmoderne. Rapport sur le savoir*. Les Éditions de Minuit.
Maasen, S. (2015). *Wissenssoziologie* (2., komplett überarb. Aufl.). transcript.
Maasen, S., Kaiser, M., Reinhart, M., & Sutter, B. (Hrsg.). (2012). *Handbuch Wissenschaftssoziologie*. Springer.
Maischatz, K. (2010). Eine Einführung in das Sozialkapital-Konzept anhand der zentralen Vertreter. In A. Fischer (Hrsg.), *Die soziale Dimension von Nachhaltigkeit – Beziehungsgeflecht zwischen Nachhaltigkeit und Benachteiligtenförderung. Berufliche Bildung und zukünftige Entwicklung* (Leuphana-Schriften zur Berufs- und Wirtschaftspädagogik, Bd. 3, S. 31–54). Schneider-Verlag Hohengehren.
Mannheim, K. (1922). *Die Strukturanalyse der Erkenntnistheorie* (Kant-Studien Ergänzungshefte, Bd. 57). Reuter & Reichard.
Mannheim, K. (1931). Wissenssoziologie. In A. Vierkandt (Hrsg.), *Handwörterbuch der Soziologie* (S. 659–680). Enke.
Mannheim, K. (1952). *Ideologie und Utopie*. Schulte-Bulmke.
Marquardt, N. (2015). *Feministische Geographie*. https://gender-glossar.de/f/item/50-feministische-geographie. Zugegriffen: 27. Nov. 2020.
Martin, G. (1973). *Platons Ideenlehre*. De Gruyter.
Massey, D. (2006). Keine Entlastung für das Lokale. In H. Berking (Hrsg.), *Die Macht des Lokalen in einer Welt ohne Grenzen* (S. 25–31). Campus.
Mathewson, K. (2009). Carl Sauer and his crititcs. In W. M. Denevan & K. Mathewson (Hrsg.), *Carl Sauer on culture and landscape. Readings and commentaries* (S. 9–28). Louisiana State University Press.
Mathewson, K. (2011). Landscape versus region. *The Wiley-Blackwell Companion to Human Geography, 16*, 130.
Mattissek, A. (2008). *Die neoliberale Stadt. Diskursive Repräsentationen im Stadtmarketing deutscher Großstädte*. transcript.
Mattissek, A. (2010). Stadtmarketing in der neoliberalen Stadt. Potentiale von Gouvernementalitäts- und Diskursanalyse für die Untersuchung aktueller Prozesse der Stadtentwicklung. In J. Angermüller & S. van Dyk (Hrsg.), *Diskursanalyse meets Gouvernementalitätsforschung. Perspektiven auf das Verhältnis von Subjekt, Sprache, Macht und Wissen* (S. 129–154). Campus.
Mattissek, A., & Wiertz, T. (2014). Materialität und Macht im Spiegel der Assemblage-Theorie: Erkundungen am Beispiel der Waldpolitik in Thailand. *Geographica Helvetica, 69*(3), 157–169.
McFarlane, C. (2011). Assemblage and critical urbanism. *City, 15*(2), 204–224.
Merker, B., Mohr, G., & Siep, L. (Hrsg.). (1998). *Angemessenheit. Zur Rehabilitierung einer philosophischen Metapher*. Königshausen & Neumann.
Merleau-Ponty, M. (1945). *Phénoménologie de la perception* (Bibliothèque des idées). Gallimard.
Merton, R. K. (1957[1949]). *Social theory and social structure* (Revised and enlarged Edition). Free Press.
Merton, R. K. (1973). *The sociology of science. Theoretical and empirical investigations*. University of Chicago Press.
Merton, R. K. (1987). The fragments from a sociologist's notebooks: Establishing the phenomenon, specified ignorance, and strategic research materials. *Annual Review of Sociology, 13*, 1–29.
Meyer, F., & Miggelbrink, J. (2018). „Der Konjuktiv ist das Problem". Zirkularität, Performativität und Reifikation in der geographischen Forschung. In F. Meyer, J. Miggelbrink, & K. Beurskens (Hrsg.), *Ins Feld und zurück – Praktische Probleme qualitativer Forschung in der Sozialgeographie* (S. 17–23). Springer Spektrum.

Michler, T., Aschenbrand, E., & Leibl, F. (2019). Gestört, aber grün: 30 Jahre Forschung zu Landschaftskonflikten im Nationalpark Bayerischer Wald. In K. Berr & C. Jenal (Hrsg.), *Landschaftskonflikte* (S. 291–311). Springer VS.

Miggelbrink, J. (2014). Diskurs, Machttechnik, Assemblage. Neue Impulse für eine regionalgeographische Forschung. *Geographische Zeitschrift, 102*(1), 25–40.

Mill, J. S. (1968 [1843]). *System der deduktiven und induktiven Logik. Eine Darlegung der Grundsätze der Beweislehre und der Methoden wissenschaftlicher Forschung* (2–4). Scientia (In Gesammelte Werke. Übersetzt unter Redaktion von Theodor Gomperz).

von Mises, L. (1927). *Liberalismus*. Verlag von Gustav Fischer.

Misik, R. (2012). *Halbe Freiheit. Warum Freiheit und Gleichheit zusammengehören* (edition suhrkamp digital). Suhrkamp.

Mittelstraß, J. (1998). Interdisziplinarität oder Transdisziplinarität? In *Die Häuser des Wissens. Wissenschaftstheoretische Studien* (S. 29–48). Suhrkamp.

Mittelstraß, J. (2004a). Szientismus. In J. Mittelstraß (Hrsg.), *Enzyklopädie Philosophie und Wissenschaftstheorie* (4, Sp – Z, unveränderte Sonderausgabe, S. 872–876). J. B. Metzler.

Mittelstraß, J. (2004b). Theoria. In J. Mittelstraß (Hrsg.), *Enzyklopädie Philosophie und Wissenschaftstheorie* (Bd. 4, S. 259–260). J.B. Metzler.

Moore, E. C. (1957). The moralistic fallacy. *The Journal of Philosophy, 54*(2), 29–42.

Moore, G. E. (Hrsg.). (1996). *Principia Ethica* (erweiterte Ausgabe). Reclam.

Moore, K., Kleinman, D. L., Hess, D., & Frickel, S. (2011). Science and neoliberal globalization: A political sociological approach. *Theory and Society, 40*(5), 505–532.

Moran, D. (2000). *Introduction to phenomenology*. Routledge.

Moran, D. (2002). *Introduction to phenomenology*. Routledge.

Morgenstern, C. (1993). *Gedichte – Verse – Sprüche*. Lechner.

Morrissey, J. (2015). Regimes of performance: Practices of the normalised self in the neoliberal university. *British Journal of Sociology of Education, 36*(4), 614–634.

Mott, C., & Roberts, S. M. (2014). *Not everyone has (the) balls. Urban exploration and the persistence of Masculinist geography*. Wiley.

Mouffe, C. (2007). Pluralismus, Diskurs und demokratische Staatsbürgerschaft. In M. Nonhoff (Hrsg.), *Diskurs – radikale Demokratie – Hegemonie. Zum politischen Denken von Ernesto Laclau und Chantal Mouffe* (S. 41–53). transcript.

Mouffe, C. (2014). *Agonistik. Die Welt politisch denken* (Bd. 2677). Suhrkamp.

Muir, R. (1998). Reading the landscape, rejecting the present. *Landscape Research, 23*(1), 71–82. https://doi.org/10.1080/01426399808706526.

Müller, A. (2017). *Planungsethik. Eine Einführung für Raumplaner, Landschaftsplaner, Stadtplaner und Architekten*. UTB.

Müller, E., & Schmieder, F. (2016). *Begriffsgeschichte und historische Semantik. Ein kritisches Kompendium*. Suhrkamp.

Müller, G. (1977). Zur Geschichte des Wortes Landschaft. In A. Hartlieb von Wallthor & H. Quirin (Hrsg.), *„Landschaft" als interdisziplinäres Forschungsproblem. Vorträge und Diskussionen des Kolloquiums am 7./8. November 1975 in Münster* (S. 3–13). Aschendorff.

Müller, M., & Schurr, C. (2016). Assemblage thinking and actor-network theory: Conjunctions, disjunctions, cross-fertilisations. *Transactions of the Institute of British Geographers, 41*(3), 217–229.

Münch, R. (2011). *Akademischer Kapitalismus. Zur politischen Ökonomie der Hochschulreform*. Suhrkamp.

Murdoch, J. (1998). The spaces of actor-network theory. *Geoforum, 29*(4), 357–374.

Murphy, A. B. (1991). Regions as social constructs: The gap between theory and practice. *Progress in Human Geography, 15*(1), 23–35. https://doi.org/10.1177/030913259101500102.

Nederveen Pieterse, J. (2005). Hybridität, na und? In L. Allolio-Näcke, B. Kalscheuer, & A. Manzeschke (Hrsg.), *Differenzen anders denken. Bausteine zu einer Kulturtheorie der Transdifferenz* (S. 396–430). Campus.

Neidhardt, F. (2010). Selbststeuerung der Wissenschaft. In D. Simon, A. Knie, S. Hornbostel, & K. Zimmermann (Hrsg.), *Handbuch Wissenschaftspolitik* (S. 280–292). VS Springer.

Nennen, H.-U., & Garbe, D. (Hrsg.). (1996). *Das Expertendilemma. Zur Rolle wissenschaftlicher Gutachter in der öffentlichen Meinungsbildung.* Springer.

Nestle, W. (1940). *Vom Mythos zum Logos. Die Selbstentfaltung des griechischen Denkens von Homer bis auf die Sophistik und Sokrates.* Kröner.

Neubert, S. (2004). Pragmatismus – thematische Vielfalt in Deweys Philosophie und in ihrer heutigen Rezeption. In L. A. Hickman, S. Neubert, & K. Reich (Hrsg.), *John Dewey. Zwischen Pragmatismus und Konstruktivismus* (Interaktionistischer Konstruktivismus, Bd. 1, S. 13–27). Waxmann.

Neun, O. (2017). Zum Verhältnis von Ludwik Flecks und Karl Mannheims Wissenssoziologie. In M. Endreß, K. Lichtblau, & S. Moebius (Hrsg.), *Zyklos 3. Jahrbuch für Theorie und Geschichte der Soziologie* (S. 71–89). Springer Fachmedien Wiesbaden.

Neurath, O. (1932). Protokollsätze. *Erkenntnis, 3*(1), 204–214. https://doi.org/10.1007/BF0 1886420.

Neurath, O., Carnap, R., & Morris, C. W. (Hrsg.). (1970). *Foundations of the unity of science. Towards an international encyclopedia of unified science.* University of Chicago Press.

Newen, A. (2018). *Analytische Philosophie zur Einführung* (3., unveränd. Aufl.). Junius.

Nicolescu, B. (Hrsg.). (2008). *Transdisciplinarity. Theory and practice.* Hampton Press.

Niemann, H.-J. (2019). Karl Poppers Spätwerk und seine ‚Welt 3'. In G. Franco (Hrsg.), *Handbuch Karl Popper* (Living reference work, S. 1–18). Springer Reference Geisteswissenschaften.

Nissen, U. (1998). *Kindheit, Geschlecht und Raum. Sozialisationstheoretische Zusammenhänge geschlechtsspezifischer Raumaneignung.* Beltz.

Nowotny, H. (2005). Experten, Expertisen und imaginierte Laien. In A. Bogner & H. Torgersen (Hrsg.), *Wozu Experten? Ambivalenzen der Beziehung von Wissenschaft und Politik* (S. 33–44). VS Verlag.

Nowotny, H., Scott, P., & Gibbons, M. (2001). *Re-thinking science. Knowledge and the public in an age of uncertainty.* Polity.

Oeing-Hanhoff, L., Kobusch, T., & Borsche, T. (2019). Individuum, Individualität. In J. Ritter (Hrsg.), *Historisches Wörterbuch der Philosophie* (Völlig neubearbeitete Ausgabe des „Wörterbuchs der philosophischen Begriffe" von Rudolf Eisler, Sonderausgabe, I – fvK, Bd. 4, S. 304–310). WBG Academic.

Olwig, K. R. (2002). *Landscape, nature, and the body politic. From Britain's Renaissance to America's new world.* University of Wisconsin Press.

Oßenbrügge, J. (2014). Zur Theoriediskussion in der Geographie und geographischen Stadtforschung. In J. Oßenbrügge & A. Vogelpohl (Hrsg.), *Theorien in der Raum- und Stadtforschung. Einführungen* (S. 24–33). Westfälisches Dampfboot.

Oßenbrügge, J., & Vogelpohl, A. (Hrsg.). (2014). *Theorien in der Raum- und Stadtforschung. Einführungen.* Westfälisches Dampfboot.

Ottmann, H. (2008). *Geschichte des politischen Denkens* (Neuzeit, Bd. 3). J. B. Metzler (Teilband 3: Die politischen Strömungen im 19. Jahrhundert).

Ottmann, H. (2012). *Geschichte des politischen Denkens* (Das 20. Jahrhundert, Bd. 4). J. B. Metzler (Teilband 2: Von der Kritischen Theorie bis zur Globalisierung).

Paasi, A. (1998). Boundaries as social processes: Territoriality in the world of flows. *Geopolitics, 3*(1), 69–88. https://doi.org/10.1080/14650049808407608.

Paasi, A. (1999). The changing pedagogies of space: Representation of the other in finnish school geography textbooks. In A. Buttimer, S. Brunn, & U. Wardenga (Hrsg.), *Text and image. Social construction of regional knowledges* (Beiträge zur Regionalen Geographie, Bd. 49, S. 226–237). Institut für Länderkunde.

Paasi, A. (2002). Place and region: Regional worlds and words. *Progress in Human Geography, 26*(6), 802–811. https://doi.org/10.1191/0309132502ph404pr.

Paasi, A. (2009). Regional geography I. In R. Kitchin & N. Thrift (Hrsg.), *International encyclopedia of human geography* (Bd. 9, S. 214–227). Elsevier.

Paris, R. (2005). *Normale Macht. Soziologische Essays*. UVK.

Parsons, T. (1951). *The social system*. Free Press.

Parsons, T. (1991 [1951]). *The social system*. Routledge.

Peirce, C. S. (1991). *Vorlesungen über Pragmatismus. Einleitung, Anmerkung und herausgegeben von Elisabeth Walther*. Felix Meiner.

Peirce, C. S. (1998). *Elements of logic. Volume 2* (58th Edition, Reprint of the 1931 Edition). Thoemmes.

Peirce, C. S. (2018 [1878]). *Über die Klarheit unserer Gedanken. How to Make Our Ideas Clear* (4., erw. Aufl.). Vittorio Klostermann GmbH.

Pennington, M. (2002). A Hayekian liberal critique of collaborative planning. In M. Tewdwr-Jones & P. Allmendinger (Hrsg.), *Planning futures. New directions for planning theory* (S. 187–205). Routledge.

Pfister, J. (Hrsg.). (2016). *Texte zur Wissenschaftstheorie*. Reclam.

Pfister, J. (2019a). *Philosophie. Ein Lehrbuch*. Reclam.

Pfister, J. (2019b). *Werkzeuge des Philosophierens*. Reclam.

Piechocki, R. (2010). *Landschaft – Heimat – Wildnis. Schutz der Natur – aber welcher und warum?* Beck.

Pieper, A. (Hrsg.). (1998). *Philosophische Disziplinen. Ein Handbuch* (Reclam-Bibliothek, Bd. 1643). Reclam.

Plessner, H. (1966). Zur Soziologie der modernen Forschung und ihrer Organisation in der deutschen Universität. In H. Plessner (Hrsg.), *Diesseits der Utopie. Ausgewählte Beiträge zur Kultursoziologie* (S. 121–142). Suhrkamp.

Popitz, H. (1992). *Phänomene der Macht* (2., stark erw. Aufl.). Mohr Siebeck.

Popper, K. R. (1959). *The logic of scientific discovery*. Harper & Row.

Popper, K. R. (1963). *Conjectures and refutations. The growth of scientific knowledge*. Routledge & Kegan Paul.

Popper, K. R. (1973). *Objektive Erkenntnis. Ein evolutionärer Entwurf*. Hoffmann und Campe.

Popper, K. R. (1979). Three worlds. Tanner lecture, Michigan, April 7, 1978. *Michigan Quarterly Review*, (1), 141–167. https://tannerlectures.utah.edu/_documents/a-to-z/p/popper80.pdf. Zugegriffen: 12. Mai 2020.

Popper, K. R. (1984). *Auf der Suche nach einer besseren Welt. Vorträge und Aufsätze aus dreißig Jahren*. Piper.

Popper, K. R. (1989). *Logik der Forschung*. Mohr Siebeck.

Popper, K. R. (1996). *Alles Leben ist Problemlösen. Über Erkenntnis, Geschichte und Politik*. Piper.

Popper, K. R. (1997). *Lesebuch. Ausgewählte Texte zur Erkenntnistheorie, Philosophie der Naturwissenschaften, Metaphysik, Sozialphilosophie* (2. Aufl.). UTB GmbH.

Popper, K. R. (2002). *The logic of scientific discovery*. Routledge.

Popper, K. R. (2003 [1945]). *Die offene Gesellschaft und ihre Feinde* (Der Zauber Platons, Bd. 1, 8. Aufl.). Mohr Siebeck.

Popper, K. R. (2010). *Die beiden Grundprobleme der Erkenntnistheorie. Aufgrund von Manuskripten aus den Jahren 1930–1933* (Karl R. Popper – Gesammelte Werke, Bd. 2, 3. Aufl., durchgesehen und ergänzt). Mohr Siebeck (Herausgegeben von Troels Eggers Hansen).

Popper, K. R. (2011 [1947]). *The open society and its enemies*. Routledge.

Popper, K. R. (2018 [1984]). *Alle Menschen sind Philosophen*. Piper (Herausgegeben von Heidi Bohnet und Klaus Stadler).

Popper, K. R. (2019 [1987]). *Auf der Suche nach einer besseren Welt. Vorträge und Aufsätze aus dreißig Jahren*. Piper.

Popper, K. R., & Eccles, J. C. (1977). *Das Ich und sein Gehirn*. Piper.

Poser, H. (2009). *Wissenschaftstheorie. Eine philosophische Einführung*. Reclam.

Poser, H. (2012). *Wissenschaftstheorie. Eine philosophische Einführung* (2., überarb. u. erw. Aufl.). Philipp Reclam jun.

Pott, A. (2007). *Orte des Tourismus. Eine raum- und gesellschaftstheoretische Untersuchung.* transcript.

Pregill, P., & Volkman, N. (1999). *Landscapes in history. Design and planning in the eastern and western traditions.* Wiley.

Putnam, H. (1995). *Pragmatism: An open question.* Blackwell Publishers.

Quante, M. (2008). *Einführung in die allgemeine Ethik* (Einführungen Philosophie, 3. Aufl.). Wissenschaftliche Buchgesellschaft.

Rademacher, C., Schroer, M., & Wiechens, P. (Hrsg.). (1999). *Spiel ohne Grenzen? Ambivalenzen der Globalisierung.* Westdeutscher.

Rapp, C. (2010). Aristoteles – Das Problem der Substanz. In A. Beckermann & D. Perler (Hrsg.), *Klassiker der Philosophie heute* (S. 38–58). Reclam.

Rau, S. (2017). *Räume. Konzepte, Wahrnehmungen, Nutzungen* (Historische Einführungen, 2., akt. Aufl., Bd. 14). Campus.

Rebay-Salisbury, K. (2013). Phänomenologie und Landschaft: der menschliche Körper in Bewegung. In R. Karl & J. Leskovar (Hrsg.), *Interpretierte Eisenzeiten. Fallstudien, Methoden, Theorie: Tagungsbeträge der 5. Linzer Gespräche zur interpretativen Eisenzeitarchäologie* (Studien zur Kulturgeschichte von Oberösterreich, Folge 37, S. 61–70). Oberösterreichisches Landesmuseum.

Redepenning, M., & Wilhelm, J. (2014). Raumforschung mit luhmannscher Systemtheorie. In J. Oßenbrügge & A. Vogelpohl (Hrsg.), *Theorien in der Raum- und Stadtforschung. Einführungen* (S. 310–327). Westfälisches Dampfboot.

Reese-Schäfer, W. (1991). *Jürgen Habermas.* Campus.

Regenbogen, A., & Meyer, U. (Hrsg.). (1998). *Wörterbuch der philosophischen Begriffe.* Meiner.

Reichenbach, H. (1938). *Experience and prediction. An analysis of the foundations and the structure of knowledge.* University of Chicago Press.

Reid, T. (2000 [1785]). *An inquiry into the human mind. On the principles of common sense.* Edinburgh University Press.

Reinhart, M. (2012). *Soziologie und Epistemologie des Peer Review* (Schriftenreihe Wissenschafts- und Technikforschung, Bd. 10). Nomos.

Richter, R. (2016). *Soziologische Paradigmen. Eine Einführung in klassische und moderne Konzepte* (2. Aufl.). UTB.

von Richthofen, F. (1886). *Führer für Forschungsreisende. Anleitung zu Beobachtungen über Gegenstände der physischen Geographie und Geologie.* Oppenheim.

Rickert, H. (1986 [1902]). *Kulturwissenschaft und Naturwissenschaft.* Reclam.

Riecke, J. (Hrsg.). (2011). *Historische Semantik.* De Gruyter.

Risse-Kappen, T. (1995). Reden ist nicht billig. Zur Debatte um Kommunikation und Rationalität. *Zeitschrift für Internationale Beziehungen, 2*(1), 171–184.

Rivera López, E. (1995). *Die moralischen Voraussetzungen des Liberalismus.* Alber.

Rodaway, P. (2014). Humanism and people-centered methods. In S. C. Aitken & G. Valentine (Hrsg.), *Approaches to human geography. Philosophies, theories, people and practices* (2. Aufl., S. 334–343). SAGE.

Rodewald, R. (2001). *Sehnsucht Landschaft. Landschaftsgestaltung unter ästhetischem Gesichtspunkt* (2. Aufl.). Chronos.

Ronneberger, K., & Vogelpohl, A. (2014). Henri Lefebvre: Die Produktion des Raumes und die Urbanisierung der Gesellschaft. In J. Oßenbrügge & A. Vogelpohl (Hrsg.), *Theorien in der Raum- und Stadtforschung. Einführungen* (S. 251–270). Westfälisches Dampfboot.

Rorty, R. (1982). *Consequences of pragmatism. Essays: 1972–1980.* University of Minnesota Press.

Rorty, R. (1989). *Kontingenz, Ironie und Solidarität.* Suhrkamp.

Rorty, R. (1991). *Objectivity, relativism, and truth.* Cambridge University Press.

Rorty, R. M. (Hrsg.). (1967). *The linguistic turn. Essays in philosophical method*. University of Chicago Press (With Two Retrospective Essays).

Rosa, H. (2013). *Beschleunigung und Entfremdung. Entwurf einer Kritischen Theorie spätmoderner Zeitlichkeit*. Suhrkamp.

Rosenberg, J. F. (2009). *Philosophieren. Ein Handbuch für Anfänger*. Vittorio Klostermann.

Rousseau, J.-J., & Rippel, P. (Hrsg.). (1998 [1755]). *Abhandlung über den Ursprung und die Grundlagen der Ungleichheit unter den Menschen*. Reclam.

Russell, B. (1967 [1912]). *Probleme der Philosophie* (2. Aufl.). Suhrkamp.

Sachsse, H. (1971). *Einführung in die Kybernetik unter besonderer Berücksichtigung von technischen und biologischen Wirkungsgefügen*. Vieweg + Sohn.

Salmon, W. C. (1983). *Logik*. Reclam.

Samers, M., Bigger, P., & Belcher, O. (2014). To build another world: Activism in the light of Marxist geographical thought. In S. C. Aitken & G. Valentine (Hrsg.), *Approaches to human geography. Philosophies, theories, people and practices* (2. Aufl., S. 344–360). SAGE.

Sarasin, P. (2016). *Michel Foucault zur Einführung* (6., erg. Aufl.). Junius.

Sauer, C. O. (1925). *The morphology of landscape* (Universitiy of California publications in geography). Universitiy of California Press.

Schaal, G. S., & Heidenreich, F. (2006). *Einführung in die Politischen Theorien der Moderne* (UTB Politikwissenschaft, Bd. 2791). Budrich.

Schaber, P. (2011). Naturalistischer Fehlschluss. In M. Düwell, C. Hübenthal, & M. H. Werner (Hrsg.), *Handbuch Ethik* (3., akt. Aufl., S. 454–456). J.B. Metzler.

Schäfer, L. (1991). Natur. In E. Martens & H. Schnädelbach (Hrsg.), *Philosophie. Ein Grundkurs* (Bd. 2, S. 467–507). Rowohlt.

Schamp, E. W. (2012). Evolutionäre Wirtschaftsgeographie. *Zeitschrift für Wirtschaftsgeographie, 56*(1–2), 121–128.

Scheler, M. (Hrsg.). (1924). *Versuche zu einer Soziologie des Wissens*. Dunker & Humbolt.

Scheler, M. (1926). *Die Wissensformen und die Gesellschaft*. Der Neue-Geist-Verlag.

Schenk, W. (2001). Kulturlandschaft in Zeiten verschärfter Nutzungskonkurrenz: Genese, Akteure, Szenarien. In Akademie für Raumforschung und Landesplanung (Hrsg.), *Die Zukunft der Kulturlandschaft zwischen Verlust, Bewahrung und Gestaltung* (Forschungs- und Sitzungsberichte, Bd. 215, S. 30–44). Selbstverlag.

Schenk, W. (2006). Der Terminus „gewachsene Kulturlandschaft" im Kontext öffentlicher und raumwissenschaftlicher Diskurse zu „Landschaft" und „Kulturlandschaft". In U. Matthiesen, R. Danielzyk, S. Heiland, & S. Tzschaschel (Hrsg.), *Kulturlandschaften als Herausforderung für die Raumplanung. Verständnisse – Erfahrungen – Perspektiven* (Forschungs- und Sitzungsberichte, Bd. 228, S. 9–21). Selbstverlag.

Schenk, W. (2011). *Historische Geographie* (Geowissen kompakt). WBG.

Schenk, W. (2017). Landschaft. In L. Kühnhardt, & T. Mayer (Hrsg.), *Bonner Enzyklopädie der Globalität* (Bd. 1, 2, S. 671–684). Springer VS.

Schenk, W., Fehn, K., & Denecke, D. (Hrsg.). (1997). *Kulturlandschaftspflege. Beiträge der Geographie zur räumlichen Planung*. Borntraeger.

Schiemann, G. (2017). Persistenz der Lebenswelt? Das Verhältnis von Lebenswelt und Wissenschaft in der Moderne. In T. Müller & T. M. Schmidt (Hrsg.), *Abschied von der Lebenswelt? Zur Reichweite naturwissenschaftlicher Erklärungsansätze* (2. Aufl., S. 181–200). Karl Alber.

Schildknecht, C., Teichert, D., & van Zantwijk, T. (Hrsg.). (2008). *Genese und Geltung. Für Gottfried Gabriel*. Mentis.

Schlottmann, A., & Miggelbrink, J. (Hrsg.). (2015). *Visuelle Geographien. Zur Produktion, Aneignung und Vermittlung von RaumBildern*. transcript.

Schlottmann, A., & Wintzer, J. (2019). *Weltbildwechsel. Ideengeschichten geographischen Denkens und Handelns* (utb Geographie, 1. Aufl.). Haupt.

Schmithüsen, J. (1959). Das System der geographischen Wissenschaft. *Berichte zur deutschen Landeskunde, 23*, 1–14.

Schmitt, C. (1933). *Der Begriff des Politischen*. Hanseatische Verlagsanstalt.

Schmitt, C. (2011 [1967]). *Die Tyrannei der Werte* (3., korr. Aufl.). Duncker & Humblot.

Schnädelbach, H. (1980). *Probleme der Wissenschaftstheorie. Studienbrief 3302 der FernUniversität Hagen*. Kurseinheit 01: Grundfragen philosophischer Wissenschaftstheorie, Hagen.

Schnädelbach, H. (1991). Philosophie. In E. Martens & H. Schnädelbach (Hrsg.), *Philosophie. Ein Grundkurs* (S. 37–76). Rowohlt.

Schnädelbach, H. (1993). *Probleme der Wissenschaftstheorie. Eine philosophische Einführung. Kurseinheit 1: Grundfragen philosophischer Wissenschaftstheorie*. Vorlesung, Fernuniversität Hagen.

Schneider, H. (2019). Essentialismus. In J. Ritter (Hrsg.), *Historisches Wörterbuch der Philosophie* (Völlig neubearbeitete Ausgabe des „Wörterbuchs der philosophischen Begriffe" von Rudolf Eisler, Sonderausgabe, D – F, Bd. 2, S. 751–753). WBG Academic.

Schneider, U. (2004). *Die Macht der Karten. Eine Geschichte der Kartographie vom Mittelalter bis heute*. Primus.

Schoeps, J. H. (1981). Konservativismus. In J. H. Schoeps, J. H. Knoll, & C.-E. Bärsch (Hrsg.), *Konservativismus, Liberalismus, Sozialismus. Einführung, Texte, Bibliographien* (Uni-Taschenbücher Politologie, Neuere Geschichte, Soziologie, Bd. 1032, S. 11–86). Fink.

Schönwälder-Kuntze, T. (2020). *Philosophische Methoden zur Einführung* (3., erw. Aufl.). Junius.

Schopenhauer, A. (2009). *Die Kunst, recht zu behalten. In achtunddreißig Kunstgriffen dargestellt*. Anaconda.

Schroeder, P. (2004). Certismus. In J. Mittelstraß (Hrsg.), *Enzyklopädie Philosophie und Wissenschaftstheorie* (1, A – G, unveränderte Sonderausgabe, S. 385–386). J. B. Metzler.

Schroer, M. (2006). *Räume, Orte, Grenzen. Auf dem Weg zu einer Soziologie des Raums*. Suhrkamp.

Schubert, H.-J., Joas, H., & Wenzel, H. (2010). *Pragmatismus zur Einführung. Kreativität, Handlung, Deduktion, Induktion, Abduktion, Chicago School, Sozialreform, symbolische Interaktion* (Zur Einführung, Bd. 382). Junius.

Schülein, J. A., & Reitze, S. (2012). *Wissenschaftstheorie für Einsteiger* (3., akt. u. erw. Aufl.). facultas wuv.

Schultz, H.-D. (2005). Zwischen fordernder Natur und freiem Willen: Das Politische an der „klassischen" deutschen Geographie. *Erdkunde, 59*(1), 1–21.

Schulz-Schaeffer, I. (2000). Akteur-Netzwerk-Theorie: Zur Koevolution von Gesellschaft, Natur und Technik. In J. Weyer & J. Abel (Hrsg.), *Soziale Netzwerke. Konzepte und Methoden der sozialwissenschaftlichen Netzwerkforschung* (Lehr- und Handbücher der Soziologie, S. 187–210). Oldenbourg.

Schurr, C., & Weichhart, P. (2020). From Margin to Center? Theoretische Aufbrüche in der Geographie seit Kiel 1969. *Geographica Helvetica, 75*(2), 53–67.

Schurz, G. (2007). Popper und das Problem der Induktion. In H. Keuth (Hrsg.), *Karl Popper: Logik der Forschung* (3., bearb. Aufl., S. 25–40). De Gruyter.

Schurz, G. (2014). *Einführung in die Wissenschaftstheorie* (4., überarb. Aufl.). WBG.

Schurz, G., & Weingartner, P. (Hrsg.). (1998). *Koexistenz rivalisierender Paradigmen. Eine postkuhnsche Bestandsaufnahme zur Struktur gegenwärtiger Wissenschaft*. Westdeutscher.

Schütz, A. (1960 [1932]). *Der sinnhafte Aufbau der sozialen Welt. Eine Einleitung in die Verstehende Soziologie* (2. Aufl). Julius Springer. (Originalarbeit erschienen 1932).

Schütz, A. (1971a [1962]). *Gesammelte Aufsätze 1. Das Problem der Wirklichkeit*. Martinus Nijhoff.

Schütz, A. (1971b). *Gesammelte Aufsätze 3. Studien zur phänomenologischen Philosophie*. Martinus Nijhoff.

Schütz, A., & Luckmann, T. (2003 [1975]). *Strukturen der Lebenswelt*. UTB.

Schützeichel, R. (2012). Wissenssoziologie. In S. Maasen, M. Kaiser, M. Reinhart, & B. Sutter (Hrsg.), *Handbuch Wissenschaftssoziologie* (S. 17–26). Springer.

Schwarz, H. (2011). Über Tabuthemen in der Wissenschaft, Programmförderung und Mainstreamforschung. *Forschung und Lehre, 18*(5), 354–355.

Schwemmer, O. (2004a). essentia. In J. Mittelstraß (Hrsg.), *Enzyklopädie Philosophie und Wissenschaftstheorie* (1, A – G, unveränderte Sonderausgabe, S. 590–591). J. B. Metzler.

Schwemmer, O. (2004b). Essentialismus. In J. Mittelstraß (Hrsg.), *Enzyklopädie Philosophie und Wissenschaftstheorie* (1, A – G, unveränderte Sonderausgabe, S. 591–592). J. B. Metzler.

Seamon, D. (2014). Lived Emplacementand the Locality of Being: A Return to Humanistic Geography? In S. C. Aitken & G. Valentine (Hrsg.), *Approaches to human geography. Philosophies, theories, people and practices* (2. Aufl., S. 35–48). SAGE.

Seiffert, H. (1996). *Einführung in die Wissenschaftstheorie 1. Sprachanalyse – Deduktion – Induktion in Natur- und Sozialwissenschaften*. C.H. Beck.

Sen, A. K. (1966). Hume's law and Hare's rule. *Philosophy, 41*(155), 75–79.

Shell, K. L. (1986). *Der amerikanische Konservatismus*. Kohlhammer.

Silvertown, J. (2009). A new dawn for citizen science. *Trends in Ecology & Evolution, 24*(9), 467–471.

Skirbekk, G. (Hrsg.). (1977). *Wahrheitstheorien. Eine Auswahl aus den Diskussionen über Wahrheit im 20. Jahrhundert*. Suhrkamp.

Sloterdijk, P. (1987). *Kopernikanische Mobilmachung und ptolemäische Abrüstung. Ästhetischer Versuch* (Edition Suhrkamp). Suhrkamp.

Smith, N. (1984). *Uneven development. Nature, capital and the production of space*. Blackwell.

Sofsky, W., & Paris, R. (1994). *Figurationen sozialer Macht. Autorität, Stellvertretung, Koalition* (Suhrkamp Taschenbuch Wissenschaft, Bd. 1135). Suhrkamp.

Soja, E. W. (1996). *Thirdspace. Journeys to Los Angeles and other real-and-imagined places*. Blackwell.

Soja, E. W. (2000). *Postmetropolis. Critical studies of cities and regions*. Blackwell.

Soja, E. W. (2003). Thirdspace – Die Erweiterung des Geographischen Blicks. In H. Gebhardt, P. Reuber, & G. Wolkersdorfer (Hrsg.), *Kulturgeographie. Aktuelle Ansätze und Entwicklungen* (Spektrum Lehrbuch, S. 269–288). Spektrum Akademischer.

Soja, E. W. (2014). *My Los Angeles. From urban restructuring to regional urbanization*. University of California Press.

Sokolowski, R. (2000). *Introduction to phenomenology*. Cambridge University Press.

Sørensen, M. L. S., & Rebay-Salisbury, K. (Hrsg.). (2012). *Embodied knowledge. Perspectives on believe and technology*. Oxbow Books.

Sorokin, P. (1959). *Social and cultural mobility*. Free Press of Glencoe.

Stegmüller, W. (1973). *Probleme und Resultate der Wissenschaftstheorie und Analytischen Philosophie. Personelle und statistische Wahrscheinlichkeit. Erster Halbband: Personelle Wahrscheinlichkeit und Rationale Entscheidung* (Bd. 4). Springer.

Stegmüller, W. (1978a). *Hauptströmungen der Gegenwartsphilosophie. Eine kritische Einführung* (Bd. 1, Nachdruck, 6. Aufl.). Kröner.

Stegmüller, W. (Hrsg.). (1978b). *Das Universalien-Problem* (Wege der Forschung, Bd. 83). Wissenschaftliche Buchgesellschaft.

Stegmüller, W. (1984). Evolutionäre Erkenntnistheorie, Realismus und Wissenschaftstheorie. In R. Spaemann, P. Koslowski, & R. Löw (Hrsg.), *Evolutionstheorie und menschliches Selbstverständnis. Zur philosophischen Kritik eines Paradigmas moderner Wissenschaft ; Referate und der Bericht über die Schlußdiskussion* (Civitas-Resultate, Bd. 6, S. 5–34). Acta Humaniora.

Stegmüller, W. (1985). *Probleme und Resultate der Wissenschaftstheorie und analytischen Philosophie* (Theorie und Erfahrung, Bd. 2). Springer (Zweiter Halbband: Theorienstrukturen und Theoriendynamik).

Stehr, N., & Meja, V. (Hrsg.). (1980) Wissenssoziologie [Themenheft]. *Kölner Zeitschrift für Soziologie und Sozialpsychologie* (22). Westdeutscher.

Steiner, C. (2009). Materie oder Geist? Überlegungen zur Überwindung dualistischer Erkenntniskonzepte aus der Perspektive einer Pragmatischen Geographie. *Berichte zur deutschen Landeskunde, 83*(2), 129–142.

Steiner, C. (2014a). *Pragmatismus – Umwelt – Raum. Potenziale des Pragmatismus für eine transdisziplinäre Geographie der Mitwelt* (Erdkundliches Wissen, Bd. 155). Franz Steiner.

Steiner, C. (2014b). Von Interaktion zu Transaktion – Konsequenzen eines pragmatischen Mensch-Umwelt-Verständnisses für eine Geographie der Mitwelt. *Geographica Helvetica, 69*(3), 171–181.

Stemmer, B. (2016). *Kooperative Landschaftsbewertung in der räumlichen Planung. Sozialkonstruktivistische Analyse der Landschaftswahrnehmung der Öffentlichkeit*. Springer VS.

Stewig, R. (Hrsg.). (1979). *Probleme der Länderkunde* (Wege der Forschung, Bd. 391). Wissenschaftliche Buchgesellschaft.

Stichweh, R. (1998). Raum, Region und Stadt in der Systemtheorie. *Soziale Systeme, 4*(2), 341–358.

Stichweh, R. (2003). Raum und moderne Gesellschaft. Aspekte der sozialen Kontrolle des Raumes. In T. Krämer-Badoni (Hrsg.), *Die Gesellschaft und ihr Raum. Raum als Gegenstand der Soziologie* (Stadt, Raum und Gesellschaft, Bd. 21, S. 93–102). Leske + Budrich.

Strawson, P. F. (2003). *Einzelding und logisches Subjekt. Ein Beitrag zur deskriptiven Metaphysik*. Reclam.

Ströker, E. (1998). Wissenschaftstheorie. In A. Pieper (Hrsg.), *Philosophische Disziplinen. Ein Handbuch* (Reclam-Bibliothek, Bd. 1643, S. 437–456). Reclam.

Stuhlmann-Laeisz, R. (1983). *Das Sein-Sollen-Problem. Eine modallogische Studie* (Problemata, Bd. 96). Frommann-Holzboog.

Sturm, G. (2000). *Wege zum Raum. Methodologische Annäherungen an ein Basiskonzept raumbezogener Wissenschaften*. VS Verlag.

Suchanek, A. (2004). Die Rolle empirischer Bedingungen für die Wirtschaftsethik. In P. Ulrich & M. Breuer (Hrsg.), *Wirtschaftsethik im philosophischen Diskurs. Begründung und „Anwendung" praktischen Orientierungswissens*. Königshausen und Neumann.

Tetens, H. (1994). *Geist, Gehirn, Maschine. Philosophische Versuche über ihren Zusammenhang* (Universal-Bibliothek, Bd. 8999). Reclam.

Tetens, H. (1999). Wissenschaft. In H. J. Sandkühler (Hrsg.), *Enzyklopädie Philosophie* (S. 1763–1773). Meiner.

Tetens, H. (2013). *Wissenschaftstheorie. Eine Einführung*. C.H.Beck.

Tetens, H. (2014). *Philosophisches Argumentieren. Eine Einführung*. Beck.

Thibaud, J.-P. (2003). Die sinnliche Umwelt von Städten. Zum Verständnis urbaner Atmosphären. In M. Hauskeller (Hrsg.), *Die Kunst der Wahrnehmung. Beiträge zu einer Philosophie der sinnlichen Erkenntnis* (S. 280–297). SFG-Servicecenter Fachverlage.

Thiel, C. (2004). Logik. In J. Mittelstraß (Hrsg.), *Enzyklopädie Philosophie und Wissenschaftstheorie* (2, H – O, unveränderte Sonderausgabe, S. 626–631). J. B. Metzler.

Thrift, N. (1991). For a new regional geography 2. *Progress in Human Geography, 15*(4), 456–466. https://doi.org/10.1177/030913259101500407.

Tilley, C. (1997). *A phenomenology of landscape. Places, paths and monuments* (Explorations in anthropology). Berg.

Tilley, C. (2005). Phenomenological archaeology. In C. Renfrew & P. Bahn (Hrsg.), *Archaeology. The key concepts* (Routledge key guides, S. 151–155). Routledge.

Toulmin, S. E. (1978). *Menschliches Erkennen I: Kritik der kollektiven Vernunft*. Suhrkamp.

Trepl, L. (2012). *Die Idee der Landschaft. Eine Kulturgeschichte von der Aufklärung bis zur Ökologiebewegung*. transcript.

Tuan, Y.-F. (1976). Humanistic geography. *Annals of the Association of American Geographers, 66*(2), 266–276. https://doi.org/10.1111/j.1467-8306.1976.tb01089.x.

Tuan, Y.-F. (1989a). *Space and place. The perspective of experience* (5. Aufl.). University of Minnesota Press.

Tuan, Y.-F. (1989b). Surface phenomena and aesthetic experience. *Annals of the Association of American Geographers, 79*(2), 233–241. https://doi.org/10.1111/j.1467-8306.1989.tb00260.x.

Tugendhat, E., & Wolf, U. (1986). *Logisch-semantische Propädeutik.* Reclam.

Vester, H.-G. (1993). *Soziologie der Postmoderne.* Quintessenz.

Vicenzotti, V. (2006). Kulturlandschaft und Stadt-Wildnis. In I. Kazal, A. Voigt, A. Weil, & A. Zutz (Hrsg.), *Kulturen der Landschaft. Ideen von Kulturlandschaft zwischen Tradition und Modernisierung* (Landschaftsentwicklung und Umweltforschung, Bd. 127, S. 221–236). Technische Universität Berlin.

Vicenzotti, V. (2011). *Der ,Zwischenstadt'-Diskurs. Eine Analyse zwischen Wildnis, Kulturlandschaft und Stadt.* transcript.

Vicenzotti, V. (2012). Gestalterische Zugänge zum suburbanen Raum – Eine Typisierung. In W. Schenk, M. Kühn, M. Leibenath, & S. Tzschaschel (Hrsg.), *Suburbane Räume als Kulturlandschaften* (Forschungs- und Sitzungsberichte, Bd. 236, S. 252–275). Selbstverlag.

Vico, G. (1966 [1725]). *Die neue Wissenschaft über die gemeinschaftliche Natur der Völker* (2. Aufl.). Rowohlt.

Viehöver, W. (2005). Der Experte als Platzhalter und Interpret moderner Mythen. Das Beispiel der Stammzellendebatte. In A. Bogner & H. Torgersen (Hrsg.), *Wozu Experten? Ambivalenzen der Beziehung von Wissenschaft und Politik* (S. 149–171). VS Verlag.

Vilsmaier, U., & Lang, D. J. (2014). Transdisziplinäre Forschung. In H. Heinrichs & G. Michelsen (Hrsg.), *Nachhaltigkeitswissenschaften* (S. 87–113). Springer Spektrum.

Voelzkow, H. (2000). Korporatismus in Deutschland: Chancen, Risiken und Perspektiven. In E. Holtmann (Hrsg.), *Zwischen Wettbewerbs- und Verhandlungsdemokratie. Analysen zum Regierungssystem der Bundesrepublik Deutschland* (S. 185–212). Westdeutscher.

Voigt, A. (2009a). ,Wie sie ein Ganzes bilden' – analoge Deutungsmuster in ökologischen Theorien und politischen Philosophien der Vergesellschaftung. In T. Kirchhoff & L. Trepl (Hrsg.), *Vieldeutige Natur. Landschaft, Wildnis und Ökosystem als kulturgeschichtliche Phänomene* (Sozialtheorie, S. 331–348). transcript.

Voigt, A. (2009b). *Die Konstruktion der Natur. Ökologische Theorien und politische Philosophien der Vergesellschaftung* (Sozialgeographische Bibliothek, Bd. 12). Steiner.

Vorwerg, C. (2013). *Raumrelationen in Wahrnehmung und Sprache. Kategorisierungsprozesse bei der Benennung visueller Richtungsrelationen* (Studien zur Kognitionswissenschaft). Deutscher Universitätsverlag.

Voss, R. (2017). *Wissenschaftliches Arbeiten … leicht verständlich!* (utb Schlüsselkompetenzen, 5. Aufl.). UVK/Lucius.

Wagner, P. L., & Mikesell, M. W. (Hrsg.). (1962). *Readings in cultural geography.* University of Chicago Press.

Wardenga, U. (1996). Von der Landeskunde zur „Landeskunde". In G. Heinritz, G. Sandner, & R. Wießner (Hrsg.), *Der Weg der deutschen Geographie. Rückblick und Ausblick* (50. Deutscher Geographentag, Bd. 4, S. 132–141). Franz Steiner.

Wardenga, U. (2001). Theorie und Praxis der länderkundlichen Forschung und Darstellung in Deutschland. In F.-D. Grimm & U. Wardenga (Hrsg.), *Zur Entwicklung des länderkundlichen Ansatzes* (Beiträge zur Regionalen Geographie, Bd. 53, S. 9–35). Selbstverlag.

Wardenga, U. (2002). Alte und neue Raumkonzepte für den Geographieunterricht. *Geographie heute, 23*(200), 8–11.

Wardenga, U. (2020). Vergangene Zukünfte – oder: Die Verhandlung neuer Möglichkeitsräume in der Geographie. Futures past – Or: The negotiation of new spaces of possibility in geography. *Geographische Zeitschrift, 108*(1), 4–22. https://doi.org/10.25162/gz-2019-0009.

Warms, C. A., & Schroeder, C. A. (1999). Bridging the gulf between science and action: The „new fuzzies" of neopragmatism. *Advances in Nursing Science, 22*(2), 1–10.

Warren, S. (1994). Disneyfication of the metropolis: Popular resistance in Seattle. *Journal of Urban Affairs, 16*(2), 89–107. https://doi.org/10.1111/j.1467-9906.1994.tb00319.x.

Wayand, G. (1998). Pierre Bourdieu: Das Schweigen der Doxa aufbrechen. In P. Imbusch (Hrsg.), *Macht und Herrschaft. Sozialwissenschaftliche Konzeptionen und Theorien* (S. 221–237). VS Verlag.

Weber, F. (2017). Widerstände im Zuge des Stromnetzausbaus – eine diskurstheoretische Analyse der Argumentationsmuster von Bürgerinitiativen in Anschluss an Laclau und Mouffe. *Berichte. Geographie und Landeskunde, 91*(2), 139–154.

Weber, F. (2018a). Ein diskurstheoretischer Zugriff auf ‚Landschaft' und ‚Stadtlandhybride' – Annäherungen zwischen Makro- und Mikroperspektive. In S. Hennecke, H. Kegler, K. Klaczynski, & D. Münderlein (Hrsg.), *Diedrich Bruns wird gelehrt haben. Eine Festschrift* (S. 122–131). Kassel University Press.

Weber, F. (2018b). *Konflikte um die Energiewende. Vom Diskurs zur Praxis.* Springer VS.

Weber, F. (2019a). Diskurstheoretische Landschaftsforschung. In O. Kühne, F. Weber, K. Berr, & C. Jenal (Hrsg.), *Handbuch Landschaft* (S. 105–117). Springer VS.

Weber, F. (2019b). Der Stromnetzausbau in Deutschland – Eine Konturierung des Konfliktes in Anschluss an Chantal Mouffe und Ralf Dahrendorf. In K. Berr & C. Jenal (Hrsg.), *Landschaftskonflikte* (S. 423–437). Springer VS.

Weber, F., & Kühne, O. (2019). Essentialistische Landschafts- und positivistische Raumforschung. In O. Kühne, F. Weber, K. Berr, & C. Jenal (Hrsg.), *Handbuch Landschaft* (S. 57–68). Springer VS.

Weber, F., Jenal, C., Roßmeier, A., & Kühne, O. (2017). Conflicts around Germany's *Energiewende*: Discourse patterns of citizens' initiatives. *Quaestiones Geographicae, 36*(4), 117–130. https://doi.org/10.1515/quageo-2017-0040.

Weber, M. (1976 [1922]). *Wirtschaft und Gesellschaft. Grundriß der verstehenden Soziologie.* Mohr Siebeck.

Weber, M. (2011 [1919]). *Wissenschaft als Beruf* (11. Aufl.). Duncker & Humblot.

Weichhart, P. (1993). How does the person fit into the human ecological triangle? From dualism to duality: The transactional worldview. In D. Steiner & M. Nauser (Hrsg.), *Human ecology. Fragments of anti-fragmentary views of the world* (S. 103–124). Routledge.

Weichhart, P. (1999). Die Räume zwischen den Welten und die Welt der Räume. In P. Meusburger (Hrsg.), *Handlungszentrierte Sozialgeographie. Benno Werlens Entwurf in kritischer Diskussion* (Erdkundliches Wissen, Bd. 130, S. 67–94). Steiner.

Weichhart, P. (2005). Auf der Suche nach der „dritten Säule". Gibt es Wege von der Rhetorik zur Pragmatik? In D. Müller-Mahn & U. Wardenga (Hrsg.), *Möglichkeiten und Grenzen integrativer Forschungsansätze in Physischer Geographie und Humangeographie* (forum ifl, Bd. 2, S. 109–136). Selbstverlag Leibniz-Institut für Länderkunde e. V.

Weichhart, P. (2006). Humangeographische Forschungsansätze. In W. Sitte & H. Wohlschlägl (Hrsg.), *Beiträge zur Didaktik des „Geographie und Wirtschaftskunde"-Unterrichts* (Materialien zur Didaktik der Geographie und Wirtschaftskunde, 4., unveränd. Aufl., Bd. 16, S. 182–198). Institut für Geographie und Regionalforschung.

Weichhart, P. (2008). *Entwicklungslinien der Sozialgeographie. Von Hans Bobek bis Benno Werlen* (Sozialgeographie kompakt, Bd. 1). Franz Steiner.

Weichhart, P. (2018a). *Entwicklungslinien der Sozialgeographie. Von Hans Bobek bis Benno Werlen* (Sozialgeographie kompakt, 2., vollst. überarb. u. erw. Aufl., Bd. 1). Franz Steiner.

Weichhart, P. (2018b [2020 erschienen]). Die Landschaft der Landschaften. *Berichte. Geographie und Landeskunde, 92*(3–4), 203–216.

Weingart, P. (1999). Neue Formen der Wissensproduktion: Fakt, Fiktion und Mode. *TATuP-Zeitschrift für Technikfolgenabschätzung in Theorie und Praxis, 8*(3–4), 48–57.

Weingart, P. (2001). *Die Stunde der Wahrheit? Zum Verhältnis der Wissenschaft zu Politik, Wirtschaft und Medien in der Wissensgesellschaft.* Velbrück Wissenschaft.

Weingart, P. (2003). *Wissenschaftssoziologie* (Einsichten). transcript.

Weingart, P. (2005). *Die Wissenschaft der Öffentlichkeit. Essays zum Verhältnis von Wissenschaft, Medien und Öffentlichkeit.* Velbrück Wissenschaft.

Weingart, P. (2008). Ökonomisierung der Wissenschaft. *N.T.M. Zeitschrift für Geschichte der Wissenschaften, Technik und Medizin, 16*(4), 477–484. https://doi.org/10.1007/s00048-008-0311-4.

Weingart, P. (2010). Wissenschaftssoziologie. In D. Simon, A. Knie, S. Hornbostel, & K. Zimmermann (Hrsg.), *Handbuch Wissenschaftspolitik* (S. 118–129). VS Springer.

Weingart, P. (2015). *Wissenschaftssoziologie*. transcript.

Weingart, P., & Winterhager, M. (1984). *Die Vermessung der Forschung. Theorie und Praxis der Wissenschaftsindikatoren*. Campus.

Weingart, P., Engels, A., & Pansegrau, P. (2008). *Von der Hypothese zur Katastrophe. Der anthropogene Klimawandel im Diskurs zwischen Wissenschaft, Politik und Massenmedien* (2., leicht veränd. Aufl.). Budrich.

Weinrich, H. (1975). System, Diskurs, Didaktik und die Diktatur des Sitzfleisches. In F. Maciejewski (Hrsg.), *Theorie der Gesellschaft oder Sozialtechnologie. Theorie-Diskussion Supplement 1* (S. 145–161). Suhrkamp.

von Weizsäcker, C. F. (2006). *Die Tragweite der Wissenschaft* (7. Aufl.). Hirzel.

Welsch, W. (1996). *Vernunft. Die zeitgenössische Vernunftkritik und das Konzept der transversalen Vernunft*. Suhrkamp.

Werlen, B. (1995). Landschafts- und Länderkunde in der Spät-Moderne. In U. Wardenga (Hrsg.), *Kontinuität und Diskontinuität der deutschen Geographie in Umbruchphasen. Studien zur Geschichte der Geographie* (Münstersche geographische Arbeiten, Bd. 39, S. 161–176). Inst. für Geographie.

Werlen, B. (1997). *Sozialgeographie alltäglicher Regionalisierungen* (Globalisierung, Region und Regionalisierung, Bd. 2, Erdkundliches Wissen Schriftenreihe für Forschung und Praxis, Bd. 119). Steiner.

Werlen, B. (2000). *Sozialgeographie. Eine Einführung*. Haupt.

Werlen, B. (2003). Kulturgeographie und kulturtheoretische Wende. In H. Gebhardt, P. Reuber, & G. Wolkersdorfer (Hrsg.), *Kulturgeographie. Aktuelle Ansätze und Entwicklungen* (Spektrum Lehrbuch, S. 251–268). Spektrum Akademischer.

van Wezemael, J., & Loepfe, M. (2009). Veränderte Prozesse der Entscheidungsfindung in der Raumentwicklung. *Geographica Helvetica, 64*(2), 106–118.

Wille, M. (2011). Die Disziplinierung des Denkens. In B. Pörksen (Hrsg.), *Schlüsselwerke des Konstruktivismus* (S. 160–174). VS Verlag.

Wiltsche, H. A. (2013). *Einführung in die Wissenschaftstheorie*. Vandenhoeck & Ruprecht.

Windelband, W. (1924). Geschichte und Naturwissenschaft. In W. Windelband (Hrsg.), *Präludien* (Bd. 2). Mohr.

Wingens, M. (1998). *Wissensgesellschaft und Industrialisierung der Wissenschaft*. Deutscher Universitätsverlag.

Wittgenstein, L. (1995 [1953]). *Tractatus logico-philosophicus. Tagebücher 1914–1916. Philosophische Untersuchungen* (Werkausgabe Bd. 1, 10. Aufl.). Suhrkamp.

Wolf, A. (2020). Landschaftskonflikte im Zuge der Energiewende: Die Windenergieanlagen von Wadgassen (Saarland). In R. Duttmann, O. Kühne, & F. Weber (Hrsg.), *Landschaft als Prozess (in diesem Band)*. Springer VS.

Wylie, J. (2005). A single day's walking: Narrating self and landscape on the South West Coast Path. *Transactions of the Institute of British Geographers, 30*(2), 234–247. https://doi.org/10.1111/j.1475-5661.2005.00163.x.

Wylie, J. (2007). *Landscape*. Routledge.

Wylie, J. (2019). Landscape and phenomenology. In P. Howard, I. Thompson, E. Waterton, & M. Atha (Hrsg.), *The Routledge companion to landscape studies* (2. Aufl., S. 127–138). Routledge.

Zima, P. V. (2004). *Was ist Theorie? Theoriebegriff und Dialogische Theorie in den Kultur- und Sozialwissenschaften*. Francke.

Zoglauer, T. (1998). *Geist und Gehirn. Das Leib-Seele-Problem in der aktuellen Diskussion* (UTB für Wissenschaft. Uni-Taschenbücher: Philosophie, Bd. 2066). Vandenhoeck & Ruprecht.

Zoglauer, T. (1999). *Einführung in die formale Logik für Philosophen.* Vandenhoeck & Ruprecht.